Sustainable Engineering Forum 2023

Held at the 2023 AIChE Annual Meeting

Orlando, Florida, USA
5-10 November 2023

ISBN: 978-1-7138-9300-4

Printed from e-media with permission by:

Curran Associates, Inc.
57 Morehouse Lane
Red Hook, NY 12571

Some format issues inherent in the e-media version may also appear in this print version.

Printed with permission by Curran Associates, Inc. (2024)

For permission requests, please contact AIChE
at the address below.

AIChE
120 Wall Street, FL 23
New York, NY 10005-4020

Phone: (800) 242-4363
Fax: (203) 775-5177

www.aiche.org

Additional copies of this publication are available from:

Curran Associates, Inc.
57 Morehouse Lane
Red Hook, NY 12571 USA
Phone: 845-758-0400
Fax: 845-758-2633
Email: curran@proceedings.com
Web: www.proceedings.com

Sustainable Engineering Forum 2023

Held at the 2023 AIChE Annual Meeting

Orlando, Florida, USA
5-10 November 2023

AIChE
The Global Home of Chemical Engineers

TABLE OF CONTENTS

ADVANCES IN ALGAE CULTIVATION, CONVERSION AND PRODUCTS

105a Assessment of the Presence of Polychlorinated Dibenzodioxins and Furans (PCDD/Fs) in Flue Gas on Microalgal Growth and Nutritional Value 1
Mahasweta Paul, Raymond Lau

105c Thermal Control of an Enclosed Photobioreactor (PBR) for Algae Cultivation 2
Tolga Pirasaci, George Philippidis, Aydin Sunol

105d Elucidating the Extraction of Fucoxanthin from Algae with Supercritical Carbon Dioxide and Ethanol as a Cosolvent for the Valorization of the Integrated Biorefinery 3
Mary Kate Lane, Elise B. Gilcher, Paul T. Anastas, Julie B. Zimmerman

105e Techno-Economic Analysis of an Integrated Approach to Produce Bio-Methanol from Micro-Algae in the UAE 4
Nasir Al Lagtah, Nizar Ashna

105f Maximizing the Biofuel Production Form Algae Using Four Element Strategy and Ultimate Economic Benefits........ 6
Fares Almomani, Ahmed M. D. Al Ketife

SUSTAINABLE ENGINEERING FORUM PLENARY SESSION (INVITED TALKS)

44b Green Ammonia for Sustainable Energy and Agriculture 7
Prodromos Daoutidis

44a Machine Learning Boosted Catalyst and Operation Design for Clean Energy 8
Fanglin Che

44c Establishing a Circular Economy in the Food Industry 9
Mariano Martin

ADVANCES IN LIFE CYCLE ASSESSMENT

68a Study on Life Cycle Assessment and Optimization of Pesticide Production Process from the Perspective of Sustainable Development........ 11
Kexuan Yang

68b Life-Cycle Assessment Integration into Scalable Open-Source Numerical Models (LiAISON) for Prospective Impact Analysis of Novel Technologies 12
Tapajyoti Ghosh, Patrick Lamers, Alberta Carpenter

68c Guiding Food Colorant Production Via Fermentation Using Techno-Economic, Life Cycle and Absolute Sustainability Assessment 14
Samir Meramo, Sumesh Sukumara, Peter Fantke, Eleonora Passuto

68d Community-Engaged Life Cycle Analysis: Ensuring Sustainability While Protecting Sovereignty........ 16
Margaret O'Connell, Jenna Trost, Kimberly Marion Suiseeya, Jennifer Dunn

68e On the Interplay of Food Packaging Design and Food Supply Chain Sustainability 17
Aurora Del Carmen Munguia Lopez, Paola Alejandra Munoz Briones, Styliani Avraamidou, Victor Zavala

BIG DATA AND ANALYTICS FOR SUSTAINABILITY

125a Dynamic Analysis for Piston-Type Work Exchanger and Network Design for Mechanical Energy Recovery 19
Abdurrafay Siddiqui, Yinlun Huang

125b Using Public Data for Predicting Industrial Chemical Pollution Source Reduction and Prevention Alternatives 20
Jose Hernandez-Betancur, Gerardo Ruiz-Mercado

125c Dynamic Sustainability Assessment in the Digital Age 21
Abdurrafay Siddiqui, Yinlun Huang

125d Modeling-Based and Data-Driven Frameworks to Promote Water Sustainability in Chemical Manufacturing Plants at Dow 22
Spyros Tsolas, Apoorva Sampat, Irfan Khan, Lin Zhao, Tim Finley, Wu Chen

125e Water Stress-Based Price for Global Sustainability: A Study Using Generalized Global Sustainability Model (GGSM) 23
Neeraj Hanumante, Yogendra Shastri, Apoorva Nisal, Urmila Diwekar, Heriberto Cabezas

125f Sustainability and Sustainizability™ Over Sets for a Multi-Species Population Model 26
Ioannis Manousiouthakis, Vasilios Manousiouthakis

PRODUCTION OF LOW-CARBON FUELS FORM BIOMASS: DESIGN, SIMULATION, AND ECONOMIC ANALYSIS

427a Net Zero and Negative Carbon Biofuels from Hydrothermal Liquefaction and Biocrude Upgrading 27
Pimphan Meyer, Shuyun Li, Lesley J. Snowden-Swan, Uisung Lee, Rodrigo Buitrago Tello, Steven Phillips

427b The Value Proposition of Bio-Intermediates as Marine Fuel for Maritime Decarbonization 28
Eric Tan, Abhijit Dutta, Kristiina Ilsa, Calvin Mukarakate

427c Techno-Economic and Life-Cycle GHG Emission Assessment for Aqueous-Phase Product Treatment and Valorization Options of Wet Waste Hydrothermal Liquefaction Process 29
Yuan Jiang, Longwen Ou, Lesley J. Snowden-Swan, Hao Cai, Shuyun Li, Karthikeyan Ramasamy, Andrew J. Schmidt, Huamin Wang, Daniel Santosa, Mariefel V. Olarte, Michael R. Thorson

427d Torrefaction of Almond Shell as a Renewable Reinforcing Agent for Plastics: Techno-Economic Analyses and Comparison to Bioethanol Process 30
Dupeng Liu, Nawa Baral, Corinne D. Scown, Ning Sun

BIOLOGICAL CONVERSIONS AND PROCESSES FOR RENEWABLE FEEDSTOCKS

221a High Temperatures and Low Soil Moisture Synergistically Reduce Switchgrass Yield and Inhibitbiofuel Production from Marginal Field Sites 31
Sarvada Chipkar, Derek J. Debrauske, Leela Joshi, Briana Cronk, Emily Burke, G. Philip Robertson, Trey Sato, Stephen Hamilton, Kurt Thelen, Rebecca Ong

221b Biochar-Assisted Acetogenic Conversion of Carbon Dioxide to C2-C6 Alcohols: A Sustainable Solution for Renewable Fuel Production 32
Rahul Thunuguntla, Hasan Atiyeh, Raymond L. Huhnke, Ralph S. Tanner

221c Lipid Production and Biofuel Yield from Wood Hydrolysates Using Oleaginous Yeast "Cutaneotrichosporon Curvatus" 33
Stephanie Ossai, Sampath Gunukula, Clayton Wheeler, Gerard Van Walsum

221e Optimal Integrated Plant for Renewable Surfactants Production 35
Sofía González Núñez, Mariano Martín, Carlos Amador

CHEMICAL AND CATALYTIC CONVERSIONS AND PROCESSES FOR RENEWABLE FEEDSTOCKS

291a Elucidating Catalytic Pyrolysis of Representative Textile Wastes Over Waste Glass-Derived Silica-Supported Transition Metal Catalysts 37
Yoke Wang Cheng, Wenming Fu, Guiying Lin, Dequan Xu, Chi-Hwa Wang

291b Elucidating the Role of Alkaline Earth Metal Promoters and Zeolite Support Acidity on Palm Oil Catalytic Upgrading: Insights from DFT and Reactive Molecular Dynamics 38
Seba Alareeqi, Daniel Bahamon, Kyriaki Polychronopoulou, Lourdes Vega

291c Catalytic Process Intensification of Bio-Renewable Surfactants Platform 40
Xiaokun Yang

291d Development and Intensification of Green Pathways for Conversion of Soybean Oil to Biodiesel 41
Aydin Sunol

CLIMATE CHANGE AND ENGINEERING SUSTAINABILITY

195a Hydrochars from Lignocellulosic Residues for Green Electronics, Environmental Remediation, Energy and Agricultural Applications 42
Andrea Landazuri, Julio Chacón, Luis M. Prócel, Yuliana Pullas Gallardo, María Fernanda Lecaro, Pierina Mirabá, Juan Fernando Posso, Lourdes Orejuela Escobar, Darío Niebieskikwiat, Marco León, Nicolás Vela, Víctor H. Guerrero, Sebastián Ponce, Herman Murillo, Victoria Alomía, Nelson Herrera, Susan K. Yépez, Katty Beltrán, Oliver Caisaluisa, Eduardo Holguín, Christian Luciani

195b Roadmap for Scale-Up of Carbon Nanotube Production for Displacement of Difficult to Decarbonize Structural Materials with Free Hydrogen Co-Production 51
Alex Dantzler, Jason Desveaux, Oliver Dewey, Arthur Sloan, Zhe Cui, Abhishek Roy, Leonardo Spanu, Glen Irvin, Matteo Pasquali

195c Synthesis of Novel Intermediate Material for Recovery of Lithium-Ion Batteries 52
Mriganka Mondal, Sergio Carrillo De Hert, Rafael Lopez-Rodriguez, Gita Singh, Tony Keene, Steven Ferguson

195d From Landfill to Cement: Developing a Negative Emission Hydrothermal Pathway to Sequester Waste Carbon ... 54
David Kenney, Shuai Wang, Nima Rahbar, Michael T. Timko, Andrew R Teixeira

195e On the Use of Low Carbon Marine Fuels with Options for on-Board Carbon Capture: A Holistic Evaluation Using Life Cycle Analysis and an Integrated Approach ... 55
Nefeli Roufogali, Antonios Kokosis

195f Sustainable Nitrogen by Design: Mechanochemically Synthesized Co-Crystal Materials for Enhanced Efficiency Fertilizer Products ... 58
Mohamed Eisa, Jonas Baltrusaitis

CO_2 CAPTURE FOR INDUSTRIAL POINT SOURCES

98a Increasing Diffusion Selectivity Via Free Volume Manipulation (FVM) and in Situ Thermal Oxidative Crosslinks of Amine-Functionalized PIM-1 Membrane ... 59
Taigyu Joo, Tae Hoon Lee, Wan-Ni Wu, Sungsool Wi, Zachary P. Smith

98b Selective CO_2 Separation by Facilitated Transport in Ionic Liquid Gel Membranes ... 60
Philip Cox, Alex Ferguson, Vincent Storhaug

98c CFD Modeling of Flue Gas Separation with Hollow Fiber Modules and Experimental Verification Via 3D Printing ... 61
Victor A. Kusuma, Lili Sun, Comfort Oluleke, Lingxiang Zhu, Nathan Diemler, Grigorios Panagakos, Glenn Lipscomb, David Hopkinson

98d CFD Modeling of High-Flux Plate-and-Frame Membrane Modules for Post-Combustion Carbon Capture ... 62
Cheick Dosso, Lingxiang Zhu, Victor A. Kusuma, David Hopkinson, Larry Biegler, Grigorios Panagakos

98e Development and Evaluation of Novel Biphasic Solvents for Post-Combustion Carbon Capture ... 63
Yongqi Lu, Hong Lu, Hafiz Salih, Paul Nielsen, Qing Ye

98f Data-Driven Design and Characterization of Aprotic N-Heterocyclic Anion (AHA) Ionic Liquids for CO_2 Capture ... 66
Austin N. Keller, Pranav J. Thacker, Louise M. Cañada, Michael Baldea, Mark Stadtherr, Joan Brennecke

98g Simulation and Uncertainty Quantification of a Cryogenic Carbon Capture System for Cement Industry ... 68
Brandon Paul, Daison Manuel Yancy Caballero, Miguel A. Zamarripa, Benjamin P. Omell, Michael S. Matuszewski

98h Continuous Carbon Capture Via Oxygen/Water Electrolysis in a Modular Solidelectrolyte Reactor ... 71
Peng Zhu, T Alan Hatton, Haotian Wang

CONCENTRATED SOLAR POWER GENERATION AND CHEMICAL PROCESSING I

186a The Important Role of Hydrogen in Solar-Assisted Synthesis of Fuels and Chemicals ... 72
Eric Miller, William Gibbons

186b Investigating Lanthanides and Transition Metal Pair Charged Compensating Doped CeO_2 Materials 73
Jayni Hashimoto, Alonzo Mendez, Christopher Muhich

186c Thermodynamic and Kinetic Characterization of Al- and Ga-Doped La-Sr-Mn Perovskites for Solar Thermochemical Water Splitting 74
Dylan McCord, Elizabeth Gager, Ximeng Wang, Simon Phillpot, Juan C. Nino, Anthony H. McDaniel, Jonathan Scheffe

186d Improving Cycle Performance of Copper Oxide Thermochemical Heat Storage Materials by Doping Cheap Metal Oxides 75
Gang Xiao

186f Thermodynamic and Computational Modeling of Thermochemical Syngas Production Via the CeO_2-CH_4-CO_2 Redox Reforming Cycle in a Tubular Reactor 77
Mario Zuber, Simon Ackermann, Aldo Steinfeld

DESIGN, ANALYSIS, AND OPTIMIZATION OF SUSTAINABLE ENERGY SYSTEMS AND SUPPLY CHAINS I

398b Designing a Li-Ion Battery Supply Chain Under Uncertainty for Nature-Positive Decisions 78
Ying Xue, Bhavik Bakshi

398c Optimization of Steel Decarbonization Strategies in the United States 81
Sydney Johnson, Emre Genã§er

398d Dynamic Systems Analysis of PET and Polyolefin Plastic Supply Chains in Circular Economy 83
Utkarsh Chaudhari, Tasmin Hossain, David W. Watkins, Damon Hartley, Anne Johnson, Kamand Sedaghatnia, Robert Handler, Barbara Reck, Vicki Thompson, Alejandra Peralta, David Shonnard

398e Designing Robust Energy Policies for Low-Carbon Technology Adoption 84
Tom Savage, Antonio Del Rio Chanona, Gbemi Oluleye

398f Systems Integration Across Multiple Processing Clusters: Design of a Cluster of Clusters 87
Mohamed Lameh, Dhabia Al-Mohannadi, Patrick Linke

398g Decision Making of Optimal Decarbonization Routes for Industry 88
Mahdi Ahmed, Joan Cordiner

ENGINEERING GEOLOGIC CARBON DIOXIDE STORAGE SYSTEMS I

457a Assessing the Potential of the St. Peter Sandstone and Everton Formation for CO_2 Storage in Illinois: A Comparison with the Mt. Simon Sandstone 89
Mansour Khosravi, Kendall Taft, Zohreh Askari

457d Illinois Basin Experiences Estimating the Area-Of-Review for Industrial-Scale Carbon Capture and Storage Projects 91
Roland Okwen, Carl Carman, Scott M. Frailey

457e The Value of 2D Seismic to Identify Regions in a Large Carbon Sequestration Project with Elevated Risk for Long Term CO_2 Storage: A Case Study from the Illinois Basin. 92
Kendall Taft

457f Supporting and Developing Commercial Carbon Storage Projects in the Western United States: The Carbon Utilization and Storage Partnership 93
Robert Balch, Brian McPherson, George El-Kaseeh

457g The Investigations of Reservoir Petrophysical Heterogeneity on the Storage Capacity and Injectivity: A Case Study from Dry Fork Carbonsafe Project, Powder River Basin, WY 96
John Jiao, Jonathan 'Fred' McLaughlin, Matthew Johnson, Ying Yu, Scott Quillinan

457h A Case Study from Wyoming Dry Fork Carbonsafe Project: The Effects of Wettability Variation on CO_2 Storage Capacity Assessment 98
Ying Yu, Sumaiya Farzana, Charles Nye, Davin Bagdonas, John Jiao, Jonathan 'Fred' McLaughlin, Scott Quillinan

FEEDSTOCK CONVERSION INTERFACE CONSORTIUM – UNDERSTANDING FEEDSTOCK VARIABILITY TO ENABLE NEXT GENERATION BIOREFINERIES (INVITED TALKS)

36a Biological Degradation and Moisture Effects on Corn Stover Comminution 99
Tiasha Bhattacharjee, Jordan Klinger, Noah Berglund, Bradley Wahlen, Vicki Thompson, Yidong Xia, Neal Yancey

36b Feed System Design Considerations for Biomass and MSW Gasification 100
Dan Dupuis

36c Experiment-Informed Virtual Physics for Granular Feedstock Preprocessing and Handling Operations: Overview of Recent Progress 101
Yidong Xia, Wencheng Jin, Ahmed Hamed, Jordan Klinger, Nepu Saha, Tiasha Bhattacharjee, Mark Small, Neal Yancey, C. Luke Williams, John E. Aston, David Thompson, Lynn Wendt, Vicki Thompson

36d Assessment of the Abrasiveness of Biomass and MSW Feedstock Particulate Materials 102
Oyelayo Ajayi, George Fenske, Jun Qu, James Keiser, Jeffrey A. Lacey, Vicki Thompson, Miranda Kuns

36e Effect of Particle Size, Moisture Content, and Tissue Fraction on Mechanical and Feeding Behavior of Milled Corn Stover and Loblolly Pine Residues 103
Nepu Saha, Jordan Klinger, Tiasha Bhattacharjee, Noah Berglund, Wencheng Jin, Yidong Xia

36f Failure Modes and Effects Analysis of Biorefinery Pathways 104
Rachel Emerson, Pralhad Burli, Lorenzo Vega-Montoto, Tiasha Bhattacharjee

FEEDSTOCK LOGISTICS FOR BIOREFINERIES

332a Variability of Sulfur Content in Loblolly Pine Residues Across Age, Soil Type, and Harvest Method 105
Jian Shi, Gary Lopez

332b Production and Characterization of Biochar from Individual Anatomical Fractions as Biorefinery Co-Products 106
Bradley Wahlen, William A. Smith, Wenqi Li, Asif Hasan Rony

332c The 1st - and nth -Plant Cost Analyses for Incorporating Advanced Biomass Fractionation into Feedstock Processing Systems 107
Yingqian Lin, Damon Hartley, David Thompson, Pralhad Burli

332d Using Municipal Wastes with Conventional Feedstocks for Biofuel Production 108
Pralhad Burli, Yingqian Lin, Damon Hartley, David Thompson

332e Advancing Forest Biorefineries Towards Commercial Applications Through Preprocessing of Biomass Wastes 109
Clayton Wheeler, Sampath Gunukula, C. Luke Williams, Sampath Karunarathne, Peter Van Walsum

FUEL AND ENERGY DECARBONIZATION

496a Analysis of Economic Impacts and Emissions Across Energy Exporters, Importers and the Shipping Industry for Alternative Low Carbon Energy Supply Chain Scenarios 110
Mohamed Lameh, Dhabia Al-Mohannadi, Gareth Burton, Joseph Rousseau, Patrick Linke

496b Dynamic Modeling and Techno-Economic Optimization of a Low-Pressure Microwave Assisted Ammonia Synthesis Process 111
Opeyemi Ogunniyan, Md Emdadul Haque, Yuxin Wang, Debangsu Bhattacharyya, Jianli Hu

496c Optimizing the Sustainable Energy Transition for Small Island Developing States Using Multiple Decision Criteria 113
Sherard Sadeek, Dhurjati Prasad Chakrabarti, Maria Papathanasiou, Keeran Ward

496d Long-Term Planning for Energy Transition Under Endogenous Uncertainty 117
Molly McDonald, Christos Maravelias

496f Decarbonization Solutions 119
Michael Hartmann

GREEN CHEMISTRY AND ENGINEERING-I

359a Scalable and Sustainable Valorization of Lignin Using Single-Step Aerosol Method............................ 120
Sujit Modi, Pratim Biswas

359b Carrier-Free Enzyme-Assembled Hydrogel Based on Synergistic Weak Interactions to Boost Enzyme-Immobilized Microflow Biocatalysis... 122
Qiang Chen, Yujun Wang, Guangsheng Luo

359c Routes to Renewable Production of Paracetamol: A Green Chemistry Analysis of Industrial Viability.. 123
Jimin Park, Caria Evans, Jacob Maier, Marta Hatzell, Stefan France, Carsten Sievers, Andy Bommarius

359d Green Solvent Exploration in Li Ion Battery Electrode Fabrication: A Sustainable Approach 124
Amrita Sarkar

359e Potential Utilization of Nitrogen Oxide from Exhausted Gas for Enhancement of Nitrogen Cycle.. 125
Isshin Ino, Yuka Sakai, Yuki Sato, Yasuki Kansha

359f Sequential Extraction of Triglycerides and Fucoxanthin with Supercritical Carbon Dioxide for the Valorization of the Integrated Biorefinery 126
Elise Gilcher, Mary Kate Lane, Paul T. Anastas, Julie B. Zimmerman

359g Preparation of Silica by Sol-Gel Methods: How Sustainable Are Current Research Methods?............. 127
Ethan Errington, Jerry Heng, Miao Guo

LIFE CYCLE ANALYSIS OF BIO-BASED FUELS, ENERGY, AND CHEMICALS

390a LCA Comparison of Char Impregnation and Direct Chemical Activation for Synthesizing Ultraporous Superactivated Hydrochars from Food Waste 129
Swarna Saha, Al Ibtida Sultana, Samira Hossain, Kawnish Kirtania, Toufiq Reza

390c Life Cycle Assessment of Hydrogen Generation from Biogenic Waste Via Anaerobic Co-Digestion Integrated with Reforming 130
Samira Hossain, Mst. Lucky Khatun, Chayan Kumer Saha, Md. Monjurul Alam, Mohammad Ismail, Toufiq Reza, Kazi Bayzid, Kawnish Kirtania

390d Regionalized Dynamic Life Cycle Assessment of Waste Biomass Utilization 131
Yuan Yao

390e A Life Cycle Analysis of the Conversion of Landfill Gas to Fischer-Tropsch Diesel by Tri-Reforming and Fischer-Tropsch Synthesis (TriFTS) 132
Tuhin Poddar, George G. Zaimes, Saurajyoti Kar, Devin M. Walker, Troy Hawkins

390f Material Flow Analysis of Bioplastics Manufacturing Using Agricultural Residues in US for Pharmaceutical Packaging 133
Apoorva Bademi, Shweta Singh

390g Technoeconomic Analysis and Life Cycle Assessment of Bio-Based Non-Isocyanate Polyhydroxyurethane and Polythiourethane Production with Monomer Recovery 135
Yasheen Jadidi

NOVEL APPROACHES TO CO_2 UTILIZATION I

514b Highly Efficient System for Direct Air Capture and Conversion of CO2 to Value-Added Chemicals 137
Codruta Loebick, Christian Junaedi, Subir Roychoudhury

514c Catalytic Conversion of CO_2 Over Cu and Ni Metals Supported on CeO_2 and La_2O_3 Via Reverse Water Gas Shift Reaction 139
Parisa Ebrahimi, Anand Kumar, Majeda Khraisheh

514d Life Cycle Analysis Tools to Evaluate Environmental Footprint of Carbon Conversion Projects 147
Sheikh Moni, Michelle Krynock, Joseph Chou

514e Catalytic Conversion of NAS-CO2 to Chemical Feedstock 149
Moumita Bhattacharya, Marty Lail, Daniel Mogollon

514f CO_2 Utilization for Production of Fuels and Chemicals Using Ereact™ 150
Peter M. Mortensen, Marené Rautenbach, Martin Oestberg, Kim Aasberg-Petersen

514g CO_2 Utilization Through Power-To-Methanol with High Temperature Heat Pump 151
Nicholas Badger, Rahim Boylu, Valentine Ilojianya, Mustafa Erguvan, Stephen Ritchie, Shahriar Amini

514h Solar Fuel Production Via Germanium Oxide-Based Thermochemical Conversion of CO2 152
Rahul Bhosale, Shelby Adams, Zachary Allen, Gabrielle Bennett, Edvinas Berezniovas, Taylor Bishop, Michael Bonnema, Sequoia Clutter, Ryan Fagan, Jordan Halabrin, Mason Hobbs, Daniel Hunt, Miguel Ivarra, Mattigan Jordan, Pooja Karunanithi, Julianna McReynolds, Valerie Ring, Sam Smith, Jonathan West

SUSTAINABLE APPROACHES FOR CHEMICAL PRODUCTION

548g Unifying CO2 Capture and Utilization Through Process Modeling: Sensitivity Analysis of Process Performance on Bifunctional Materials in a Single Bed Reactor 164
Abdullah Al Moinee, Ali Rownaghi, Fateme Rezaei

548a Techno-Economic Analysis and Life Cycle Assessment of Hydroxylamine Eco-Manufacturing Via Wastewater Electrochemical Reduction 165
Manish Mosalpuri, Mark Mba Wright, Wenzhen Li

548b Moving to a Zero Carbon Emissions World While Managing Key Transitions in the Food and Energy Supply Chains. 167
John Starr

548c Process Simulation Coupled with Economic, Energy and Environmental Indicators as an Eco-Design Tool for Sustainable Products 168
Andrea Mio, Elena Barbera, Alessandro Massi Pavan, Alberto Bertucco, Maurizio Fermeglia

548d Towards the Development of Safe and Sustainable by Design Chemicals, Materials and Products 169
Dimosthenis Sarigiannis, Fotini Nikiforou, Achilleas Karakoltzidis, Antonis Gypakis, Spyros Karakitsios

548e Electrochemical Systems and Machine Learning for Cost-Effective Wastewater Treatment to Maximize Renewable Energy 172
Abdullah Maghfuri Sr., Manish Mosalpuri Sr., Wenzhen Li Sr., Mark Mba Wright

548f Grain Storage Goes Green: Innovations in Sustainable Processes and Products 173
Vinod Amar

POSTER SESSION: SUSTAINABILITY SCIENCE AND ENGINEERING, BIOREFINERIES, AND ENERGY

86a Selective and Efficient Valorization of Industrial Wastes Via Carbon Mineralization 174
Ning Zhang, Danny Huang, Aaron Moment, Ah-Hyung Park

86b Unique Synergies Between Carbon Mineralization and Biomass Conversion: Alkaline Thermal Treatment for Combined Carbon Capture and Hydrogen Production 175
Jonah Williams, Ah-Hyung Park

86c Evaluation of Bioenergy Production from Advanced MFC-Fermentation Technology 176
Kundan Kumar, Ming-Hsun Cheng

86e Chemical Sector Risk Management Agency Resources in Support of a Sustainable and Resilient Chemical Sector 177
Janine Mason, Ashley Pennington, Cheryl Louck, Jacob Mehl

86g Landfill Gas Converted into Renewable Natural Gas: A Case Study in Texas 178
Helen Lou, Elizabeth Hastings, Jian Fang, Jessica Ledwith, Samantha Adelan

86h Synthesis and Tribological Properties of Bio-Based Lubricants from Vegetable Oils 179
Paulo Roberto C. F. Ribeiro Filho, Francisco Murilo Tavares De Luna, Celio Cavalcante Jr.

86i Materials for Direct Air Capture of CO_2 Via Particle Molecular Layer Deposition 180
Hailey Loehde-Woolard, Bergen Evans, Kent J. Warren, Alan Weimer

86j Integration of Sustainable Extraction and Recovery of Energy-Relevant Metals Using Aldoxime-Based Ligands Systems from Unconventional Resources 182
Amanda Whai Shin Ooi, Hunter B. Vibbert, Ah-Hyung Park

86k Guidelines for Designing Amine-Based CO_2 Capture Systems Onboard LNG Fuelled Ships 183
Anikesh Kumar, Preethi Sridhar, Farooq Shamsuzzaman, Iftekar Karimi

86m A Novel Integration of Solar Chimney Power Plant with Wind Farm for Green Hydrogen and Distilled Water Production 185
Fares Almomani, Jihad Eriekat, Hamza Alnawafah, Emad Abdelsalam

86n Electrochemical Performance of Hybrid Ferrite/(Mn, Ti)-Oxide Electrode for Asymmetric Supercapacitor 186
Khang Huynh, Bharathkiran Maddipudi, Rajesh Shende

243c System Level Analysis of Intermitent Photocatalytic Processes for the Production of Liquid Fuels 187
Luisdomingo Guzman, Juan Manuel Restrepo-Florez

PROCESS DESIGN: INNOVATION FOR SUSTAINABILITY

399a Planning of Electricity Unit Commitment in Synergy with Nature's Ability to Mitigate Carbon Dioxide and Criteria Air Pollutants 188
Xinyu Zhen, Bhavik Bakshi

399b Economic and Environmental Analyses of Optimal Integrated Processes to Produce Chloromethanes Production from Wastes. 191
Diego Santamaria, Borja Hernández, Mariano Martin

399c Simulation of the Conversion of Japanese Waste Biomass to Light Olefins: A Comparative Revision of the Effect of Biomass Chemical Composition in the Yield of Olefins in PRO/II 193
Gabriel Talero, Yasuki Kansha

399d Process Operability Analysis of the Recovery of Rare Earth Elements from Coal Fly Ash 195
Beatriz Dantas, Deniz Talan, Fernando Lima

399e Evaluation of Polyol Ester Oils as Absorbents in a Hybrid Absorption Refrigeration Cycle 197
Aaron Liebold, Paul Yelvington, David Sykes, Ted Amundsen

399f Spectroscopic Monitoring of Dissolved Carbon Dioxide for Ccus 219
Richa Sharma, T. S. Ramakrishnan

REACTOR ENGINEERING FOR BIOMASS FEEDSTOCKS

466a Effective Combination of Reactors for Biomass Feedstock for High-Efficiency Power Generation from Manure 220
Yukihiko Matsumura, Mengli Zhang, Takayuki Ichikawa, Wookyung Kim, Yutaka Nakashimada, Keiya Nishida

466b Ash Adhesion on a Metal Mesh During Combustion of Bamboo Powder with Inorganic Additives 221
Ken-Ichiro Tanoue, Nao Umehara, Masayuki Fujita Masayuki Fujita, Yoshihiko Ninomiya Yoshihiko Ninomiya

466c Exhaust Gas Characteristics and NOx Formation Mechanism on Spark Ignition Engine Using by Bio-Syngas 222
Hossain Zakir, Naoya Kishi, Yukihiko Okumura

466d Bio-Derived Gamma-Butyrolactone Production from Succinic Acid with Bio-Derived Solvents 223
Nanami Hirano, Ryu Ukawa-Sato, Ryuta X. Suzuki, Joerg Freitag, Hiroshi Hori, Masafumi Kono, Chihiro Fushimi

RESILIENT AND SUSTAINABLE SUPPLY CHAINS AND PRODUCT SYSTEMS

428a Indicator-Based Supply Stability Analysis of Pharmaceutical Excipients: Method and Application 226
Kosuke Nemoto, Yusuke Hayashi, Hirokazu Sugiyama

428b Ecological Network Analysis and Information Theory to Evaluate the Performance of Circular Economy Designs 227
Farshid Nazemi, Brian D. Fath, Bhavik Bakshi

428c Design of Resilient Supply Chain Networks for Pharmaceutical Products 229
Arun Sankar Eenhakkattu Mana, Smrithi Prakash, Venkat Venkatasubramanian

428d An Agent-Based Simulation Model Towards Mitigating Domino Effects of Maritime Disruptions in LNG Supply Chains 230
Venkataramanan Swaminathan, Rajagopalan Srinivasan

428e Midstream Infrastructure Cybersecurity, Physical Security and Safety 233
Helen Lou, Tianxing Cai

428f Chemical Sector Risk Management Agency Resources for Enhanced Supply Chain Resilience and Security 234
Ashley Pennington, Cheryl Louck, Janine Mason, Jacob Mehl

SUSTAINABLE ENERGY: GENERATION AND STORAGE

565b Low Desolvation Energy Tailored Electrolyte for Enhanced Subzero Temperature Performance of Lithium-Ion Batteries 235
Ethan Adams, Jason Packard, Vilas Pol

565c Solid State Electrolytes with Low-Temperature Plasticizer Optimization 236
Nestor Levin, Annie Sun, Ethan Adams, Soohwan Kim, Vilas Pol

565e Can Fusion Help Decarbonize the Power Sector? 238
Amanda Farnsworth, Emre Genã§er

565f Electricity Generation from Hydraulic Fracturing Wastewater Using a Reverse Electrodialysis Stack 240
Arash Emdadi, Jamie Hestekin, Lauren F. Greenlee, Bruce E. Logan

565g Theoretical Analysis of Salt Selection for a Reverse Electrodialysis System 241
Arash Emdadi, Jamie Hestekin, Lauren F. Greenlee, Bruce E. Logan

SUSTAINABLE MANAGEMENT AND USES OF POST-CONSUMER MATERIALS AND WASTE

444b Rare-Earth Elements Recovery from Electronic Waste: Techno-Economic and Life Cycle Analysis 242
Esther Sanchez Moran, Mark Mba Wright

444d Development of a Predictive Model for in-Situ Resource Utilization of Space Exploration Waste 243
Ray Pitts, Joel Olson, Anne Meier

444e Optimization of Advanced Oxidation Processes for Removal of 1,4-Dioxane from Water Influents 244
Serenity Broussard, William E. Holmes, Rafael Hernandez, Daniel Gang, Mark E. Zappi

444f Respirator Cartridge Performance Testing for Chemicals of Potential Concern 245
Satish Nune, Michael Minette, Angela Melville, Stephen, D Davidson, Thomas Brouns, Michael Zabel, Eugene Morrey, Kevin D. Miller, Kenneth J Way

DESIGN FOR CIRCULAR ECONOMY

286a Designing Chemical Value Chains for Net-Zero Emissions and Nature Positive Decisions 246
Amrita Sen, Ying Xue, Bhavik Bakshi

286b Material Flow Assessment and Circularization of Waste in the Acetaminophen Production Network Through Integration of a New Industrial System 249
Abhimanyu Raj Shekhar, Shweta Singh

286c Developing a Nitrogen Circular Economy by Recovering and Utilizing N_2o as a Feedstock for Phenol Production 251
Chayse Lavallais, Jennifer Dunn

286d Integrated Anaerobic Membrane Bioreactor Platform for Resource Recovery from Swine Wastewater- Evaluation of Life Cycle Economic and Environmental Impacts 252
Prathap Parameswaran, Madison Kratzer, Vikas Khanna

286e Towards a Circular Economy for PET Bottles in the US - a System-Dynamics Approach 253
Tapajyoti Ghosh, Alberta Carpenter

286f Supply Chain Design and Optimization for Plastics at the End-Of-Life: City-Based Study 254
Baibhaw Kumar, Natalia Cano-Londoño, Gerardo Ruiz-Mercado, Heriberto Cabezas, Csaba Deak

286g Computational Solvent-Based Extraction of Additives for Polymer Recycling and Upcycling 255
Pooja Bhalode, Sean Najmi, Pavel Kots, Montgomery Baker-Fales, Dionisios Vlachos

ENGINEERING GEOLOGIC CARBON DIOXIDE STORAGE SYSTEMS II

484a Relative Permeability of $scCO_2$ and Brine in Core from Potential CO_2 Reservoirs 256
Dustin Crandall, Thomas Paronish, Magdalena Gill, Johnathan Moore, Scott Workman

484c CCS-Analytics: Application of Artificial Intelligence for Carbon Dioxide Storage Systems 257
Shahab Mohaghegh, Vyacheslav Romanov, Guoxiang Liu, Derek M. Vikara

484d Enabling Gigaton-Scale CO_2 Mineralization and Utilization in Reactive Reservoirs 259
Ruoshi Cao, Katherine Muller, Quin Miller, Allie Nagurney, Emily Nienhuis, Eusebius Kutsienyo, Pejman Rasouli, Herbert T. Schaef

484e Geologic Characterization and Fluid-Flow Modeling for Commercial-Scale CO_2 Injection and Storage in the Mt. Simon Sandstone Storage Complex 260
Oladipupo Babarinde, Fang Yang

484f Benefits of Injecting Carbon Dioxide into Coal Seams 262
John Pope

484h Imaging the Dissolution of Chalk During HCl Injection 263
Peter Winkel Rasmussen, Anders Nymark Christensen, Nicolas Bovet, Benaiah Anabaraonye

DESIGN, ANALYSIS, AND OPTIMIZATION OF SUSTAINABLE ENERGY SYSTEMS AND SUPPLY CHAINS II

438a Roadmap Towards Zero-Emission Shipping: Global Long-Term Planning of Green Corridors 265
Hanchu Wang, Prodromos Daoutidis, Qi Zhang

438b the Role for Synthetic and Biofuel Pathways in an Integrated, Net-Zero Energy System...................... 266
Jun Wen Law, Dharik Mallapragada

438c Multi-Period Joint Capacity Expansion Planning of Renewable Electricity and Hydrogen Production: An Application to South Korea 267
Yechan Choi, Jay Hyung Lee, Seongmin Heo

438d Spatial Trade-Offs in Designing a Sustainable Cellulosic Biofuel Supply Chain: Insights from High Resolution Modelling 269
Eric O'Neill, Caleb Geissler, Christos Maravelias

438e Industry Perspective on Green Ammonia: Current Status and Challenges in Synthesis, Transportation and Utilization 270
Shantanu Patwardhan, Kiran Chaudhari

438f Comprehensive Analysis of Various Types of Hydrogen Refueling Stations : Economics, Environmental, and Safety 271
Ga-Young Lee, Joo-Sung Lee, Chul-Jin Lee

438g Economic and Environmental Potential of Green Hydrogen Carriers (GHCs) Produced Viareduction of Amine-Captured CO_2............................ 272
Changsoo Kim, Younggeun Lee, Hyunjoo Lee, Ung Lee, Kyeongsu Kim

NOVEL APPROACHES TO CO_2 UTILIZATION II

533a In-Situ CO_2 Capture-Conversion in Methane Dry Reforming for Direct Synthesis of Hydrogen Over Bifunctional Structured $Ce_{1-X}Co_x$nio/Ca perovskites-Type Oxides Monoliths...................... 274
Khaled Baamran, Shane Lawson, Ali Rownaghi, Fateme Rezaei

533b Tandem CO2 Electrolysis System for Chemical Production 275
Feng Jiao

533c Production of Carbonates from Industrial Waste Via Carbon Mineralization Towards a Circular Economy............................ 276
Ning Zhang, Danny Huang, Aaron Moment

86f Techno-Economic and Life Cycle Assessment of Reactive CO2 Capture to Renewable Natural Gas 277
Alvina Aui, Hannah M. Goldstein, Nathan C. Ellebracht, Wenqin Li, Simon H. Pang

533f Feasibility Study of Alkaline Metal Oxide Modified Cu-ZnO-Al_2O_3 for Reactive Capture of CO_2 to Methanol 278
Chae Jeong-Potter, Martha A. Arellano-Treviño, Wilson McNeary Iv, Anh To, Daniel Ruddy

533g Accelerated CO2 Mineralization of Acid Mine Drainage Waste 280
Hamid Radfarnia, Kourosh Zanganeh, Laleh Dashtbankenari, Sanaz Mosadeghsedghi, Konstantin Volchek

CONCENTRATED SOLAR POWER GENERATION AND CHEMICAL PROCESSING II

264a Dynamic Model of a Solar Fuel System for the Thermochemical Production of Syngas from H_2o and CO_2 281
Remo Schäppi, Aldo Steinfeld

264b Continuum Model for Transport Phenomena and Thermochemical Reactions in Oxidation Reactors for High-Temperature (>1000 °C) Heat Extraction 282
David Korba, Michael Hayes, Philipp Schimmels, Kelvin Randhir, James F. Klausner, Nick Auyeung, Like Li

264c Thermal Energy Transport Using the Sorption-Assisted Boudouard Reaction 283
Alexandre Yokochi, Riley Choquette, Benjamin Phillips, Annette Von Jouanne

264d Heat Transfer in Directly-Irradiated High-Temperature Particle–Gas Flows for Solar Particle Receiver Applications 285
Jingjing Chen, Apurv Kumar, Joe Coventry, Wojciech Lipinski

264e Challenges in Using Particulate Materials as Heat Transfer Fluid Or Reactant in Solar-Chemical Processes 286
Martin Roeb

264f Development and Testing of a Novel Combined Moving-Fluidized Bed Oxidation Reactor for Efficient High-Temperature Discharge of Thermochemical Energy Storage Particles 288
Juvenal Ortiz-Ulloa, Lucas Freiberg, Fuqiong Lei, Owen Ramsey, David Korba, Michael Hayes, Philipp Schimmels, Kelvin Randhir, Like Li, Joerg Petrasch, James F. Klausner, Nick Auyeung

CONCENTRATED SOLAR POWER GENERATION AND CHEMICAL PROCESSING III

325a Experimental Demonstration of Heat Recovery in a Solar Redox Reactor with an Integrated Thermal Energy Storage 291
Alon Lidor, Sven Hellmüller, Philipp Haueter, Aldo Steinfeld

325b Experimental Dataset and Numerical Model Validation of a Lab Scale Solar Volumetric Receiver for High Temperature Industrial Heating 292
Aysha Melhim, Fathya Salih, Insha Shaikh, Konstantinos Kakosimos

325c Solar-Driven Biomass Chemical Looping Gasification for Production of Syngas and H_2: Thermodynamic Analysis and Proof-of-Concept Experiment 293
Dequan Xu, Bo Wang, Xian Li, Yanjun Dai, Chi-Hwa Wang

325d Techno-Economic Assessment of Heat Integration from Concentrated Solar Thermal Energy into Solid Oxide Electrolysis for Green Hydrogen Production 294
Martin Janse Van Vuuren

325e Self-Controlled Electric Power Supply System from Solar Cells Using Hydrogen Storage 296
Satoshi Wada, Katsuhiko Tsuno, Takeharu Murakami, Takayo Ogawa, Katsushi Fujii

325f Potential of Monetizing Stranded Natural Gas Through Process Intensification and Green Energy 306
Abdullah Al-Aboosi, Fadhil Al-Aboosi, Mahmoud El-Halwagi, Wei Zhan

Author Index

105a Assessment of the Presence of Polychlorinated Dibenzodioxins and Furans (PCDD/Fs) in Flue Gas on Microalgal Growth and Nutritional Value

Mahasweta Paul, Raymond Lau

Microalgae is known to be one of the most sustainable food sources for the future and have gained much attention as a food supplement and replacement. They can be "harvested" in large quantities perennially and can minimally provide three-fold higher protein yield than plant-based protein sources, including wheat, pulse legumes, and soybean. Their growth rate can be increased by feeding high volumes of CO_2 to promote photosynthesis. Waste-to-Energy (WTE) incinerators are the most concentrated and readily available source of CO_2 (~12%) and can be used for microalgae cultivation. Microalgae has been reported to capture 1.83 kg of CO_2 / kg biomass produced, showing promising prospect in carbon fixation. The biomass produced can then be fed into various manufacturing processes. WTE incinerator power plants utilize municipal solid waste as a fuel source for power generation, reducing the dependence on fossil fuels, while diverting wastes from landfills, indirectly reducing CH_4 emission. However, incineration of such waste, which often includes significant quantity of plastics and other halogenated compounds, will produce PCDD/Fs. Therefore, to economically use CO_2 from WTE flue gas for microalgae production, assessment of the presence of the dioxins on the microalgae is an important consideration. *Chlorella sp.* (a freshwater species) is chosen for the work due to its extremely high protein content (50-60%). To facilitate the understanding, Chlorella is exposed to different concentrations of the dioxins, followed by the analysis of change in several physiological parameters such as cell growth, cell morphology, nutritional content of protein (amino acid profiles), lipids and carbohydrates as well as Chlorophyll a. These measurements would then be used as the evaluation criteria to provide a more comprehensive view on what steps need to be included or improved for the enhanced scrubbing of the PCDD/Fs from the incinerator flue gas such that a sustainable food production cycle from waste can be generated, creating a self-sustaining ecosystem.

105c Thermal Control of an Enclosed Photobioreactor (PBR) for Algae Cultivation

Tolga Pirasaci, George Philippidis, Aydin Sunol

Microalgae and cyanobacteria are promising sources of not only biofuels but also many nutraceutical, cosmetic and pharmaceutical value-added bioproducts, thanks to their rapid growth rates and significantly higher oil yields compared to terrestrial oil crops.

Closed photobioreactors (PBRs) and open channel pools (RWPs) have been used extensively for growing microalgae. Although closed PBRs have significantly higher biomass volumetric productivity compared to open RWPs, they have several disadvantages that limit their use on a commercial scale.

Temperature control is one of the most important factors. For the proper production the temperature inside the PBR should be kept constant despite changing climatic conditions. Various solutions – using heat exchangers, direct immersion of the PBR in a thermo-regulated pool and water spraying on the PBR surface - have been suggested in the literature.

This study focused on water spraying technique. In this study numerical heat transfer model of a closed PBR was developed and validated with the experimental data. The effect of the operation parameters (cooling system on and off conditions, mass flow rate etc) on the PBR temperature was investigated during a year for the various cultivation periods.

The findings were consistent across pilot-scale reactor simulations. However, caution should be exercised when using the presented data for scaling purposes, as commercial scale experimental data must be collected and fitted to our numerical model.

105d Elucidating the Extraction of Fucoxanthin from Algae with Supercritical Carbon Dioxide and Ethanol as a Cosolvent for the Valorization of the Integrated Biorefinery

Mary Kate Lane, Elise B. Gilcher, Paul T. Anastas, Julie B. Zimmerman

Integrated biorefineries that produce a variety of chemicals, fuels, and products from renewable biomass sources are key in the move towards a sustainable future not dependent on fossil fuels. The integrated biorefinery's success is tied to its economic viability, creating high-value products along with lower-value biofuels. This work examines the extraction of high-value nutraceuticals from algae biomass as a way to economically enable the integrated biorefinery using green chemistry approaches. Green solvents, such as supercritical carbon dioxide and pressurized carbon dioxide with ethanol as a cosolvent, are deployed to extract high-value nutraceuticals such as fucoxanthin from the algae biomass, Phaeodactylum tricornutum. There has been growing interest in fucoxanthin as a nutraceutical from algae sources, but experimental studies do not agree on an optimized condition for maximum fucoxanthin recovery in extractions with supercritical carbon dioxide or pressurized carbon dioxide with ethanol as a cosolvent. This work analyzes the proposed optimal conditions for fucoxanthin extraction in current literature and statistical trends are examined for variables that increase fucoxanthin yield. Statistical analysis indicates that higher polarity achieved by either higher operating pressures in supercritical carbon dioxide or the addition of ethanol as a co-solvent correlates with higher yields of fucoxanthin, but it is difficult to untangle the effect of the algae (as many different species of algae are used, both macro- and microalgae) and other variables from these studies. Here, the solubility of fucoxanthin in supercritical carbon dioxide and the role of ethanol as a co-solvent in extracting fucoxanthin is fully investigated through extractions on pure fucoxanthin standard, which are fit using the Chrastil model of solubility in supercritical carbon dioxide. These experiments are then compared with extractions on the microalgae, Phaeodactylum tricornutum, to elucidate the effect of the algae matrix on fucoxanthin recovery. In microalgae extractions, a significant finding was that in mildly-polar high pressure (50 MPa) supercritical carbon dioxide extractions, there was evidence of selectivity of the desired product, fucoxanthin, over the undesired polar pigment co-product, chlorophyll. This selectivity was not seen in more polar extractions with ethanol, indicating an additional benefit of using supercritical carbon dioxide solvent. This study aims to impart a greater understanding of the extraction of the high-value nutraceutical, fucoxanthin, in order to enable the integrated biorefinery.

105e Techno-Economic Analysis of an Integrated Approach to Produce Bio-Methanol from Micro-Algae in the UAE

Nasir Al Lagtah, Nizar Ashna

Global economic growth and energy demand go hand in hand. With future energy demands predicted to rise between 17% to 50% by 2040, heavy reliance on fossil fuels to meet these demands can pose drastic consequences, where the global atmospheric temperature is likely to rise above 3.6 °C, causing detrimental impacts on ecological communities and alarming rise in greenhouse gas emissions. This necessitates development of renewable and sustainable energy technologies. To control and limit the effects of climate change the Paris Agreement, established during the 21st Conference of Parties (COP21), and the United Nations Framework Convention on Climate Change (UNFCCC) encourage the application of bio- based energy sources.

Micro-algae is one of the most promising non-food crop feedstocks. Due to its ability to be processed into both bioenergy and valuable co-products, and to fix CO_2 using sunlight at ten times terrestrial plant efficiency, micro-algal cultivation is one of the most promising processes of carbon capture today. The United Arab Emirates has a sizable potential for algal biofuel production owing to its vast deserts, long shorelines, year-round sunny days, flue gas, seawater and wastewater availability, and the wealth of native adapted algal species with salinity tolerance twice as high as that in the gulf waters. As UAE is hosting the upcoming UN Conference of Parties (COP28), more research is required to investigate the potential of micro-algae production and conversion.

In this proposed research, an integrated approach is taken to produce bioenergy from micro- algae, by modelling bio-methanol production through gasification of biomass using Aspen Plus. The key stages of the process are gasification, syngas post treatment, methanol synthesis followed by purification. The developed model performance is then evaluated for three different types of micro-algal species: (1) *S. Acuminatus* (2) *N. Oculata* and (3) *S. Almeriensi*. The performance of the micro-algal species was also compared with a plant biomass (pine). The simulation results were found to be comparable to experimental data available in literature.

Amongst the three feedstocks, S. *Acuminatus* was found to produce the greatest amount of bio-methanol and the greatest HHV value under identical conditions. A sensitivity analysis was conducted to evaluate the optimum gasification temperature for obtaining maximum bio-methanol yield, for each of the feedstocks. The optimum temperature was found to be 860; 880; 860 and 840 °C for S. *Almeriensi*, N. *Oculata*, S. *Acuminatus*, and Pine Biomass respectively.

This model was then used to evaluate the economic feasibility of the proposed plant. The profitability indicators used in this case were IRR (%), ROI (%), NPV and Payback time. Since obtaining specific raw material costs for each micro-algal species was difficult, a general feedstock of *Chlorella vulgaris* is considered. A sensitivity analysis was conducted for wide range of bio-methanol selling prices and two *Chlorella vulgaris* feedstock prices ($269 and $45 per ton). The average bio-methanol selling price is $770 per ton. However, when applied as a shipping fuel, the cost of bio-methanol increases manifold to $3,336 per ton. Hence, the best scenario is when methanol is sold at the bunker prices. An IRR of 525% was recorded even when the raw material price was as high as $269 per ton.

The worst case was identified to be when the bio-methanol price was $693 per ton. Minimum profitability was obtained with a payback time of 14 years (without 3-year construction) with a plant lifespan of 20 years. An IRR of only 2.6 % was obtained.

Considering the standard average price of \$770 per ton, an ROI of 2.6% and 13.2% was noted when raw material prices were \$269 and \$45 per ton, respectively. Therefore, the plant can be classified as feasible.

Given the current price of conventional methanol however, the bio-methanol is not yet commercially competitive. But, with emergence of the new legislature and regulations, the current focus on decarbonization and looking for more sustainable fuel alternatives, this should gradually reverse the trend. The sudden boom in the bio-methanol price for use as a shipping fuel for example is a sign pointing in this direction. Keeping sustainability at its core, biofuels are directly in line with the sustainable development goals of the United Nations and the commitments of COP27.

105f Maximizing the Biofuel Production Form Algae Using Four Element Strategy and Ultimate Economic Benefits

Fares Almomani, Ahmed M.D. Al Ketife

The Four Element Strategy (FES) was used to maximize biofuel production from algae strains and assess the associated economic benefits. The FES is based on considering the type of algal strains, concentration of input CO_2 ($C_{c,g}$), light intensity (L_I) energy, and nutrients requirements (Nitrogen (N) to phosphorus (P) ratio, $R_{N/P}$) achieve the maximum benefits . Two algal strains was used in the current study, encompassing slain algae *Nannochloropsis sp.* (*Ns*), and fresh water strain *chlorella vulgaris* (*Cv*). Response Surface Methodology (RSM) and Box Behnkin Design (BBD) combined with the biokinetic model was used to investigate the relation between $C_{c,g}$ in the range0.03-20 %, $R_{N/P}$ in the 1- 11) and L_I 100-400 μE m^{-2} s^{-1}) and the corresponding CO_2 gas fixation rate (R_{CO2}, g L^{-1} d^{-1}), biomass accumulation (X, g L^{-1}), and algal growth rate (μ, d^{-1}). The results revealed that the optimum productivity conditions are; L_I 100-270 μE m^{-2} s^{-1}, $C_{c,g}$ 0.03-15%, $R_{N/P}$ 3.5 -11. The corresponding to a maximum R_{CO2} , X and μ were 1 g L^{-1} d^{-1},.6 g L^{-1}, and 1.6 d^{-1}, respectively. The optimized conditions using the desirability function support that a R_{CO2} of 1.2 g L^{-1} d^{-1}, X of 1.8 g $L^{-1}$and μ of 1.7 d^{-1}, when the $R_{N/P}$,Cc,g , and L_I were 11, 4.5% and 200 μE m^{-2} s^{-1}respectively. The N and P uptake rates for *NS* were reported at 85 mg L^{-1} d^{-1} and 17.1 mg L^{-1} d^{-1} compared with only 60.15 mg L^{-1} d^{-1} and 2.1 mg L^{-1} d^{-1} for *Cv,* respectively. The maximum B_{YP} of 1.47 g L^{-1} d^{-1} was achieved based on the optimum MCT conditions of $C_{c,g}$ in the range 4-16%); L_I in the range 200-250 μE m^{-2} s^{-1}, and $R_{N/P}$ in the range 8-11 respectively. The total cost required to produce 1 L of biofuel was found to be 0.629 $, and the required cost to cover this cost is 0.138 $ and 0.375 $ for *NS* and *Cv,* respectively. The current study provide the guide way for large-scale process for MCT.

44b Green Ammonia for Sustainable Energy and Agriculture

Prodromos Daoutidis

Ammonia is the basis of synthetic nitrogen fertilizer production and is thus the backbone of modern agriculture. Its large-scale production using the classic Haber-Bosch process, with hydrogen obtained from fossil fuels, is responsible for 1-2% of both global energy consumption and CO 2 emmissions. Reducing the carbon intensity of ammonia production is critical to a more sustainable agricultural system. Furthermore, ammonia's relative ease of storage and transportation (compared to hydrogen), and its potential use as a power/heat generation or transportation fuel with no associated carbon emissions at the point of use, give it considerable promise as a sustainable energy vector, provided it can be produced in a low emissions manner. For these reasons, green ammonia, i.e., ammonia synthesized from hydrogen obtained from water electrolysis and nitrogen obtained from air using renewable electricity, is currently
receiving world-wide attention from governments, industry, non-governmental organizations, and private investors. Green ammonia requires more distributed siting and smaller-scale production facilities than traditional ammonia due the inherently localized nature of renewable resources and the relative lack of economies of scale for electrolysis technologies. Optimal design and deployment are critical factors for its economic competitiveness.

In this talk, I will describe recent advances in the production, deployment, and utilization of green ammonia that address these challenges. These include:

* The concurrent design and scheduling of ammonia synthesis processes that operate dynamically, as a framework for accounting for the intermittency of the main feedstock (wind or solar irradiance) to minimize the levelized cost of production.

* A supply chain optimization framework for optimal placement of distributed, modular, ammonia production facilities to meet local fertilizer demands.

* The optimal design of a green ammonia-based system where ammonia is used as fertilizer, grain drying fuel, tractor fuel, and energy storage medium on a farm, as an illustration of the potential benefits from the synergistic use of ammonia across multiple sectors.

* The optimal design, scheduling, and operation of islanded renewable microgrids which utilize ammonia for energy storage.

The results I will discuss document the transformative potential of green ammonia toward sustainability in the agriculture and energy sectors. I will also address some outstanding challenges on deploying green ammonia production at scale, as well as adopting its utilization across multiple sectors.

44a Machine Learning Boosted Catalyst and Operation Design for Clean Energy

Fanglin Che

Developing advanced theory, modeling, and data science approaches for understanding field-enhanced catalysis has extensive applications in clean energy technologies, particularly for plasma catalysis, electrostatic catalysis, interfacial electrocatalysis, programmed catalysis, and fuel cells. Theoretically determining and quantifying the physics and chemical understanding of electrified interfacial structure and field-dipole interactions on controlling the activity and selectivity of chemical processes and then integrating these physics/chemical rules to establish deep collaborations between interpretable, physics-informed machine learning and electrified interfacial chemical processes is crucial for rationally designing these electrified modular systems for energy storage and sustainable chemical production.1,2 This presentation will focus on several examples, including organic-inorganic interface and its impact on electrocatalysis of carbon dioxide3, field-dipole interaction effects on sustainable ammonia synthesis,1 non-thermal plasma catalysis for ammonia decomposition and synthesis and so on.

Reference

1. M. Wan, H. Yue, J. Notarangelo, H. Liu, F. Che*, "Deep-Learning Assisted Electric Field-Accelerated Ammonia Synthesis", JACS Au, Accepted, 2022, ACS Editor's Choice.

2. Mou, T.; Pillai, H. S.; Wang, S.; Wan, M.; Han, X.; Schweitzer, N. M.; Che, F.; Xin, H., Bridging the complexity gap in computational heterogeneous catalysis with machine learning. Nat. Catal. 2023, 6 (2), 122-136.

3. Wan, M.; Gu, Z.; Che, F., Hybrid Organic-Inorganic Heterogeneous Interfaces for Electrocatalysis: A Theoretical Study of CO2 Reduction to C2. ChemCatChem 2021, (14), e202101224

44c Establishing a Circular Economy in the Food Industry

Mariano Martin

With the increase in the generation of waste from the daily operation of society in the production of food, process systems engineering offers the tools to evaluate the optimal management of the residues and the recovery of added value products to develop and
implement a circular economy. We will focus on three types of products, added -value products, utilities and fertilizers. A methodology is presented [1] to identify the product or set of products to recover and applied to various cases of interest such as the coffee [2] , olive oil [1] , wine [3] and orange industries [4] , as well as the production of sweeteners [5] . In addition, we will delve into the production of fertilizers to close the production cycle [6] . The biorefinery problem is extended into the process and product design problems to select the best biomass to produce a product or a set of products. In this way it is expected to improve the economic, environmental, and
social aspects simultaneously. Interesting products are not only produced to improve the economy of plants, such as food supplements from waste, but also with a view to improving the energy efficiency of production plants through the production of essential services for their operation, water vapor and electricity. Thus, highly integrated facilities [7] that using only waste are designed to avoid the use of external chemicals and limit the need for external utilities towards the production of added value products and commodities, creating a circular economy within the process. The concept is extended towards a circular economy for the food industry presenting a case of meat production within Spain. [8]

References:

[1] Guerras, L.; Sengupta, D.; Martín, M., El-Halwagi, M. (2021) Multi-layer approach for product portfolio optimization: Waste to added value products. Acs. Sust. Chem. Eng. 9, 18, 6410–6426

[2] Taifouris, M.; Corazza, M.; Martín, M., (2021) Integrated design of biorefineries based on spent coffee ground Ind. Eng. Chem. Res. 60 (1) 494-506

[3]. Taifouris, M., Martín, M (2023) Methodology for estimating a country's production of biomethane from agricultural residues: a multiscale and holistic approach Comp. Aided. Chem Eng. Vol 51 33rd European Symposium on Computer Aided Process Engineering (ESCAPE33), June 18-21, 2023, Athens,Greece Antonis Kokossis, Michael C. Georgiadis, Efstratios N. Pistikopoulos (Eds.) 2197-2202

[4].- Criado, A., Martín, M., (2020) Integrated multiproduct facility for the production of chemicals, food and utilities from oranges. Ind. Eng. Chem. Res. 59, 16, 7722-7731

[5] Galán, G.; Martín, M., Grossmann, I.E. (2021) Integrated Renewable Production of Sorbitol and Xylitol from Switchgrass. Ind. Eng. Chem Res. 60,15, 5558-5573

[6] Martín-Hernández, E, Martín, M., Ruiz-Mercado, G.J. (2021) A geospatial environmental and techno-economic framework for sustainable phosphorus management at livestock facilities. Resources. Cons. Recyl. 175 105843

[7] Roldán San Antonio, J.E., Martin, M (2023) Optimal integrated plant for biodegradable polymers production. ACS Sust Chem Eng. 11, 6, 2172–2185 10.1021/acssuschemeng.2c05356

[8] Taifouris, M., Martín, M. (2022) Integrating intensive livestock and cropping systems: sustainable design and location. Agri. Systems. 203, 103517

68a Study on Life Cycle Assessment and Optimization of Pesticide Production Process from the Perspective of Sustainable Development

Kexuan Yang

With the growing interest in sustainability management of the pesticide industry, coupled with the incompleteness of the life cycle assessment (LCA) framework in pesticides, it was imperative to conduct process-based life cycle assessment to support environmentally friendly optimization. Herein, the cradle-to-gate LCA of glyphosate production was executed as a case study to investigate its determinants and multiple environmental impacts during the production processes. Results showed that methanol had the highest environmental damage in toxicity with a normalized value of 85.7×10^{-8}, followed by coal-fired power in energy consumption with a value of 6.00×10^{-8}. Based on these results, the optimized LCAs were further performed including energy improvement and wastewater targeted conversion to reduce the environmental impact of the original process. Between them, the latter was more efficient with the normalized value decreasing by 21.10×10^{-8} while that of the former only decreased by 6.50×10^{-8}. Moreover, the reliability of the data was verified via sensitivity analysis (SA) and uncertainty analysis (UA). Between them, the results of SA showed a high linear correlation as the correlation coefficients were all above 0.98, while the UA demonstrated significantly higher accuracy since the coefficient of variations based on the Monte Carlo simulation were far lower than those in the existing pesticide-LCAs. This study ultimately provided an integrated LCA framework of pesticide production from upstream control, offer ideas for optimized measures based on the processes, and further facilitate waste resource utilization and sustainable development.

68b Life-Cycle Assessment Integration into Scalable Open-Source Numerical Models (LiAISON) for Prospective Impact Analysis of Novel Technologies

Tapajyoti Ghosh, Patrick Lamers, Alberta Carpenter

Decarbonizing the industrial sector is a significant challenge in achieving a net-zero greenhouse gas (GHG) emissions economy by 2050[1] and the Paris Agreement, i.e., a global climate change mitigation target of achieving a maximum average temperature change potential of 1.5°C or less by 2100 with respect to pre-industrial levels.[2] In the United States (US), the industrial sector accounts for 23% of total GHG emissions[3] and is home to a number of hard-to-electrify activities. The chemicals subsector has the single largest subsector emissions profile after direct emissions from fossil fuel combustion and leakage from fossil fuel distribution systems.[3] Within the chemicals subsector, many processes depend on hydrogen or ammonia precursors. Decarbonizing these two commodities would contribute significantly to decarbonizing the industrial sector as hydrogen could also be used for low carbon steel production (e.g., hydrogen-based direct reduction of iron) and other industrial applications.

Emerging technologies require the application of prospective life cycle assessment (LCA)[4], which can account for technology (foreground) scaling and process improvements via learning-by-doing, among others. In many cases, the future system context (background) in which the technologies are assumed to operate in is equally relevant.[5] Background scenarios generated by integrated assessment models (IAM) can coherently incorporate potential future dynamics of the energy-climate-human-land system. Further, IAM scenarios are harmonized across socioeconomic and climate change mitigation pathways[6], which facilitates the comparability of prospective LCAs using different IAMs.

We introduce an open source prospective LCA framework, the Life-cycle Assessment Integration into Scalable Open-source Numerical models (LiAISON), to analyze the non-linear relationships between technology foreground and the future energy system background across a series of midpoint and resource use metrics The integration of LCA and IAM data is achieved using prospective environmental Impact assessment (PREMISE)[7]. We showcase it by assessing two Power-to-Hydrogen (PtH_2) processes, namely Solid Oxide Electrolysis (SOE) and Polymer Electrolyte Membrane Electrolysis (PEME). We compare the technologies to a baseline of hydrogen production via natural gas-based Steam Methane Reforming (SMR) in a US context of multiple energy system and climate change mitigation futures. Besides providing an analysis that specifies the LCA results ranges with temporal and geospatial explicitness across the two technologies, metrics, and impact assessment methods, this research also aims to establish a base framework that can be expanded to use other IAM generated scenarios and US open-source life cycle inventory (LCI) databases. We find that the temporal environmental performance of either technology or their difference to SMR is directly influenced by the underlying background dynamics. Additionally we compare our results by linking two other prospective models with LiAISON - GCAM(Global Change Assessment Model) and ReEDS (Regional Energy Deployment System) to analyze the effect of changing background scenarios using varying predictions in life cycle analysis.

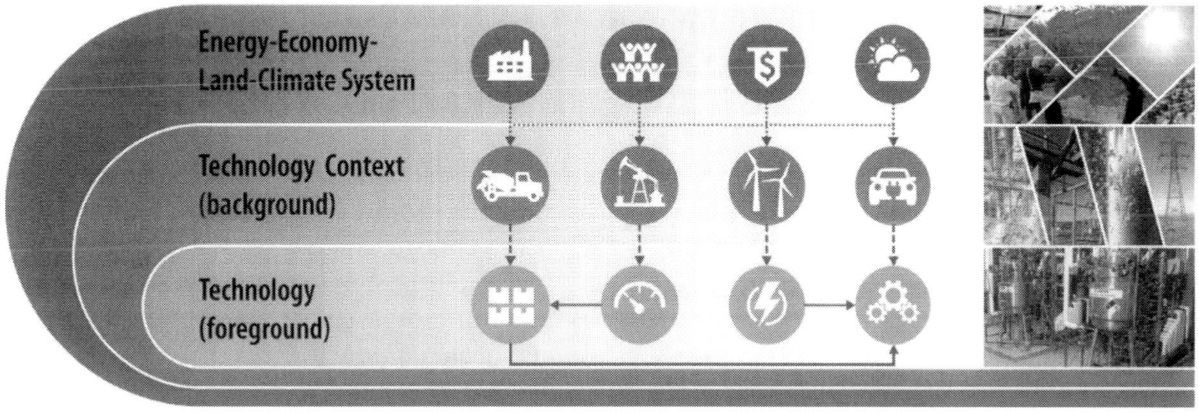
Energy-Economy-Land-Climate System
Technology Context (background)
Technology (foreground)

68c Guiding Food Colorant Production Via Fermentation Using Techno-Economic, Life Cycle and Absolute Sustainability Assessment

Samir Meramo, Sumesh Sukumara, Peter Fantke, Eleonora Passuto

In the past decade, biotechnology progress has been accelerated by higher investment flows to the growth of bio-based innovations. The impact becomes higher, considering that biotechnology has applications in the pharma, food, and chemical industries. This growth has resulted in more opportunities to develop bio-based products based on the biological transformation of renewable resources. As a result, many people take the sustainability of bio-based products for granted without proper sustainability assessment. This could lead to unexpected outcomes and innovations that face challenges towards commercialization. With this motivation, the authors developed the sustainability assessment toolbox (presented at the AIChE last annual meeting) to guide product innovation based on sustainability assessment from the inception point. The toolbox provides an efficient framework to identify sustainability hotspots and seek potential improvement areas and optimize the technology earlier.

The sustainability toolbox integrates process model, Techno-economic Assessment (TEA), and Life Cycle Assessment (LCA) to quantify the economic and environmental sustainability performance of biotechnology innovations. The LCA is a standardized tool to quantify environmental impacts over a product or technology life cycle. This tool estimates potential environmental impacts at different technology readiness levels. The toolbox uses two relevant features of LCA, which are the hotspots analysis, prioritizing improvement areas, and compare different alternatives, to identify best-in-class solutions. Similarly, TEA has been extensively applied in chemical process design to predict the economic viability of processes at different technology maturity levels. TEA results allow for benchmarking the financial performance of scaled-up innovations with respect to the standard process in industrial practice. Therefore, the toolbox uses the system perspective to overcome the limits of biomanufacturing or bioprocess level by inclusion of broader impacts (for instance, biomass production). The toolbox includes a combined TEA and LCA module to show the tradeoffs between process alternatives. It uses the single-score approach to provide a combined sustainability assessment score. In addition, the toolbox integrates the Absolute Environmental Sustainability Assessment (AESA) model to bring the relative environmental sustainability performance to absolute terms. Subsequently, the impacts are quantified using the Planetary Boundary-Life Cycle Assessment (PB-LCA) methodology. Applying the ASEA methodology requires allocating the safe operating space for the functional unit, where we first downscale the planetary boundaries to the individual level and then upscale it to the product, sector, or country level. Different sharing principles are evaluated to show the sensitivity of this parameter.

A case study based on fermentation-based colorant production is assessed using the sustainability assessment toolbox. The case study is a representative example of low-maturity biotechnology innovation (considering a Technology readiness level of 3) with the potential to become both environmentally and economically sustainable in the near future. A functional unit of 1 kg of colorant is set as a common function (for TEA and LCA) to guarantee that inventory flows (and then the results) in LCA and TEA are taken under the same reference. As the AESA focuses on consumption-based impacts and absolute terms, the function is adapted to the market and consumption-based data on food colorants. Therefore, the functional unit for the AESA is connected to the final consumption expenditure of this product. As the toolbox includes different module dashboards, detailed economic results are illustrated, describing the cost breakdown and financial variables like internal rate of return, return on investment, and payback period. Similarly, for LCA and integration modules.

Finally, the toolbox will show AESA results and transgression levels for each planetary boundary, indicating the different sharing principles used.

68d Community-Engaged Life Cycle Analysis: Ensuring Sustainability While Protecting Sovereignty

Margaret O'Connell, Jenna Trost, Kimberly Marion Suiseeya, Jennifer Dunn

Decarbonization of the transportation sector is critical for reducing greenhouse gas emissions and mitigating climate change. Often, transportation decarbonization technologies – such as electric vehicles – require minerals like lithium, cobalt, copper, nickel, and more. Increased decarbonization may therefore result in increased mining operations in order to acquire these minerals. It is critical that these minerals are acquired in ways that do not cause more environmental problems than those they mitigate through their use in decarbonization technologies, and it is critical that life cycle analysis (LCA) frameworks assess the impact of these operations in a holistic manner. Given the history of extractive industry operations located on Indigenous lands, it is also imperative that LCA frameworks incorporate Indigenous worldviews and knowledge systems. Current LCA frameworks almost entirely rely on scientific ecological knowledge (SEK), excluding traditional ecological knowledge (TEK) and other ways of knowing. As sovereign Peoples, Indigenous Peoples have a right to assess its sustainability based on the criteria and knowledge systems they see fit. LCA frameworks centered on SEK alone limit who gets to define sustainability and risk violating the rights of Indigenous Peoples to determine their own priorities using their own institutions.

Collaborating with the Ojibwe Nations, we expand on a co-developed community-engaged LCA (CE-LCA) framework to assess the potential impacts of a proposed copper-nickel mine in Minnesota. CE-LCA aims to be a framework inclusive of diverse worldviews and knowledge systems, in this case rooted in the Four Orders (Physical, Plant, Animal, and Human) central to Ojibwe lifeways. CE-LCA centers community priorities, values, and culture to ensure that LCA results are locally-relevant. In this case, we highlight the relationships between the Four Orders, assessing how mining operations impact the ecological, social, and cultural health that underpin these relationships. Existing work conducted by Ojibwe tribes and the Great Lakes Indian Fish & Wildlife Commission identifies members of the Four Orders most at risk from climate change impacts. Using both SEK and TEK, we build on this work to illustrate the risk factors to each of the relationships between these species and their relation to Ojibwe health and sovereignty. In contrast to existing, separate environmental and social LCA frameworks, the results from the CE-LCA process provide a holistic, integrated assessment of impacts on environmental, human, and cultural health and the interconnectedness between them. Ultimately, the results will identify what impacts are most critical for mining operations to address to ensure a sustainable supply of the minerals needed for decarbonization.

68e On the Interplay of Food Packaging Design and Food Supply Chain Sustainability

Aurora Del Carmen Munguia Lopez, Paola Alejandra Munoz Briones, Styliani Avraamidou, Victor Zavala

Nearly 40% of all plastic produced worldwide is used for packaging, and projections suggest that this amount can increase in the future [1,2]. Plastic packaging, in particular, is essential in helping preserve, extend shelf-life, and maintain the quality and safety of diverse food products. As a result, plastic packaging plays a fundamental role in food supply chain design and operations (from production to storage to transportation to retail and consumption). Food packaging represents nearly 66% of the total packaging waste by volume [3]. Notably, most of this packaging becomes waste after a single use; moreover, a fraction of plastic packaging is lost at each stage of the food supply chain.

There are strong and interesting trade-offs between the economic and environmental impacts of plastic packaging and of food supply chains. For instance, packaging helps avoid food waste by extending food shelf life, and increases food accessibility as food can travel longer distances, but this comes at the expense of using materials that are complex and difficult to recycle (e.g., multilayer plastic films, polystyrene, and plastic-coated paper cartons). Moreover, current food production trends, such as food processing and marketing strategies, can lead to overpackaging, resulting in increasing amounts of non-biodegradable waste [4,5]. These trends exacerbate the environmental impacts of food supply chains such as carbon emissions, land-use change, and plastic pollution [6]. The impact of plastic packaging is commonly overlooked in food supply chain studies (which often focus on the boundary of the food system) [7-11]. As such, there is an urgent need for holistic frameworks to study interconnected impacts of plastic packaging and food supply chains [3].

In this work, we quantify the economic and environmental impacts of different plastics for food packaging (e.g., multilayered fossil-based films, bio-based materials, and compostable materials) [12,13]. We integrate techno-economic analysis and life cycle assessment with a plastic pollution analysis into a computational framework to assess the hidden impacts of plastic packaging, including its production and end-of-life disposal. By considering the plastic leakages into the environment, we aim to provide a framework for a fair comparison between food packaging options. Furthermore, we analyze the trade-offs (food waste reduction against plastic pollution leading to food insecurity) and flexibility (transport and shelf-life) that each option provides. We apply our framework to the supply chain and packaging of cheese and coffee, which have significant environmental and socio-economic impacts.

[1] Geyer, R., Jambeck, J.R. and Law, K.L., 2017. Production, use, and fate of all plastics ever made. Science advances, 3(7), p.e1700782.

[2] Stegmann, P., Daioglou, V., Londo, M., van Vuuren, D.P. and Junginger, M., 2022. Plastic futures and their CO2 emissions. Nature, 612(7939), pp.272-276.

[3] Marsh, K. and Bugusu, B., 2007. Food packaging—roles, materials, and environmental issues. Journal of food science, 72(3), pp.R39-R55.

[4] Molina-Besch, K., Wikström, F. and Williams, H., 2019. The environmental impact of packaging in food supply chains—does life cycle assessment of food provide the full picture?. The International Journal of Life Cycle Assessment, 24, pp.37-50.

[5] Georgakoudis, E.D. and Tipi, N.S., 2021. An investigation into the issue of overpackaging-examining the case of paper packaging. International Journal of Sustainable Engineering, 14(4), pp.590-599.

[6] Persson, L., Carney Almroth, B.M., Collins, C.D., Cornell, S., de Wit, C.A., Diamond, M.L., Fantke, P., Hassellöv, M., MacLeod, M., Ryberg, M.W. and Søgaard Jørgensen, P., 2022. Outside the safe operating space of the planetary boundary for novel entities. Environmental science & technology, 56(3), pp.1510-1521.

[7] Williams, H. and Wikström, F., 2011. Environmental impact of packaging and food losses in a life cycle perspective: a comparative analysis of five food items. Journal of Cleaner Production, 19(1), pp.43-48.

[8] Boulay, A.M., Verones, F. and Vázquez-Rowe, I., 2021. Marine plastics in LCA: current status and MarILCA's contributions. The International Journal of Life Cycle Assessment, 26(11), pp.2105-2108.

[9] Chitaka, T.Y. and von Blottnitz, H., 2021. Development of a method for estimating product-specific leakage propensity and its inclusion into the life cycle management of plastic products. The International Journal of Life Cycle Assessment, 26, pp.1431-1438.

[10] Beitzen-Heineke, E.F., Balta-Ozkan, N. and Reefke, H., 2017. The prospects of zero-packaging grocery stores to improve the social and environmental impacts of the food supply chain. Journal of Cleaner Production, 140, pp.1528-1541.

[11] Baratsas, S.G., Pistikopoulos, E.N. and Avraamidou, S., 2021. A systems engineering framework for the optimization of food supply chains under circular economy considerations. Science of the Total Environment, 794, p.148726.

[12] Dilkes-Hoffman, L.S., Lane, J.L., Grant, T., Pratt, S., Lant, P.A. and Laycock, B., 2018. Environmental impact of biodegradable food packaging when considering food waste. Journal of Cleaner Production, 180, pp.325-334.4.

[13] Alias, A.R., Wan, M.K. and Sarbon, N.M., 2022. Emerging materials and technologies of multi-layer film for food packaging application: A review. Food Control, p.108875.

125a Dynamic Analysis for Piston-Type Work Exchanger and Network Design for Mechanical Energy Recovery

Abdurrafay Siddiqui, Yinlun Huang

The fundamental concept of work exchanger network (WEN) and basic system analysis were first introduced by Huang and Fan in 1996. [1] Today, WEN design for mechanical energy recovery has become an active area of research in process systems engineering. [2] The process units used for WEN are of two types: the flow work exchanger which can be called the direct work exchanger, and the compressor-turbine type which can be termed indirect work exchanger. Amini-Rankouhi and Huang developed a thermodynamic modeling based method for predicting the maximum amount of recoverable mechanical energy prior to WEN synthesis, and reported a basic analysis for the design of a direct work exchanger. [3],[4]

In this paper, we extend our research on piston-type direct work exchanger to the dynamic domain and on the dynamic-behavior-incorporated WEN design. In dynamic study, CFD simulation is performed to reveal the dynamic behavior in the pressurization-depressurization cycle under different operational scenarios, such as in adiabatic, isothermal, and non-isothermal conditions within a direct work exchanger. In simulation, thermal behavior of process streams is also observed, which provides critical information for unit design. The dynamic characterization of work exchanger is incorporated in the thermodynamic-model-based heat-integrated WEN design. We will show how mechanical energy can be cost effectively recovered in a direct work exchanger based network as compared with those using other types of work exchangers. We will also discuss a number of factors for operational stability and safety in WEN.

References

[1] Huang, Y. L. and L. T. Fan, 'Analysis of Work Exchanger Networks,' Industrial & Engineering Chemistry Research, 35, 3528-3538, 1996.

[2] Yu, H., C. Fu, and T. Gundersen, “Work Exchange Networks (WENs) and Work and Heat Exchange Networks (WHENs): A Review of the Current State of the Art,” Industrial & Engineering Chemistry Research, 59, 507-525, 2020.

[3] Amini-Rankouhi, A. and Y. Huang, 'Prediction of Maximum Recoverable Mechanical Energy via Work Integration: A Thermodynamic Modeling and Analysis Approach,' 63(11), 4814-4826, 2017.

[4] Amini-Rankouhi, A. and Y. Huang, “Modeling and Simulation of a Piston-Type Work Exchanger for Mechanical Energy Recovery,” Industrial & Engineering Chemistry Research, 58(39), 18292-18300, 2019.

125b Using Public Data for Predicting Industrial Chemical Pollution Source Reduction and Prevention Alternatives

Jose Hernandez-Betancur, Gerardo Ruiz-Mercado

Some commercial chemicals have inherent hazardous characteristics that may affect human health and the environment. Thus, environmental releases and exposure scenarios need proper identification and assessment. Chemical pollution generation is critical due to the uncertainty of tracking chemicals across the use and end-of-life supply and management chains. The green chemistry and engineering principles advise preventing and minimizing pollution generation than treating and controlling it. Nonetheless, due to several decision-making factors like chemical properties, environmental regulation, economic behavior, and industrial production size, it could be difficult to determine a proper source reduction activity/method to prevent hazardous chemical pollution generation from industrial activities. In addition, the appearance of new chemicals in the market, new applications for existing chemicals, and data unavailability for decision-making increase the complexity of selecting adequate strategies for preventing and reducing the impact of chemicals of concern. This contribution moves forward on creating data engineering for integrating scattered data with chemical-structured-based machine learning models to predict potential pollution source reduction activities during the manufacturing of chemicals. This contribution supports environmental practitioners, stakeholders, and decision-makers in exploring pollution source reduction and prevention activities and designing strategies for the safer use of chemical substances.

125c Dynamic Sustainability Assessment in the Digital Age

Abdurrafay Siddiqui, Yinlun Huang

Industries are in the midst of significant, compelling transformation regarding technology innovation and the ways products are manufactured. This is largely impacted by Industry 4.0, which is mainly featured by digitalization. New and relatively low-cost technologies for smart sensing and operation, fast data communication among sensor node, database, internet, and cloud services in manufacturing sites have started to be used. These provide a variety of opportunities for advancing engineering sustainability research and practice. In industry, sustainability assessment is commonly conducted annually and improvement strategies are developed accordingly. It has been questioned whether sustainability problems could be identified, assessed, analyzed and predicted in time so that solutions could be derived and actions be taken quickly. With abundant data accessible in the digital age, traditional sustainability assessment could be performed dynamically. This renders a need to study methods for dynamic sustainability assessment.

In this paper, we introduce a concept of dynamic sustainability assessment and discuss data accessibility, data generation using Digital Twin technology, and data spatial-temporal classification and utilization. Then we analyze the features of sustainability indicators in the economic, environmental, and social categories and classify them into groups that demand different data acquisition frequencies. We will also describe what sustainability aspects should be assessed more often as compared with others for annual evaluation. The method for dynamic sustainability assessment has been applied to a study on electroplating system sustainability. The assessment results will be compared with those by conventional annualized sustainability assessment. It shows a potential to open new areas of study for industrial sustainability as a whole.

125d Modeling-Based and Data-Driven Frameworks to Promote Water Sustainability in Chemical Manufacturing Plants at Dow

Spyros Tsolas, Apoorva Sampat, Irfan Khan, Lin Zhao, Tim Finley, Wu Chen

Freshwater availability is key to Dow's continued operations. Recognizing that water is a shared and finite resource, Dow is working to accelerate water stewardship in its operations and to innovate water-saving products. Through the World-Leading Operations Performance 2025 Sustainability Goal, Dow has identified six of its manufacturing sites as key water-stressed sites where it is committed to reducing the freshwater intake intensity by 20 percent [1]. At these sites, we are leveraging data driven techniques to develop freshwater reduction strategies and advanced forecasting techniques to mitigate the impacts of drought events.

In this talk, we will present two such efforts. The first effort presents a framework on leveraging open-source and site-specific water availability data to identify current water supply risk and predict future water availability scenarios. We have developed a digital dashboard that identifies the short-, medium-, and long-term risk of water availability to the site. We will also present a framework to leverage timeseries forecasting techniques to predict future water supply and guide drought contingency plans for the site. These modeling capabilities can also be used in guiding infrastructure investment decisions to develop strategies that mitigate the impact of drought events.

In the second sub-topic, we will present a framework to assess and improve the cooling strategy efficiency in existing sites. Inefficient cooling can affect the nominal plant operation and may give rise to (i) excessive water withdrawal from external reservoirs (or make up), and (ii) subpar asset productivity due to cooling bottlenecks. Water quality also affects the condition of the pipeline which in turn affects productivity due to the arising need for more frequent maintenance. Finally, in extreme cases seasonal hot periods may pose an additional strain on cooling efficiency since less external water may be available to cool down the units and on top of that with higher incoming temperature. To this end, we present a framework that (i) assesses the current cooling water requirements of a plant in terms of cooling duty and water flows, (ii) evaluates the economic effect of cooling efficiency on productivity, and (iii) offers alternative cooling strategies to minimize strain on external water, maximize cooling efficiency and minimize implementation capital.

<u>References:</u>

[1] Recycling Water in Water-Stressed Communities. *Dow Corporate homepage* [online accessed: March 27th, 2023]. Available at: https://corporate.dow.com/en-us/science-and-sustainability/2025-goals/world-water-day/recycling-water-20.html.

125e Water Stress-Based Price for Global Sustainability: A Study Using Generalized Global Sustainability Model (GGSM)

Neeraj Hanumante, Yogendra Shastri, Apoorva Nisal, Urmila Diwekar, Heriberto Cabezas

Water is a resource that is critical for human survival. Recently, water consumption across the world has been increasing significantly, leading to a rapid rise in water stress; that is, water demand exceeds the supply. The state of California in the U.S. and South Africa are two examples of regions with high water stress. One way of addressing this situation is increasing the supply of water through the development of more water-related infrastructure, such as desalination plants, dams, and canals. The other way is water demand management, that is, altering the water demand to reduce water stress. Water pricing is a tool that can be used for water demand management. Globally, flat rate pricing was a popular model used for water pricing. More recently, other models such as uniform (volumetric) rate, block or tire rate, and complex rate have been utilized for water price-setting. However, conventionally, the effect of water stress on the water price is incorporated as a constant factor. Hence, examination of the suitability of the water stress-based price as a tool for water demand management becomes the need of the hour.

Water is consumed for agriculture, energy, industry, and municipal applications. The relationship between these sectors with each other is complex; for instance, biomass produced by the agricultural sector can be used to produce energy, and the agricultural sector uses electricity-powered tools for farming processes. Because of such inter-dependencies, integrated models capable of capturing these relations, that is, the Food-Energy-Water nexus, are well suited to analyze the effect of policies implemented for water demand management. One such model, Generalized Global Sustainability Model (GGSM), was developed by Nisal et al. (2021) and Hanumante et al. (2022). GGSM, as shown in Figure 1, is a global model with integrated economic and ecological dimensions. Its ecological dimension has multiple trophic levels with intra-trophic level diversity. At the lowest trophic level, the plants are the primary producers that consume the natural resources from the Resource Pool. The higher trophic levels depend on lower ones for survival. Since there are multiple compartments at each trophic level, these interactions give rise to a complex food web. These interactions are represented using Lotka-Volterra-type equations. Some compartments of the ecological dimension are human-controlled; for instance, P1 represents the agricultural sector. On the right side, we have the human households HH, industrial IS, and energy production EP sectors. The industrial sector operates by taking resources from the ecological dimension and the energy from the Energy Production sector. This is where the ecological and economic dimensions integrate. At the end of their life, the used industrial goods are transferred to the Inaccessible Resource Pool, imitating the landfills. This GGSM is parameterized to reflect global realities. It also incorporates a global water system, where the water consumption in the world is represented in six regions (Africa, Asia, Europe, North America, Oceania, and South America) and three sectors (municipal, industrial, and agricultural). The historical consumption of water by these regions and sectors has been captured in this model.

However, GGSM has a key limitation in the form of an assumption that irrespective of the water stress, the water consumption patterns would not change. As a result, even if the water stress reaches a value close to 100%, the water consumption rate does not change. In reality, lower availability of water is expected to limit the water demand; it may be because water access becomes harder or water becomes costlier. For instance, in water-starved regions, the supply of water for municipal use is carried out via tankers; this might change consumption patterns. Even the higher water price might persuade the consumer to use water more judiciously. Hence, feedback on water stress is a critical component for the realistic modeling of the FEW nexus. Here, we examine the utility of water-stress-based prices from a global sustainability standpoint.

We develop a water-stress-based water price model and a feedback mechanism that facilitates the manifestation of the effects of the price change on water consumption and hence, water stress. We first model the water price with a base price and a variable component which is a linear function of the water stress. Thus, instead of a constant environmental factor, we develop a dynamic model for water price, which is based on water stress. Then we model the water stress feedback through the demand elasticity of water price. The demand elasticity of water price represents the ratio of the percent change in demand to the percent change in water price. This change in the water demand then results in a change in the state variable of the GGSM.

We parameterize the water price model and the feedback model for six regions (Africa, Asia, Europe, North America, Oceania, and South America) and three sectors (municipal, industrial, and agricultural). For each of these combinations (region and sector), for the water price model, we determine the base price and multiplier for the water stress, that is, the water stress sensitivity of water price; and for the water stress feedback model, we determine the demand elasticity of water price. The parametrization mainly depends upon two regional factors, regional affluence, and regional water availability. We assume that, for the municipal sector, affluent regions have a higher base price, higher price sensitivity, and higher elasticity. For the industrial sector, on the other hand, owing to a well-developed water market and regulations, affluent regions are expected to have lower elasticity. However, the base price and price sensitivity for the affluent regions would be higher. The agricultural sector is expected to be inelastic till the water price crosses a threshold value. Agricultural practices that govern the regional water use pattern are developed to conform with regional water availability, and hence, the demand elasticity of water price for the agricultural sector is determined as a function of the regional availability of water. We determine the limits of the parameter values from the literature and use GDP per capita (from the World Bank and UN) and water availability per capita (from the works of Hanumante et al. (2022)) as a proxy for regional affluence and regional water availability.

We carried out simulations using this modified GGSM to analyze various scenarios. Simulation results indicated that the proposed water price model could capture the regional realities through relative changes in price and demand. The change in water price can alter the regional water demand and hence the water stress. Thus, a water stress-based price model can be used as a policy instrument. The simulation results indicate that, in the absence of a water stress-based price, the African continent may experience increasing water stress well beyond 100%. With the implementation of a water stress-based price, the water stress can be controlled; however, it may lead to a reduction in regional food production by about 26%. Because of the trade-off between regional food production and water stress, cooperation between various regions could help in reducing the impact of the impending water and food crisis. North America and Europe may produce surplus food products and play a pivotal role in alleviating the critical situation in Africa.

To summarize, water-stress-based water prices can be used as a tool for water demand management. The analysis identifies the advent of a potential food crisis in the African continent owing to increased water stress. This model will be used to carry out further simulation studies to compare policy options and determine ways and means to achieve specific policy targets.

References:

Nisal, Apoorva, et al. 'Integrated model for food-energy-water (FEW) nexus to study global sustainability: The main generalized global sustainability model (GGSM).' *PloS one* 17.5 (2022): e0267403.

Hanumante, Neeraj, et al. 'Integrated model for Food-Energy-Water (FEW) nexus to study global sustainability: The water compartments and water stress analysis.' *Plos one* 17.5 (2022): e0266554.

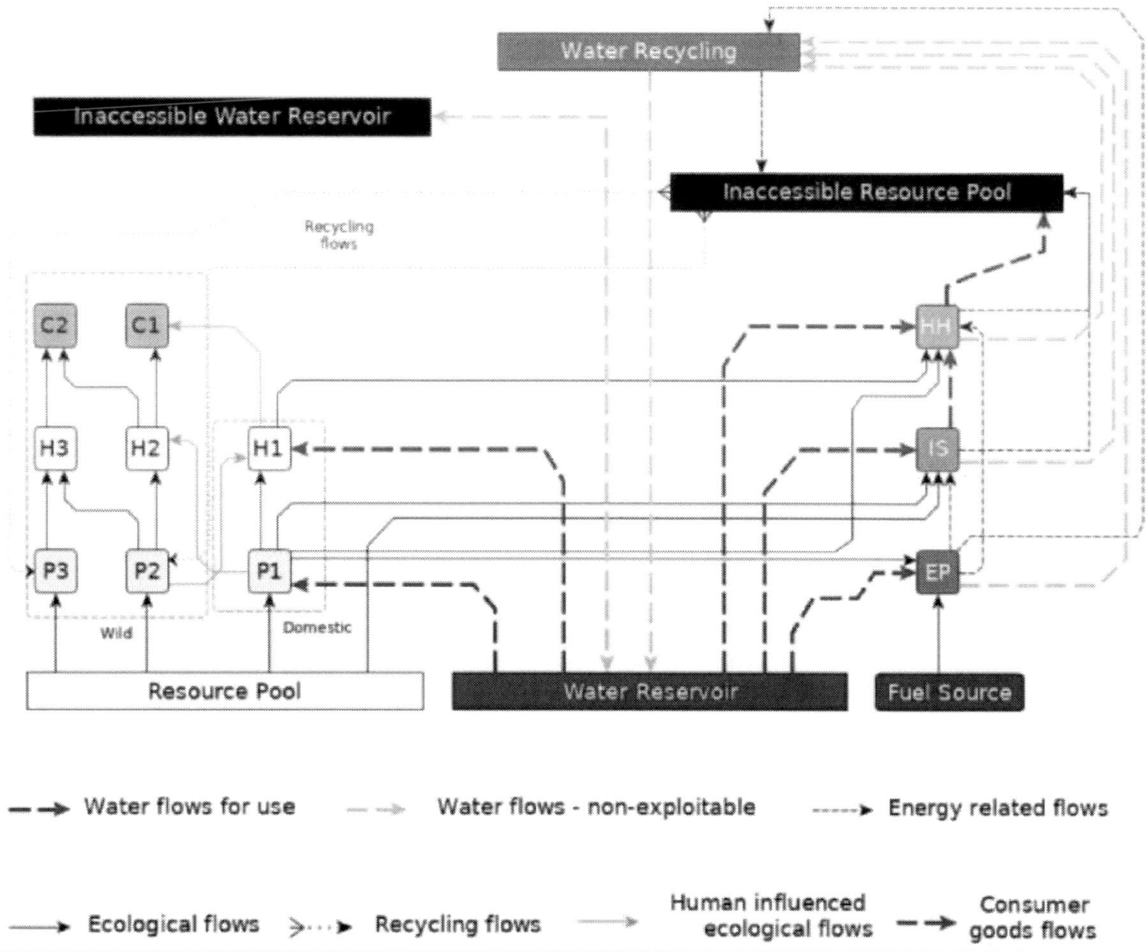

Figure 1: shows the Generalized Global Sustainability Model (GGSM). Higher trophic level depends on the lower one for its survival. For example, herbivores depend on plants and carnivores depend on herbivores. These ecological compartments can be feral or controlled by humans. Agriculture and livestock are represented by P1 and H1 respectively. P2 represents the open access grasslands. The human households HH are shown on the right side along with energy production EP and industrial sector IS. The water supply to human controlled compartments emanate from the Water Reservoir. The connections and dependencies of the agricultural sector, energy sector and the water sector create the Food-Energy-Water nexus.

125f Sustainability and Sustainizability^{TM} over Sets for a Multi-Species Population Model

Ioannis Manousiouthakis, Vasilios Manousiouthakis

The Gause-Lotka-Volterra (GLV) population model has been the subject of many studies. In this work, two theorems are proved for the general, n-species GLV model, establishing necessary and sufficient sustainability and sustainizability^{TM} conditions over rectangular sets. A case study on the May-Leonard, 3-species, GLV model is employed to illustrate the power of the two established Theorems.

427a Net Zero and Negative Carbon Biofuels from Hydrothermal Liquefaction and Biocrude Upgrading

Pimphan Meyer, Shuyun Li, Lesley J. Snowden-Swan, Uisung Lee, Rodrigo Buitrago Tello, Steven Phillips

Hydrothermal liquefaction (HTL) is a thermochemical pathway used to convert wet biomass to biocrude material which can be further upgraded to transportation fuel via hydrotreating process. HTL process is relatively simple and unique because it can process wet biomass feedstock without drying. Experimental data from the HTL and biocrude upgrading flow reactors at Pacific Northwest National Laboratory was used in base case study. Techno-economic analysis (TEA) and life cycle assessment (LCA) were performed to identify process parameters significantly driving biofuel production cost and process carbon intensity. LCA results suggested that the biofuel produced from HTL and biocrude upgrading processes could be near net zero carbon fuel when waste organic feedstocks such as wastewater treatment sludge, animal manure and food waste were used. In addition, case studies of utilizing low carbon heat and power such as renewable resources were performed. Negative carbon fuel could be produced when renewable heat and electricity were applied. TEA also suggested that renewable heat could significantly increase the production cost. On the other hand, renewable electricity did not give a significant impact on the biofuel production cost.

427b The Value Proposition of Bio-Intermediates As Marine Fuel for Maritime Decarbonization

Eric Tan, Abhijit Dutta, Kristiina Iisa, Calvin Mukarakate

Marine biofuels have the potential to play an important role in decarbonizing the shipping industry and reducing its impact on the environment. They are an important solution for reducing greenhouse gas emissions in the shipping industry, which is responsible for a significant portion of global emissions. Biofuels are produced from renewable sources and can be used as a direct replacement for traditional fossil fuels. In addition to reducing greenhouse gas emissions, marine biofuels also offer other benefits, such as reducing air pollution and improving energy security by diversifying fuel sources. However, there are also challenges associated with the use of marine biofuels, such as the availability of sustainable feedstocks and the potential for increased competition with food production. It is important to carefully consider the environmental and social impacts of biofuel production and use and to ensure that sustainability criteria are met.

Catalytic fast pyrolysis (CFP) bio-oil is a bio-intermediate derived from biological sources, such as woody biomass and forest waste materials, and non-food crops, which can help reduce the reliance on fossil fuels and create a more circular economy. CFP bio-oil can be further processed into a wide range of value-added chemicals, including sustainable aviation and marine fuels, plastics, and other materials. One of the advantages of CFP bio-oil is that it is a renewable and sustainable drop-in fuel intermediate. Additionally, bio-oil typically has significantly lower life cycle greenhouse gas emissions compared to fossil marine fuels, as it captures carbon from the atmosphere and stores it in the bio-oil and bio-char.

While the fast pyrolysis raw bio-oil is highly acidic and unstable, making it difficult to store and transport, CFP bio-oil exhibits more favorable fuel properties. Moreover, we have demonstrated that only a relatively mild upgrading of the bio-intermediate is required to enable its blending with commercial very low sulfur fuel oil (VLSFO). Consequently, this presentation will discuss the value proposition of using bio-intermediate derived from woody biomass via catalytic fast pyrolysis as marine fuel for maritime decarbonization, lying in its potential to provide a sustainable and low-cost alternative to traditional fossil fuels.

References:

1. Li S, Tan ECD, Dutta A, *et al.* Techno-economic Analysis of Sustainable Biofuels for Marine Transportation. *Sci. Technol.* 56(23), 17206–17214 (2022).
2. Tan ECD, Hawkins TR, Lee U, *et al.* Biofuel Options for Marine Applications: Technoeconomic and Life-Cycle Analyses. *Sci. Technol.* 55(11), 7561–7570 (2021).

427c Techno-Economic and Life-Cycle GHG Emission Assessment for Aqueous-Phase Product Treatment and Valorization Options of Wet Waste Hydrothermal Liquefaction Process

Yuan Jiang, Longwen Ou, Lesley J. Snowden-Swan, Hao Cai, Shuyun Li, Karthikeyan Ramasamy, Andrew J. Schmidt, Huamin Wang, Daniel Santosa, Mariefel V. Olarte, Michael R. Thorson

Waste-to-fuel via hydrothermal liquefaction, a promising technology with high fuel yield and carbon efficiency, can simultaneously decarbonize the transportation sector and reduce the landfill demand. However, a significant amount of carbon, and nitrogen in the feedstock (23.2%, and 50.4%, respectively) goes to the aqueous phase product, and results in an aqueous-phase stream with high chemical oxygen demand. Treatment is required before disposal, of which the cost, utility consumption, and emission can be high when using conventional activated sludge process in the state-of-art design for hydrothermal liquefaction. In this work, four alternative treatment and valorization options were investigated for the hydrothermal liquefaction aqueous-phase product, including two catalytic hydrothermal gasification technologies using different catalysts, steam-phase catalytic reduction of wastewater and ambient-pressure aqueous-phase catalytic upgrading. Techno-economic and life-cycle assessments were conducted based on continuous-flow experimental data and rigorous process models developed in Aspen Plus V11. The results suggest that a treatment option producing valuable by-products will simultaneously improve the economic performance and reduce the life-cycle greenhouse gas emission of the entire process. Among all treatment options investigated, the ambient-pressure aqueous-phase catalytic upgrading technology achieved the lowest minimum fuel selling price (13% lower than the state-of-art) and lowest life-cycle greenhouse gas emission (39% lower than the state-of-art). It is because this technology requires less severe operation conditions (less energy consumption and capital investment) and less expensive catalyst to enable deep chemical oxygen demand reduction and production of fuel gas and fertilizer as by-products.

427d Torrefaction of Almond Shell As a Renewable Reinforcing Agent for Plastics: Techno-Economic Analyses and Comparison to Bioethanol Process

Dupeng Liu, Nawa Baral, Corinne D. Scown, Ning Sun

The U.S. state of California alone produced nearly 3.5 billion kilograms of almonds accounting for approximately 84% of the world's almond production. This generates about 2.58 million metric tons of almond residues. The almond shells are currently either burned to generate power or disposed of in the landfill. Valorizing almond shells and hulls provides an opportunity to replace petroleum-derived products and divert organic material from landfills. Here we demonstrate a detailed techno-economic analysis of an almond shell torrefaction process capable of utilizing the 520,000 metric tons of almond shells produced annually in California. Our process also includes preprocessing the torrefied biomass to exploit it as a reinforcing agent for plastics. We further compared the revenue generated from the torrefied biomass and bioethanol derived from the same amount of almond shells. We considered three different torrefaction facility scales to evaluate trade-offs between economies of scale at the facility and trucking costs to deliver almond shells. A facility that takes in 200,000 metric tons (MT)/year of almond shells results in lower per-unit-output basis capital and operating cost relative to other smaller scales of torrefaction facility, including 10,000 MT/year and 50,000 MT/year, considered for analysis in this study. The large-sale facility results in the minimum selling price of the torrefied biomass of \$311.4/MT. An analogous techno-economic analysis on converting almond residues into bioethanol is also investigated. The minimum selling price of almond shells derived ethanol (\$1.71/kg) is higher than corn (\$0.48/kg) or cellulosic biomass (\$0.88/kg) derived ethanol. Compared to the bioethanol route, the torrefied almond shells result in 3 times more revenue if it is utilized as a reinforcing agent for plastics.

221a High Temperatures and Low Soil Moisture Synergistically Reduce Switchgrass Yield and Inhibitbiofuel Production from Marginal Field Sites

Sarvada Chipkar, Derek J. Debrauske, Leela Joshi, Briana Cronk, Emily Burke, G. Philip Robertson, Trey Sato, Stephen Hamilton, Kurt Thelen, Rebecca Ong

Marginal field sites are defined based on their unsuitability for food production, and for this reason are considering promising locations for production of bioenergy crops, which have lower input requirements. However, the effect of drought in some marginal sites may be more severe than in arable cropland. This has implications for both bioenergy crop yields and downstream conversion, as we have previously shown that severely drought-stressed switchgrass completely inhibited downstream yeast fermentation. The objective of this project was to understand how marginal field sites with various soil characteristics under simulated drought conditions influenced switchgrass yields, composition, and conversion to biofuels. An experiment was designed at five marginal sites with diverse soil characteristics, spanning a latitudinal gradient in Michigan and Wisconsin. Rainout shelters were installed in three replicate field plots at each location to simulate drought stress. During the first year of installation (2018), the rainout shelters reduced the rainfall by 60%, causing a statistically significant reduction in switchgrass fermentability harvested at the Wisconsin Central - Hancock field site compared to the ambient samples, which was not observed at the other locations. Among the five sites, the Hancock site had a unique Entisol soil type that led to comparatively low soil moisture and high soil temperatures during the growing season, which was exacerbated by the simulated drought. This reduction in fermentability was not expected based on either the biomass or hydrolysate composition, which were comparable between the control and treatment, indicating that more research is needed before we can predict the impact of drought on biofuel production.

221b Biochar-Assisted Acetogenic Conversion of Carbon Dioxide to C2-C6 Alcohols: A Sustainable Solution for Renewable Fuel Production

Rahul Thunuguntla, Hasan Atiyeh, Raymond L. Huhnke, Ralph S. Tanner

Globally, carbon dioxide (CO_2) emissions have exceeded 36 billion tons annually, contributing to a rise in greenhouse gas (GHG) emissions. Development of carbon capture and conversion technologies is crucial to reduce the global carbon footprint. Current technologies for the capture and direct conversion of CO_2 into various products have the potential to improve process economics and reduce GHG emissions. However, electrochemical and thermochemical processes are energy-intensive, suffer from catalyst poisoning, and have low conversion efficiencies. In contrast, microbial CO_2 fixation and transformation into biofuels and value-added products using autotrophic microorganisms have gained attention for their ability to occur at near ambient temperature and pressure. Additionally, biochar has emerged as a promising material for carbon sequestration and soil improvement, which can complement the benefits of microbial CO_2 conversion. This carbon-rich material has a high surface area and pH buffering capacity and contains minerals and metals that can replace expensive medium components. In this study, biochar produced from switchgrass and poultry litter at 350°C and 700°C was investigated for its effects on the conversion of CO_2 into C2-C6 alcohols and acids using *Clostridium ragsdalei* strain P11, *C. carboxidivorans* strain P7, and *C. muellerianum* strain P21. Fermentations were performed in 250 ml serum bottles with 50 ml working volume using H_2:CO_2:N_2 (60:20:20) at 37°C and 125 rpm. Results showed that medium supplemented with the switchgrass biochar produced at 350°C (SGB350) resulted in more C2-C6 alcohols with the three strains compared to control and other biochar media. The medium supplemented with poultry litter biochar made at 350°C (PLB350) also enhanced growth and formation of C2-C6 fatty acids with the three strains compared to other media. Strain P11 in the SGB350 medium produced ethanol (6.4 g/L), which was 3- and 2-fold more than with strains P7 and P21 in the same medium. Strain P11 in the PLB350 medium produced acetic acid (9.1 g/L), which was 21% and 55% more than with strains P7 and P21, respectively, in the same medium. Strain P21 produced over 2-fold more butyric acid and butanol in the PLB350 and SGB350 media, respectively, than with strain P7. Only strain P21 produced hexanol. The amounts of CO_2 converted by strain P11 in the SGB350 medium were 35% and 4% higher than with strains P7 and P21, respectively. The type of biochar and the temperature at which it was produced affected the three strains differently. These findings demonstrate the potential for SGB and PLB biochar to be used as a medium substitute for expensive nutrients to enhance the production of C2-C6 alcohols and fatty acids from CO_2, increasing the feasibility of the process. Furthermore, the produced C2-C6 alcohols from CO_2 could serve as intermediates for making renewable jet fuels, which are in high demand.

221c Lipid Production and Biofuel Yield from Wood Hydrolysates Using Oleaginous Yeast "*Cutaneotrichosporon Curvatus*"

Stephanie Ossai, Sampath Gunukula, Clayton Wheeler, Gerard Van Walsum

Problem:

The United Nations' sustainable development goals seven and thirteen focus on improving the global energy mix and promoting net-zero global greenhouse gas emissions by 2050. Producing sustainable aviation fuels (SAF) is one of the energy applications that are most difficult to wean away from fossil fuels. Non-edible lignocellulosic biomass is the most abundant renewable bioresource in the ecosphere and can be used to generate SAF. It comprises polysaccharides (cellulose and hemicellulose) and a natural polymer (lignin).

One approach to producing SAF from sustainable lignocellulose is thermal deoxygenation (TDO). This process converts cellulose-derived levulinic acid to high-energy-density hydrocarbon liquids, which can be hydrotreated into molecules suitable for jet fuel applications. A current challenge of this conversion process includes reducing the amount of pentosan sugars feeding into the initial hydrolysis/dehydration producing the levulinic acid. In the strong acid environment of the hydrolysis/dehydration process, pentosan sugars convert to resinous materials that glaze over the equipment surfaces. One approach to reducing pentosan sugars in the feed stream is adding a pre-conditioning step that enables the preemptive removal of some hemicellulose sugars. These sugars could then be converted into other biochemical or biofuel products.

Current studies in microbial biofuel production have gained attention due to the deteriorating amount of fossil fuels. In this study, we explore the use of fermentable sugars produced from dilute sulfuric acid pre-conditioning of white pine wood chips for the production of biofuel precursors. Oleaginous yeast is suggested as a viable lipid producer to promote a more sustainable biofuel industry. Using oleaginous yeasts to consume unwanted sugars has various advantages, such as a fast growth rate, high lipid accumulation that does not compete with human food, ease of handling, and no environmental or climate effects. Studies have shown that *Cutaneotrichosporon Oleaginosus* has the highest lipid content and biofuel yield among other yeast strains. We explore sustainable and efficient renewable sources such as wood pre-hydrolysates for biofuel production.

Methods: White pine chips were used as the starting biomass material. Mild acid (0.25% H2SO4) pre-conditioning of the wood chips was carried out at Idaho National Lab using a ton-per-day plug screw feeder with a reactor operating at different operating conditions, including different feed rates, particle sizes, temperatures, and feedstock concentrations. The pre-hydrolysate was analyzed using HPLC (HPX-87H and HPX-87P columns) and contained glucose, xylose, mannose, arabinose and galactose, and some degradation compounds. Batch fermentation was used to culture the oleaginous yeast strain, *C. Oleaginosus,* using wood hydrolysate as the principal substrate with additional defined nutrients. Production and accumulation of microbial lipids in the cells are enabled by nitrogen-limited stagnant growth in the presence of abundant sugars. Degradation compounds such as hydroxymethylfurfural (HMF), acetic acid, and furfural can significantly impact the growth of microorganisms, reducing their metabolic rate and limiting their lipid yield. Detoxification of hydrolysates using over-liming and activated carbon adsorption reduced this hazard. Hydrothermolysis (HTL) processing of the whole oleaginous yeast or their extracted lipids will be performed with resulting products characterized by GC-MS for applicability to biofuel applications.

Results: HPLC analysis of the pine pre-hydrolysis showed that most solubilized sugars were monomer sugars, and several degradation products were also present. Glucose and mannose were the sugars most abundant in the pre-hydrolysate. Growth of the oleaginous yeast *C. Oleaginosus* showed consumption of all the types of sugars, though glucose and mannose were the most sugars completely consumed. Glucose was the most rapidly consumed sugar. The effects of over-liming and activated carbon adsorption on wood hydrolysate were investigated, with activated carbon appearing more effective at reducing identified contaminants. Acid hydrolysis and solvent extraction of the disrupted cells and lipid recovery will be investigated. HTL of the whole cells and extracted lipids will be carried out, and the final products will be analyzed via GC-MS.

Implications: The TDO process with hydrotreatment has been shown to be effective at converting cellulose to SAF. Work from this project seeks to improve the operability of the acid hydrolysis/dehydration reaction that initiated the TDO process and increase the conversion of raw materials through fermentation of byproduct sugars and conversion of microbial lipids via HTL.

221e Optimal Integrated Plant for Renewable Surfactants Production

Sofía González Núñez, Mariano Martin, Carlos Amador

The sustainable development goals pose a number of challenges to the traditional industry. The sustainable production and consumption from the UN requires a complete transformation of the chemical and consumer goods industry. Chemicals traditionally produced from crude fractions must find another raw material that is renewable and alternative production routes. Surfactants are specialty chemicals whose purpose is to wet, emulsify, help solubilize or soften species. They are used in laundry as well as in other industries such as gas recovery from fracture and, so far, they have been typically produced from crude oil. Washing clothes or dishes means that sooner or later the species end up in a water body. Even if the waste water treatment plants are capable of processing and removing surfactants from water, including the two major ones such as linear alkylbenzene sulphonates (LAS) and the alkyl phenol ethoxylates (APE), that are aerobically degraded and partially absorbed to the sewage sludge, biodegradable ones can certainly help in the sustainability of products such as detergents. There are already synthetic paths for sustainable biosurfactants (Kumar et al., 2022; Reis et al 2013) In most linear processes that transform biomass into the biosurfactants are presented. Surfactants consist of two sections, a hydrophilic head group and a hydrophobic alkyl chain. Alternatively, it is possible to generate both fractions from biomass to build alkyl poluglucosides, APG's, (Rivai et al 2017) and that can be produced from manure and CO_2.

This study presents the conceptual design for the production of sustainable surfactants from waste, including CO_2 and manure. An integrated facility is designed to process the waste employing it to grow algae and produce biogas. From the algae, intermediates such as glucose and lipids are obtained. From biogas, hydrogen is produced. Next, alcohols are produced via lipids hydrogenation so as to synthesize the APG's from them and the glucose. No further raw material is required, at the expense of a large investment. The design problem is formulated as a large NLP for the selection of processing route of the intermediates and final product as well as the operating conditions. Such formulation allows for the selection of the algae growing and its composition for the limited use of additional utilities and chemicals. An economic analysis and process scale down has also been performed.

The algae composition recommended consists of 60% lipids and 27% carbohydrates. The yield of the facility reaches 0.47kg/kg algae (0.08kg/kg manure). For the production of 252 kt/yr of APG, the facility consumes 17.7 MW of thermal energy and steam and 7.9 MW of electricity, absorbing 788 kt/yr of CO_2 after discounting the emissions related to thermal and power consumption. The investment adds up to 196 M€ for a production cost of 0.17€/kg_{APG}. The large amount of manure to be processed, corresponding to 300k cows, suggests a scale down study to evaluate the effect on the investment and production costs. Current market price of APG can be achieved processing the waste of over 2.5k animals.

References

Kumar, S., Kumar Kansal, S., Pandey, A., Saravanamurugan, S. (2022) Chapter 14 - Polyalkylglycosides: sustainable production of nonionic biosurfactants from lignocellulosic biomass, 373-390. In Biomass, Biofuels, Biochemicals Biochemicals and Materials Production from Sustainable Biomass Resources (Li, H., Pandey, A., Saravanamurugan S., Elumalai S edas. Elsevier Oxford.

Reis, R.S., Pacheco, G.J., Pereira A.G. , Freire D.M.G. (2013) Biosurfactants: Production and

Applications DOi: 10.5772/56144

Rivai, M., Astuti, Y., Hambali, E., Permadi, P., Suryani, A., & Sutanto, A. (2017). Selection of PTSA Catalyst Concentration for the Synthesis Alkyl Polyglycosides from Fatty Alcohol (C16) and Glucose Syrup. 11

291a Elucidating Catalytic Pyrolysis of Representative Textile Wastes over Waste Glass-Derived Silica-Supported Transition Metal Catalysts

Yoke Wang Cheng, Wenming Fu, Guiying Lin, Dequan Xu, Chi-Hwa Wang

In practice, the textile wastes are incinerated to ash before dumped in landfill whereas the non-incinerable glass wastes persistently occupy the landfill. At present, the global recycling rates of textile wastes (12%) and glass wastes (21%) are dismal, which aggravate the shortage issue of landfill sites. Unlike highly pollutive waste incineration, the pyrolysis is a relatively cleaner thermochemical valorisation approach that empower the co-production of syngas, char, and oil from textile wastes under inert atmosphere. As textile wastes fall under municipal solid wastes (MSW), the pyrolysis of textile wastes is often benchmarked against other segregated MSW components, mixed/simulated MSW, and refused derived fuels. Withal, the actual composition of textile wastes is affected by spatial, temporal, and spatio-temporal variations. Since pyrolytic behaviour of different textile materials remain nebulous, six representative textile wastes (such as cotton, polyester, cotton-polyester, denim, polyacrylic, and wool-plant fibre) are selected as the pyrolysis feedstock to unveil their pyrolytic behaviour. To date, none of the textile pyrolysis studies incorporated heterogeneous catalysis, wherein the so-called "catalysts" are ostensibly chemical activating agent that pre-impregnated onto or physically mixed with the feedstocks. Herein, various commercial silica-supported transition metal (Ni, Co, Fe, Cu, or Mn) catalysts are used to hasten the textile pyrolysis process or modulate the product distribution in favour of desired pyrolytic products. Lastly, the waste glass-derived silica-supported transition metal catalysts are synthesized and benchmarked against their commercial counterparts for catalytic performance in textile pyrolysis. This study imparts some insights on the reutilization of textile wastes and glass wastes by repurposing them as pyrolysis feedstock and silica feedstock, respectively.

291b Elucidating the Role of Alkaline Earth Metal Promoters and Zeolite Support Acidity on Palm Oil Catalytic Upgrading: Insights from DFT and Reactive Molecular Dynamics

Seba AlAreeqi, Daniel Bahamon, Kyriaki Polychronopoulou, Lourdes Vega

Over the past decades, a shift to biomass conversion technologies has been observed to counteract the negative effects of a fossil fuel-based economy [1]. Palm oil, an abundantly present agricultural biomass, holds the potential of producing biofuels that can be used as drop-in alternatives to fossil fuel hydrocarbons [2][3]. Among the various biomass conversion routes, fast pyrolysis is regarded as an economical biomass-to-fuel method with up to 70% wt. produced liquid product, known as bio-oil [4]. However, a major drawback of bio-oil direct use is its highly oxygenated nature, causing its lower energy density compared to fossil fuels [5]. Thus, O-removal using a HDO process is a promising route to stabilize long-chain bio-oils for commercial applications as biodiesel fuels [3], or as feedstocks for the production of added-value products [6][7].

A variety of heterogenous microporous heterogeneous catalysts have been investigated for the HDO of modeled palm oil compounds [3,8–10]. Among such structures, tuned aluminosilicates incorporated with metal sites, have attracted great attention towards the decarboxylation/decarbonylation (deCOx) and hydrodeoxygenation (HDO) of bio-oils catalytic upgrading [8,11–14]. Despite recent experimental studies, the upgrading of long-chain palm oil compounds has not yet been explored computationally. Thus, it remains unclear how the competitive mechanism proceeds to take place on the catalyst active-sites.

In this work, the upgrading of oleic acid and palmitic acids, two representative modelled palm oil molecules, have been examined using DFT calculations and ReaxFF simulations to study the role of (1) alkaline earth metal promoters on Ni(111) surface and examine the (2) activity of Brønsted–acid sites incorporated into (001) and (100) BEA-zeolite support. The Results from DFT reflected the effect of Mg^{2+} on improving the bio-oils' adsorption energy, reducing the O–M separation distance, and elongating the C–O bond length towards HDO upgrading. Notably, correlating the periodic number of the ionic promoter active-sites with the palmitic acid adsorption energy and O–M separation provided a linear relationship with R^2=0.988 and R^2=0.987. Such derived linear correlations are valuable in further anticipating the reactivity trends of transition metal-based modified catalysts. Moreover, ReaxFF simulations demonstrated the greater deCOx reactivity of BEA(100) compared to (001), revealing the molecular-level interactions governing the aggregation of oleic acid and palmitic acid on both facets. Aside from the effect of facets, different Si/Al ratios (all-silica, Si/Al=7, and 15), reaction temperatures (300K, 700K, and 1200K), and feedstocks (bio-oil/dodecane) were tackled. The protonated sites induced the zeolite activity as depicted by the higher number of CO/CO_2 and H_2O produced, also signified by elevated $O_{OleicAcid}$-H_{BEA} $g(r)$ peaks and reduced separation distances. Besides, the presence of dodecane gave rise to the bio-oils conversion and hydrocarbon yields quantitatively examined by the ratio of $C_{feedstock}/C_{hydrocarbon\text{-}fuels}$ production. Although increasing the temperature from 700K to 1200K enhanced the palmitic acid and oleic acid conversions, a reduction in n-C_{16} and n-C_{18} yields was found coupled with higher ethylene production. This contribution deepens the fundamental understanding, at the molecular level, of the role of modified nickel surfaces and tuned BEA-zeolite supports to pinpoint the optimum factors governing the catalytic upgrading of long-chain bio-oil feedstocks towards value-added hydrocarbons and potential biofuels.

This work has been partially financed by Khalifa University of Science and Technology under the Research and Innovation Center on CO_2 and Hydrogen (RICH) and the Catalysis and Separation Center (CeCaS) (projects RC2-2019-007 and RC2-2018-024, and AARE19-223).

References:

- [1] B.B. Uzoejinwa, X. He, S. Wang, A. El-Fatah Abomohra, Y. Hu, Q. Wang, Energy Convers.

Manag. 163 (2018) 468–492.
- [2] Y. Makkawi, Y. El Sayed, M. Salih, P. Nancarrow, S. Banks, T. Bridgwater, Renew. Energy 143 (2019) 719–730.
- [3] B. Yoosuk, P. Sanggam, S. Wiengket, P. Prasassarakich, Renew. Energy 139 (2019) 1391–1399.
- [4] S. Kim, E.E. Kwon, Y.T. Kim, S. Jung, H.J. Kim, G.W. Huber, J. Lee, Green Chem. 21 (2019) 3715–3743.
- [5] T.P. Vispute, H. Zhang, A. Sanna, R. Xiao, G.W. Huber, Science (80-.). 330 (2010) 1222–1227.
- [6] P.L. Cruz, E. Montero, J. Dufour, Fuel 196 (2017) 362–370.
- [7] K.S. Arias, M.J. Climent, A. Corma, S. Iborra, Energy Environ. Sci. 8 (2015) 317–331.
- [8] O.B. Ayodele, H.F. Abbas, W.M.A.W. Daud, Energy Convers. Manag. 88 (2014) 1111–1119.
- [9] S. Xing, Y. Liu, X. Liu, M. Li, J. Fu, P. Liu, P. Lv, Z. Wang, Appl. Catal. B Environ. 269 (2020) 118718.
- [10] M. Peroni, I. Lee, X. Huang, E. Baráth, O.Y. Gutiérrez, J.A. Lercher, ACS Catal. 7 (2017) 6331–6341.
- [11] K.N. Papageridis, N.D. Charisiou, S. Douvartzides, V. Sebastian, S.J. Hinder, M.A. Baker, S. AlKhoori, K. Polychronopoulou, M.A. Goula, Renew. Energy 162 (2020) 1793–1810.
- [12] I. Hita, T. Cordero-Lanzac, G. Bonura, C. Cannilla, J.M. Arandes, F. Frusteri, J. Bilbao, J. Ind. Eng. Chem. 80 (2019) 392–400.
- [13] M. Hellinger, H.W.P. Carvalho, S. Baier, D. Wang, W. Kleist, J.D. Grunwaldt, Appl. Catal. A Gen. 490 (2015) 181–192.
- [14] J. Xing, L. Song, C. Zhang, M. Zhou, L. Yue, X. Li, Catal. Today 258 (2015) 90–95.

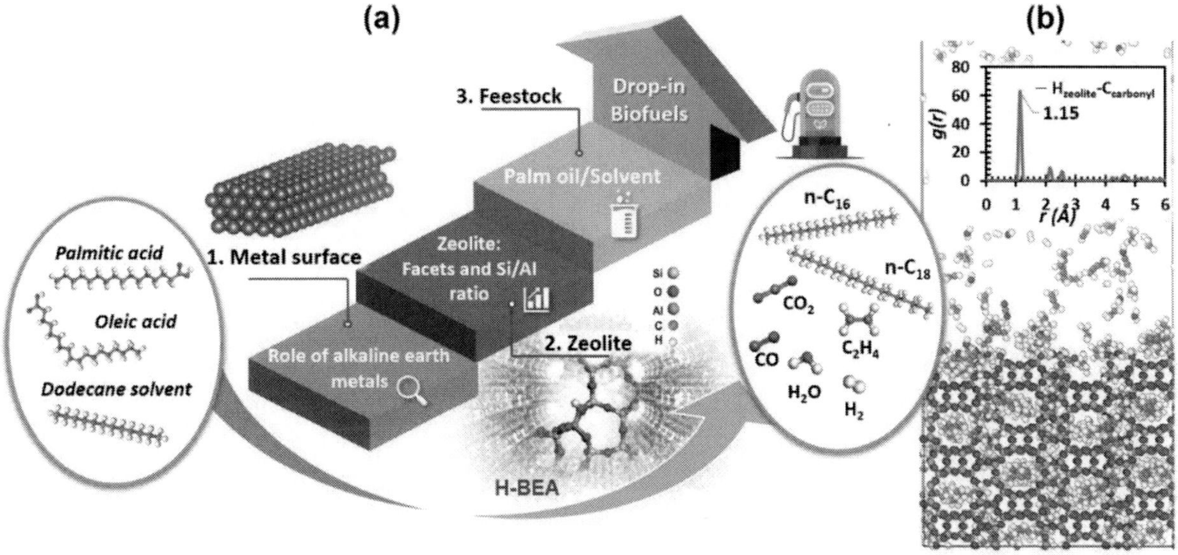

Figure. (a) Schematic illustration of the modelled systems and investigated effects in this work. (b) Optimized configuration of Oleic acid/BEA(100) structure following 1ns at 1200K. Radial distribution function plot inserted in (b) signifies the strong interactions between protonated BEA(100) sites and the oleic acid carbonyl functional group.

291c Catalytic Process Intensification of Bio-Renewable Surfactants Platform

Xiaokun Yang

The global surfactants market size is projected to reach USD 52.4 billion by 2025. Surfactants are the key active ingredient in cleaning and personal care products and are one of many petrochemical products that could be supplemented with biomass-derived materials. *Sironix Renewables LLC.* has invented a new class of surfactants, called Oleo-Furan Surfactants (OFS), which utilize the unique properties of biomass by linking the function of bio-based furan building blocks with natural oils to provide 100% bio-based multifunctional and eco-friendly cleaning and personal care products. This project is helping *Sironix Renewables* de-risk their process scale-up for new product development and maturation. The joint research project with *Sironix Renewables* is designed to leverage the catalyst development resources and catalytic reaction engineering of the ChemCatBio consortium with the surfactant's platform of *Sironix Renewables* to accelerate the DOE-invented and DOE Small Business Innovation Research-funded technology toward market commercialization. This presentation will cover the key technical goals accomplished including the development of new catalysts and reaction pathways of various furan-based structures to lower production costs with environmentally benign materials, process improvements to achieve the efficient scale-up strategy of existing surfactants, and techno-economic analyses to guide experimental efforts and measure market potential.

291d Development and Intensification of Green Pathways for Conversion of Soybean Oil to Biodiesel

Aydin Sunol

Experience and key learnings from experimental, piloting and modeling work that was undertaken for the past dozen years for transesterification of soybean and other oils in supercritical alcohols will be presented, focusing on supply chain sustainability and process intensification.

The two projects highlighted will be the process for novel reactive extraction at supercritical conditions and novel heat integrated 100,000 gallon/yr plant with an unique reactor design.

The supply chain implications of the novel distributed production approach and optimization of facility sizes will be discussed.

Hydrochars from Lignocellulosic Residues for Green Electronics, Environmental Remediation, Energy and Agricultural Applications

Andrea C. Landázuri[1,2,3,4*], Julio Chacón[5,6], Luis Miguel Prócel[7], Yuliana Pullas[1], María Fernanda Lecaro[1], Pierina Mirabá[7], Juan Fernando Posso[7], Katty Beltrán[7], Oliver Caisaluisa[7], Lourdes Orejuela[1,2,3,4], Darío Niebieskikwiat[8], Marco León[2,9], Nicolás Vela[2,10], Víctor H. Guerrero[11], Sebastián Ponce[2,12], Herman Murillo[12], Victoria Alomía[13], Nelson Herrera[7], Susan Yépez[1], Eduardo Holguín[7,14], Christian Luciani[8]

[1] *Universidad San Francisco de Quito USFQ, Colegio de Ciencias e Ingenierías, Applied Circular Engineering & Simulation Group (GICAS), Chemical Engineering Department, Diego de Robles y Vía Interoceánica, P.O. Box 17-0901, Quito, Ecuador*

[2] *Universidad San Francisco de Quito USFQ, Institute for Energy and Materials, Diego de Robles y Vía Interoceánica, P.O. Box 17-0901, Quito, Ecuador*

[3] *Universidad San Francisco de Quito USFQ, Instituto de Investigaciones Biológicas y Ambientales (Biósfera), Diego de Robles y Vía Interoceánica, P.O. Box 17-0901, Quito, Ecuador*

[4] *Universidad San Francisco de Quito USFQ, Instituto IBioMed, Diego de Robles y Vía Interoceánica, P.O. Box 17-0901, Quito, Ecuador*

[5] *Yachay Tech University, School of Physical Sciences and Nanotechnology, 100119-Urcuquí, Ecuador*

[6] *Department of Physics, Freie Universität Berlin, Arnimallee 14, 14195 Berlin, Germany.*

[7] *Universidad San Francisco de Quito USFQ, Instituto de Micro y Nanoelectrónica, Diego de Robles y Vía Interoceánica, P.O. Box 17-0901, Quito, Ecuador*

[8] *Universidad San Francisco de Quito USFQ, Colegio de Ciencias e Ingenierías, Departamento de Física, Diego de Robles y Vía Interoceánica, P.O. Box 17-0901, Quito, Ecuador*

[9] *Universidad San Francisco de Quito USFQ, Colegio de Ciencias e Ingenierías, Departamento de Ingeniería Mecánica, Diego de Robles y Vía Interoceánica, P.O. Box 17-0901, Quito, Ecuador*

[10] *Universidad San Francisco de Quito USFQ, Colegio de Ciencias e Ingenierías, Departamento de Ingeniería Ambiental, Diego de Robles y Vía Interoceánica, P.O. Box 17-0901, Quito, Ecuador*

[11] *Department of Materials, Escuela Politécnica Nacional, Quito, 170525, Ecuador*

[12] *Universidad San Francisco de Quito USFQ, Colegio de Ciencias e Ingenierías, Chemical Engineering Department, Diego de Robles y Vía Interoceánica, P.O. Box 17-0901, Quito, Ecuador*

[13] *Universidad San Francisco de Quito USFQ, Colegio de Ciencias e Ingenierías, Departamento de Ingeniería en Agronomía, Diego de Robles y Vía Interoceánica, P.O. Box 17-0901, Quito, Ecuador*

[14] *ISEP-Institut Supérieur d'Electronique de Paris, 10 rue de Vanves*

*Corresponding author: alandazuri@usfq.edu.ec

Abstract

The use of residues to obtain high-value materials, compounds, and products has gained attention in the last decade. Examples of lignocellulosic residues that are relevant to the Ecuadorian reality are those related to banana, cocoa, moringa, mango, and rice crops, among others. For instance, the banana, cocoa, and mango agribusinesses in Ecuador are recognized worldwide for having high-quality products that are attractive to consumers. However, their waste usually does not enter the value chain of industries, and millions of tons of waste are generated annually without relevant industrial reuse. Another example is the blackberry (Rubus niveus), which is considered one of the worst invasive plants in the Galapagos Islands, causing serious problems for local biodiversity and agriculture. Control methods such as manual eradication, application of herbicides, and biological control have been tried without significant success. There is also the added problem that agro-ecosystems in the islands are affected by low productivity caused, among other things, by the lack of local fertilizers. The goal of this project is to physically, chemically, and electrically characterize lignocellulosic residues in their native and hydrochar forms to suggest potential electrical, agricultural, and environmental applications that promote a circular engineering approach along with sustainable and green technological methods.

Introduction

Hydrothermal carbonization (HTC) is a promising method for recovering biomass waste and efficiently using biological resources, which aligns with the principles of Circular Bioeconomy and Circular Engineering [1]. HTC produces hydrochar which can be used as a nanostructured material for bio nano adsorbents, bio nanocatalysts, and nanoparticle immobilization during wastewater treatment, which enables the removal of organic and inorganic compounds [2], [3]. It can also be used for oxygen capture in agricultural applications and soil conditioning, which improves soil quality. HTC is an environmentally friendly method of improving biomass characteristics that eliminates the need for toxic reagents. Hydrochars can be characterized physically, chemically, and electrically to better understand their properties. HTC temperature, processing times, and activation methods influence the suitability of hydrochars for specific applications.

Besides abundant crops such as the banana, cocoa and mango crops, agribusiness of introduced crops such as the moringa agribusiness, has increased exponentially in Ecuador since it was introduced in the coastal region in 2012, it is a recognized superfood. Their residues (husks and pressed cotyledons) have been vastly studied for water treatment applications [4]. Rice is another important crop that feeds more than 50% of the world's population [5] and efforts to use rice husk as source of renewable energy and water treatment have also been made [6]. Regarding blackberry species, this is also of particular interest since it presents high adaptability to new environments, displacing native vegetation that may alter the ecological balance. For example, blackberry is considered an invasive species in the Galapagos Islands and solutions are necessary to be implemented in situ to avoid its propagation. Hence, an extraordinary number of lignocellulosic residues can be found worldwide from which high value-added materials, compounds and fine chemicals can be produced and extracted.

Lignocellulosic residues are rich in nutrient content and possess their own the chemical/phytochemical composition (identity) and with this in mind, the goal is to suggest which materials and under what conditions are suitable for i. electronic purposes (as dielectric materials or semiconductive ones), ii. soil amendments due to amount of nutrients present in hydrochars, iii. water treatment applications, iv. energy applications such as biofuel production.

Methodology

Materials characterization is performed to all native and hydrochar samples via X-ray Diffraction (XRD) (for determination of crystallographic structure), Thermogravimetric Analyses (TGA) (for evaluation of temperature effects), Fourier Transform Infrared (FTIR) spectroscopy (for material composition), Scanning Electron Microscopy (for morphology observation), Brunauer-Emmett-Teller (BET) surface area analysis, Raman spectroscopy (vibrational modes of the samples), and X-ray photoelectron spectroscopy (XPS) (for determination of the chemical state of the surface). Inductively coupled plasma - optical emission spectrometry (ICP-EOS) is also used for metal/metalloid quantification. In parallel, pellets for electrical characterization using a probe-station connected to a Semiconductor Analyzer is used for the determination of relative permittivity of the samples.

Results and Discussion

For this, results from the cocoa and banana residues are presented as an example. Relevant results showed that cocoa husk hydrochars (CHH) depict less variability in relative permittivity values with respect to ripe banana peels hydrochars (BPH) with approaching values between 8.25 to 10.29 for HTC temperatures between 150°C to 210°C (Figure 1). Hence, it is concluded that these materials possess dielectric capacities which imply their suitability for consumer electronics applications such as their use in capacitors. Loss tangent values, reported previously in another work [7] also show that these materials fall into the category of "low-loss" dielectric materials meaning a tendency towards insulating capacities as well. The corresponding tangent loss values varied between 0.034 to 0.041 (CHH) and 0.046 to 0.091 (BHH). In both cases the HTC process decreased the tangent loss, therefore improving dielectric properties. In general, the relative permittivity of the materials decreases with temperature and treatment time until approaching a limit value.

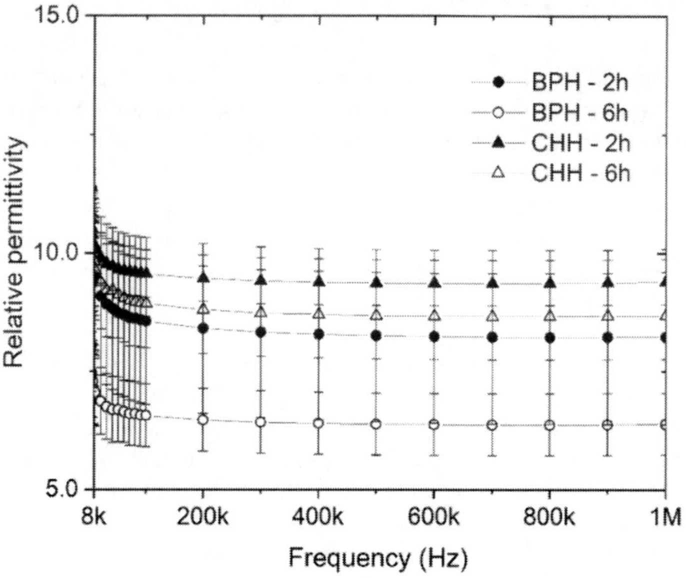

Figure 1. Relative permittivity of CHH and BPH for 210°C HTC treatments during 2 h and 6 h.

In addition, according to N_2 adsorption – desorption isotherms, as temperature and time increases the specific surface area represented as m^2/g of sample, increase as well. These materials are classified as non-porous or microporous materials (type II) and specific surface areas for this set of experiments were 0.071 (150°C) – 6.41 (210°C) m^2/g for BPH and 1.09 (150°C) – 35.8 (210°C) m^2/g for CHH. More detail can be found elsewhere [7]. CHH present higher areas at higher temperatures compared to BPH. Surface morphology (Figure 2) of the samples show that the BPH particles are slightly larger than CHH particles with an average length of 630 μm for 230°C HTC treatments (Figure 3).

The particle surface of native banana and BHH (Figure 2 a, b) are different to native cocoa husk and CHH (Figure 2c, d). Native banana (Figure 2a) and BPH particle morphology (Fig. 2b.) are quite similar. These particles are rough, with rugged surfaces, however the BPH particles showed a slightly rougher surface. In the case of cocoa husk particle surfaces, differences are observed between native and CHH particles. The native particles are rough (Figure 2c), and the hydrochar particles are porous (Figure 2d). Comparing native cocoa particles with banana particles, these are less rough and flatter.

From XRD (Figure 4) and XPS (Figure 5) results it is found that graphitic material is present in all the samples, and the absence of the characteristic (002) reflection may indicate that graphite is in turbostratic form, such that the disordered stacking of the carbon layers would kill the intensity of this peak. The only exception would be the BPH produced at 210°C, where the (002) reflection of graphite is clearly observed through XRD analyses. XPS reveals the presence of C, O, N, P, K and Ca, being C, the largest contribution due to the presence of lignin and cellulose molecules within the structures. Also, it important to notice that N content was not affected, therefore these hydrochars can be considered as N-doped hydrogenated carbon nanostructures. On top, BPH C1s spectra clearly resembles a more graphitic structural system [8] when compared to the CHH. BPH generates a more crystalline structure compared to CHH. This is due to the relative carbon/oxygen concentration which is higher than the pristine materials and CHH. Additionally, the shape of the C1s peak is more similar to a graphitic material for this case. BPH generates a more crystalline structure compared to CHH. This is due to the relative carbon/oxygen concentration which is higher than the pristine materials and CHH. Additionally, the shape of the C1s peak is more similar to a graphitic material for this case.

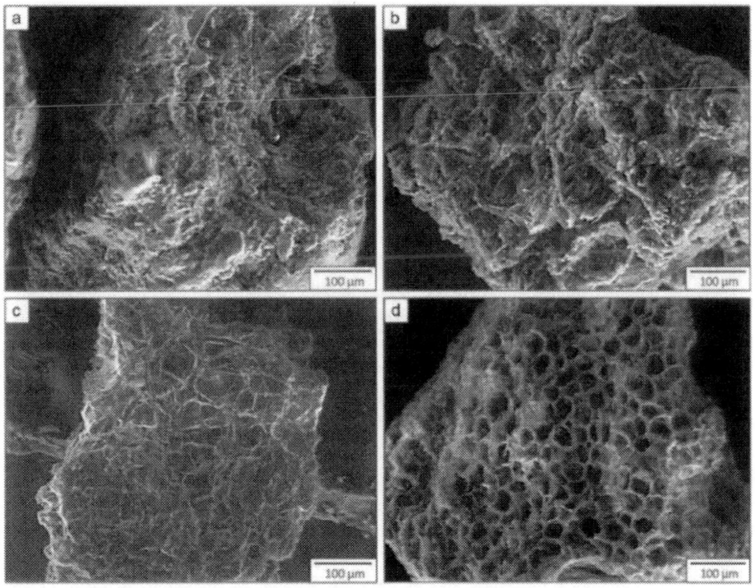

Figure 2. SEM micrographs at 200x magnifications of: a) native banana peel particles; b) BPH particles; c) native cocoa husk particles; d) CHH particles.

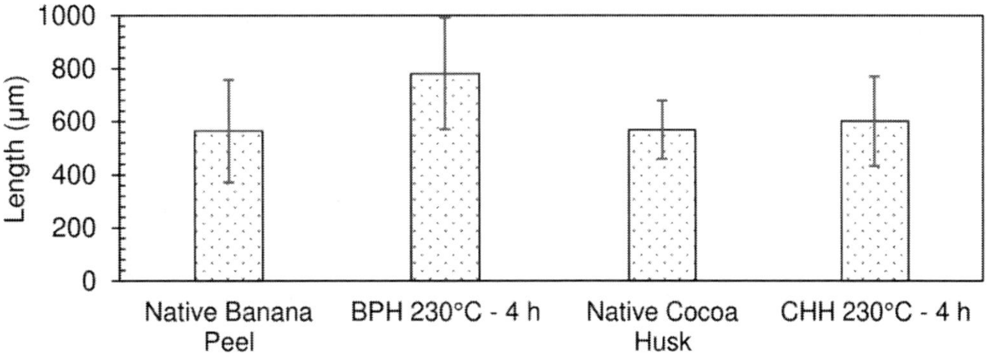

Figure 3. Average length of hydrochar particles (e.g. native banana peels and cocoa husks, and hydrochars produced at 230°C and 4 h).

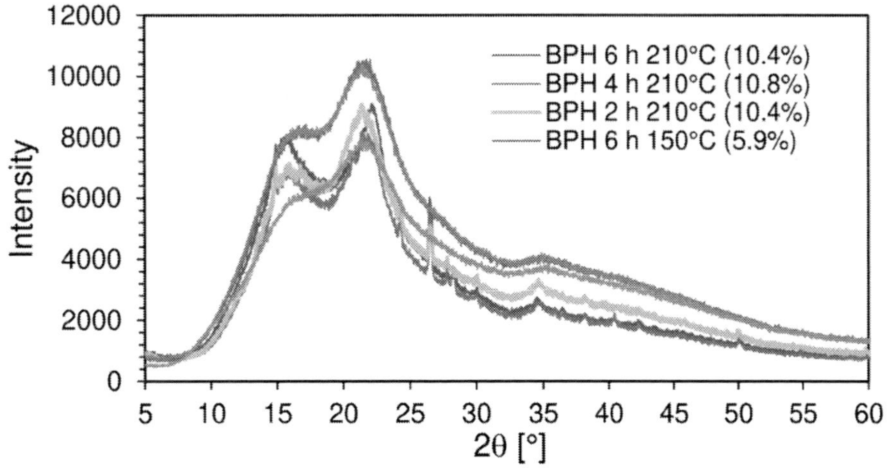

Figure 4. X-ray diffractograms of BPH and CHH treated at 150°C and 210°C. In general, as temperature and HTC time increases, so does the crystallinity (values in parentheses) of the samples.

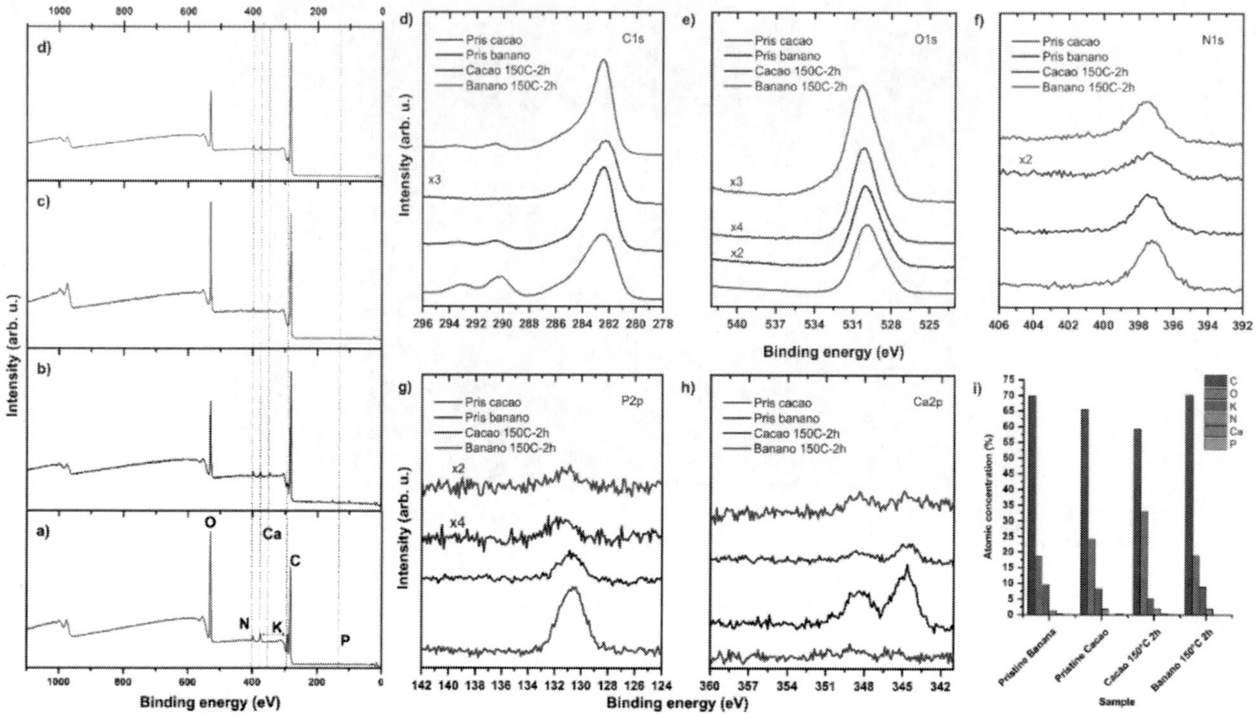

Figure 5. X-ray photoelectron spectroscopic analyses. Survey spectrum of a) native cocoa husk, b) native banana peel, c) CHH 150°C 2h and d) BPH 150°C 2h revealing the presence of oxygen, nitrogen, calcium, potassium, carbon and phosphorous elements present in the samples. e) Atomic concentration analysis extracted from the a) to c) survey spectrum specific to C, O, K, N, Ca and P atoms. f to j) High resolution spectral analyses from the C1s, O1s, N1s, Ca2p, and P2p regions disclosing the chemical state modifications of the native samples (wine, black lines) with respect to the thermal treated ones at 150°C (blue top lines). A deconvolution spectral analysis was conducted in BPH 150°C 2h C1s and O1s regions to confirm the evolution of the chemical states in the samples.

The metal content can vary depending on the feedstock and process conditions used during HTC, and can have implications for the potential applications of the resulting hydrochar. For example, metal/metalloid from blackberry bushes, mango endocarps and their hydrochars (Figure 6) present considerable amounts of Ca, K, Mg, P.

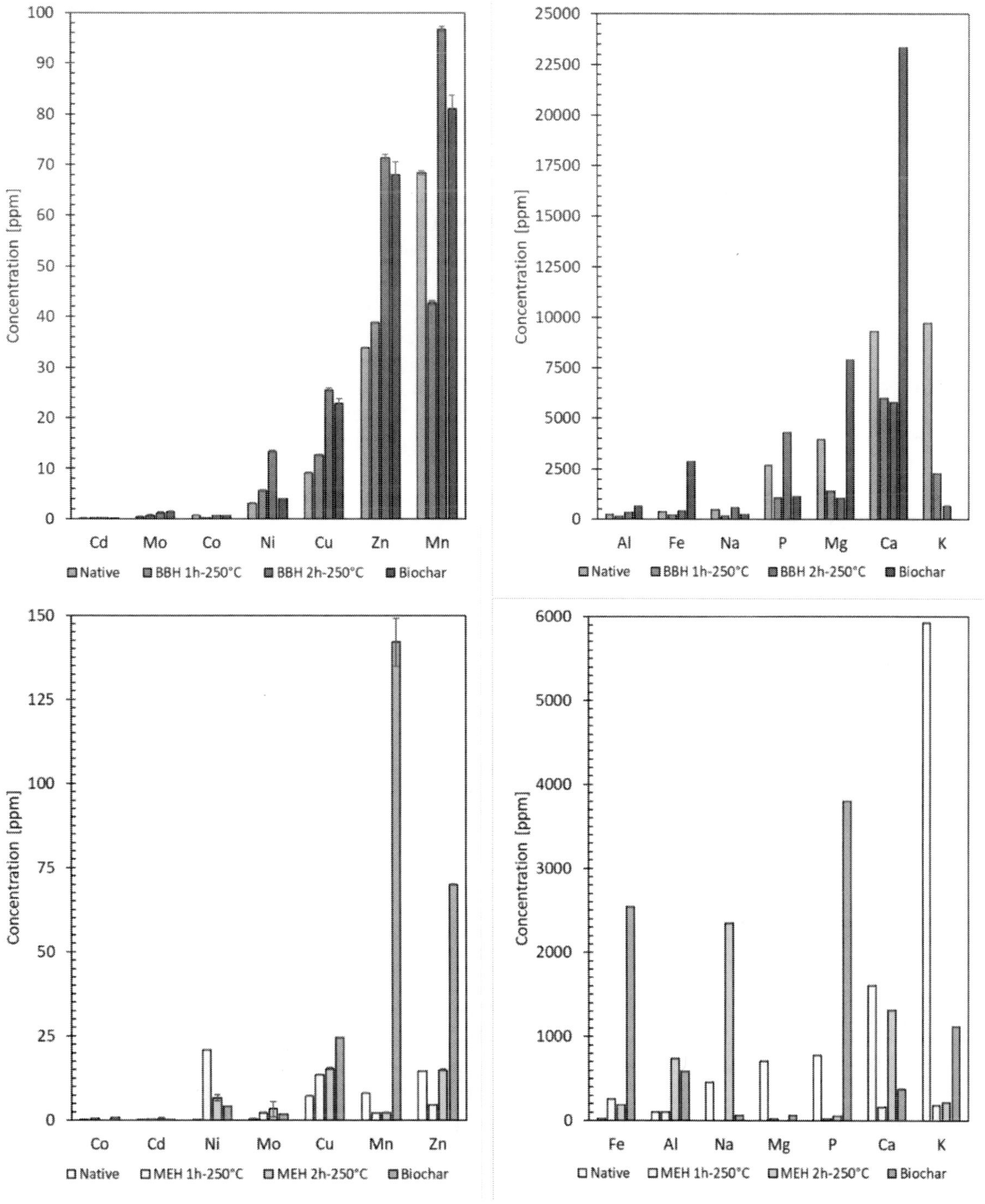

Figure 6. Metal and metalloid content present in (a) native blackberry bushes from the Galápagos, blackberry bushes hydrochars (BBH) processed for 1h and 2 h at 250°C, and pyrolyzed blackberry bushes (biochar), (b) mango endocarps and mango endocarps hydrochars (MEH) processed for 1h and 2h at 250°C and pyrolyzed mango endocarps (biochar). Note that data is sorted in ascending order of concentrations based on native materials.

Discussion

Hydrochars from lignocellulosic residues have potential applications in electronic fields, fuel production, agriculture and in environmental fields such as water treatment [9], nutrient recovery [10], soil improvement [11], and greenhouse gas emissions reduction [12]. Overall, this project depicts the potential applications of abundant and exotic lignocellulosic residues based on their physical, chemical and electrical properties. These materials can be tailored and activated to enhance a desired effect. For example, these carbon-based materials can be chemically treated for improving their adsorptive capabilities. The use of chemical agents such as acid (HCl), base (NaOH, KOH), and metal oxides would increase their specific surface area in conjunction with a surface functionalization [13]. Thus, a faster and selective adsorption process can be performed. On the other hand, HTC can be carried out towards fuel production. Lignocellulosic residues, especially those from the agricultural sector can be used as fuels; however, their direct combustion is not suitable from an environmental point of view. These residues usually show significant contents of fuel-bound nitrogen, which increases NO_X concentrations; high volatile matter contents that promote the release of products of incomplete combustion such as carbon monoxide, volatile organic compounds, and polycyclic aromatic hydrocarbons (PAHs); and high ash contents rising the inorganic particulate matter emissions. In this sense, agro-forest-based hydrochar fuels in form of pellets have been demonstrated to successfully reduce CO and particulate matter emissions in domestic heating stoves [14]. In the case of NO_X emissions, modifying the operating conditions by using catalysts during the HTC process seems to be a proper way to obtain hydrochars showing lower nitrogen contents [15], [16] thus reducing NO_X concentrations during combustion.

Hydrochars have shown good adsorption properties for a variety of contaminants, including heavy metals, organic pollutants, and dyes. This makes them a promising material not only for water treatment applications but also for the recovery of nutrients from wastewater. For example, hydrochars produced from sludge and lignocellulose mixtures [12] have been shown to have good potential for nitrogen recovery. Hydrochars can also be used to improve soil properties, such as water retention, nutrient availability, and pH. This can have implications for agriculture and land remediation. As for metal and metalloid content in native materials and in their hydrochars, this is an important property to consider for both environmental and valorization reasons. It has been shown that residues such as mango endocarps and blackberry residues possess important amounts of Ca, K, P, and Mg. These are important nutrients for plant growth and development in agricultural purposes. Mass transfer of these compounds into the liquid phase (liquor) is to be investigated. These nutrients are considered essential elements that plants must have in order to grow properly[11].

The characterization of hydrochars is crucial in order to propose downstream applications. It has already been pointed out that crops such as the banana, cocoa, and mango crops are globally relevant, but tons of their residues do not enter the industrial value chain. Therefore, it is important to find alternative uses for these residues that could lead to novel and sustainable applications in diverse fields, such as consumer electronics. HTC is a promising tool for the valorization of these wastes. The dielectric characteristics of hydrochars produced from ripe banana peels and cocoa pod husks were explored in a previous study [7] where HTC produced highly stable materials with low loss tangent values, making them good candidates for insulation and possessing good dielectric properties that can be further processed for downstream applications in the electronics industry. Similar analyses for other residues such as moringa husks, mango endocarps, rice husks, palm and mangosteen rinds are being carried out along with nutrients determination mainly for soil conditioning, metal recovery [9], immobilization in contaminated soils[17], emerging organic pollutants removal [18], and the correlation between all evaluated properties. In parallel, the liquid fractions from the HTC treatments will also be explored.

Conclusions

This work focuses on Sustainable Development Goals: 6 ,7, 9, 12 and 13, mainly, to contribute towards ensuring clean water, affordable energy, responsible consumption and production, while promoting innovation to industrial applications with less impacts to climate change. HTC gives greater added value to the agro-industrial residues to develop the Circular Bioeconomy in the country and find new industrial applications with a sustainable approach. Future work involves the use of all characterization data as input for AI algorithms that can help predict the evolution of lignocellulosic materials, other properties and even predict the HTC process itself.

References

[1] L. M. Orejuela-Escobar, A. C. Landázuri, and B. Goodell, "Second Generation Biorefining in Ecuador: Circular Bioeconomy, Zero Waste Technology, Environment and Sustainable Development – The Nexus," *Journal of Bioresources and Bioproducts*, 2020.

[2] B. A. Omran and K.-H. Baek, "Control of phytopathogens using sustainable biogenic nanomaterials: Recent perspectives, ecological safety, and challenging gaps," *J Clean Prod*, vol. 372, p. 133729, 2022, doi: https://doi.org/10.1016/j.jclepro.2022.133729.

[3] W. Liang *et al.*, "Hydrothermal carbonization of forest waste into solid fuel: Mechanism and combustion behavior," *Energy*, vol. 246, p. 123343, 2022, doi: https://doi.org/10.1016/j.energy.2022.123343.

[4] J. S. Villarreal, J. R. Gándara, D. Navarrete, M. L. Bejarano, and A. C. Landázuri, "Lead (Pb2+) adsorption by means of pristine and prewashed residual Moringa oleifera Lam. seed husk biomass for water treatment applications," *International Journal of Sustainable Engineering*, vol. 00, no. 00, pp. 1–13, 2020, doi: 10.1080/19397038.2020.1862350.

[5] D. Portalanza *et al.*, "Potential Impact of Future Climates on Rice Production in Ecuador Determined Using Kobayashi's 'Very Simple Model,'" *Agriculture (Switzerland)*, vol. 12, no. 11, Nov. 2022, doi: 10.3390/agriculture12111828.

[6] W. I. Mortada, R. A. Mohamed, A. A. A. Monem, M. M. Awad, and A. F. Hassan, "Effective and Low-Cost Adsorption Procedure for Removing Chemical Oxygen Demand from Wastewater Using Chemically Activated Carbon Derived from Rice Husk," *Separations*, vol. 10, no. 1, p. 43, Jan. 2023, doi: 10.3390/separations10010043.

[7] A. C. Landázuri *et al.*, "Valorization of ripe banana peels and cocoa pod husk hydrochars as green sustainable 'low loss' dielectric materials," *J Clean Prod*, p. 139044, Oct. 2023, doi: 10.1016/j.jclepro.2023.139044.

[8] A. Fujimoto, Y. Yamada, M. Koinuma, and S. Sato, "Origins of sp3C peaks in C1s X-ray Photoelectron Spectra of Carbon Materials," *Anal Chem*, vol. 88, no. 12, pp. 6110–6114, Jun. 2016, doi: 10.1021/acs.analchem.6b01327.

[9] S. Li, B. Ji, and W. Zhang, "A review on the thermochemical treatments of biomass: Implications for hydrochar production and rare earth element recovery from hyperaccumulators," *Chemosphere*, vol. 342, p. 140140, 2023, doi: https://doi.org/10.1016/j.chemosphere.2023.140140.

[10] L. Wu, W. Wei, D. Wang, and B.-J. Ni, "Improving nutrients removal and energy recovery from wastes using hydrochar," *Science of The Total Environment*, vol. 783, p. 146980, 2021, doi: https://doi.org/10.1016/j.scitotenv.2021.146980.

[11] D. Skrzypczak *et al.*, "Innovative fertilizers and soil amendments based on hydrochar from brewery waste," *Clean Technol Environ Policy*, 2023, doi: 10.1007/s10098-023-02557-5.

[12] M. Ebrahimi *et al.*, "Effects of hydrochar derived from hydrothermal treatment of sludge and lignocellulose mixtures on soil properties, nitrogen transformation, and greenhouse gases emissions," *Chemosphere*, vol. 307, p. 135792, 2022, doi: https://doi.org/10.1016/j.chemosphere.2022.135792.

[13] M. S. Soffian, F. Z. Abdul Halim, F. Aziz, M. A. Rahman, M. A. Mohamed Amin, and D. N. Awang Chee, "Carbon-based material derived from biomass waste for

wastewater treatment," *Environmental Advances*, vol. 9. Elsevier Ltd, Oct. 01, 2022. doi: 10.1016/j.envadv.2022.100259.

[14] H. A. Murillo, L. A. Díaz-Robles, R. E. Santander, and F. A. Cubillos, "Conversion of residual oat husk and pine sawdust by co-hydrothermal carbonization towards biofuel production for pellet stoves," *Ind Crops Prod*, vol. 174, Dec. 2021, doi: 10.1016/j.indcrop.2021.114219.

[15] J. Massaya, G. Pickens, B. Mills-Lamptey, and C. J. Chuck, "Enhanced Hydrothermal Carbonization of Spent Coffee Grounds for the Efficient Production of Solid Fuel with Lower Nitrogen Content," *Energy & Fuels*, vol. 35, no. 11, pp. 9462–9473, Jun. 2021, doi: 10.1021/acs.energyfuels.1c00870.

[16] B. Zhang, J. Wang, Z. Xu, S. Wu, R. Luque, and H. Zhang, "Sewage sludge valorisation by hydrothermal carbonization: A new method to enhance nitrogen removal in hydrochar catalyzed with Ni–Mg–Al layered double oxides," *J Clean Prod*, vol. 386, Feb. 2023, doi: 10.1016/j.jclepro.2023.135880.

[17] Y. Chen, J. Chen, S. Chen, K. Tian, and H. Jiang, "Ultra-high capacity and selective immobilization of Pb through crystal growth of hydroxypyromorphite on amino-functionalized hydrochar," *J Mater Chem A Mater*, vol. 3, no. 18, pp. 9843–9850, 2015, doi: 10.1039/C5TA01011J.

[18] U. A. Qureshi, B. H. Hameed, and M. J. Ahmed, "Adsorption of endocrine disrupting compounds and other emerging contaminants using lignocellulosic biomass-derived porous carbons: A review," *Journal of Water Process Engineering*, vol. 38, Dec. 2020, doi: 10.1016/j.jwpe.2020.101380.

195b Roadmap for Scale-up of Carbon Nanotube Production for Displacement of Difficult to Decarbonize Structural Materials with Free Hydrogen Co-Production

Alex Dantzler, Jason DesVeaux, Oliver Dewey, Arthur Sloan, Zhe Cui, Abhishek Roy, Leonardo Spanu, Glen Irvin, Matteo Pasquali

Manufacturing structural materials makes-up an enormous amount CO_2 emissions, with steel alone constituting ~8% of global CO_2 annually. Structural materials like metals and cement are required in large volumes to support our infrastructure, and the manufacturing process is CO_2 intensive and considered difficult-to-decarbonize because of stoichiometric emissions from high-temperature reduction steps. High-performance materials like carbon fiber seek to address these issues by requiring less material to meet the same strength requirements; however, half of the mass of fossil-based feedstock is lost converting to carbon fiber, so the process is too inefficient to achieve a cost and supply able to meet structural material demands, beyond niche applications in aerospace and recreation. Carbon nanotubes (CNT) and their macrostructures (e.g., fibers, yarns, woven mats, etc.) have much higher predicted strength and conductivity than carbon fiber but have a simpler manufacturing process that suggests a low floor for production cost and environmental footprint. Research on CNT to-date has largely focused on property improvement, but with reported tensile strength of CNT fibers already an order of magnitude higher than steel (on a mass basis) and demonstrated doubling of fiber properties every three years (Taylor et al. Improved Properties, Increased Production, and the Path to Broad Adoption of Carbon Nanotube Fibers, Carbon 171 (2021) 689-694) the question of scalability becomes interesting. Here, we propose a techno-economic model for the scale-up of CNT production with H_2 co-production using lab-scale reactor conditions like reaction conversion, selectivity, and temperature to create a roadmap for researchers to push the field towards scalable and high-performance CNT at low environmental costs as measured in CO_2 intensity, embodied energy, and process efficiency. Preliminary analysis of this roadmap suggests that CNT production has the potential to be a carbon sequestering process and can ultimately compete with steel given focused research and investment to push the material along its learning curve. Process simulation (Aspen Plus) confirms the tractable model, and work is ongoing to explore the implications of various hydrogen separation schemes on the techno-economics.

195c Synthesis of Novel Intermediate Material for Recovery of Lithium-Ion Batteries

Mriganka Mondal, Sergio Carrillo De Hert, Rafael Lopez-Rodriguez, Gita Singh, Tony Keene, Steven Ferguson

Lithium Ion Batteries (LIBs) are widely used for energy storage in portable devices, electric vehicles, and storage as most countries aim to be environmentally friendly. The use of LIBs is lucrative due to their high specific energy density, high rated voltage, and low self-discharge rate. Primarily LIBs contain cathode-active materials, such as Lithium Cobalt Oxide ($LiCoO_2$), Lithium Nickel Manganese ($LiNi_xMn_yCo_zO_2$, $x+y+z = 1$), a graphite anode, a copper cathode, an aluminum cathode collector, PVDF binder, and the electrolyte. With their increase in consumption, recycling and disposal of LIBs pose a significant challenge for human health and the environment as they contain a high percentage of heavy metals and toxic electrolytes. In recent years, a lot of effort has been made in order to recover different components of the batteries, however, recovery of spent cathode material poses a significant future for the circular economy. The common method to extract the cathode metallic components (such as $LiCoO_2$, LCO) is pyrometallurgical, hydrometallurgical, mechanical, and mechano-chemical processes. The pyrometallurgical method is energy intensive and low efficient as most of the valuable metal sources are obtained in alloy and lithium is lost in slag. The hydrometallurgical method typically involves the use of inorganic acids and the use of electrochemical processes leading the process to be energy intensive and high material cost. The mechanical and mechano-chemical processes have the benefit of using minimum energy but the downstream processing can be very complex and the use of heavy machinery can make it expensive.

Thus, in the present study, a novel solvothermal method is proposed for the reactive crystallization of an intermediate novel precursor material (PM1), $[Li(C_2O_4)]_2[Co_5(OH)_8]$. PM1 has a global structure of heterometallic layered double hydroxide with an asymmetric unit that can be represented in Figure 1 (a) and the visual representation can be made from Figure 1(b) which shows the AB layers containing lithium oxalate and cobalt hydroxide that are present in the PM1 structure. Additionally, cobalt can be octahedrally coordinated or tetrahedrally coordinated where systematic depilation takes place.

Lithium cobalt oxide contains a layered structure similar to what has been mentioned in the case of PM1. This enables the recovery of LCO at much lower temperatures $T = 300\text{-}350^{\circ}C$ compared to the aforementioned conventional process, making it more energetically favorable. The spent LCO from a battery waste is primarily digested in an organic acid (typically oxalic acid can be used) at elevated temperature (>80°C) resulting in Lithium oxalate and cobalt oxalate dihydrate. Later, by adjusting the stoichiometric ratio of cobalt and lithium oxalate to 1:1 in the pH range of 8-10, the PM1 can be synthesized in an aqueous medium at $T = 80\text{-}100^{\circ}C$. Powder X-ray diffraction indicates this new synthetic route is successful in synthesizing LCO with high phase purity at temperatures far below what has been reported previously. However, the presence of excess cobalt in the PM1 lattice leads to the production of an LCO and cobalt oxide (II, III) mixture (CO_3O_4) upon calcination along with CO_2 and H_2O. The separation of CO_3O_4 from the calcined material becomes crucial with a number of novel separation operations developed and characterized to this end. Furthermore, the method has been demonstrated to be extendable to NMC chemistries using an analogous intermediate PM2. In this study, a novel process utilizing these novel intermediates and associated plant designs is developed and compared with industrial processes to demonstrate their economic and industrial feasibility for lithium-ion battery recycling.

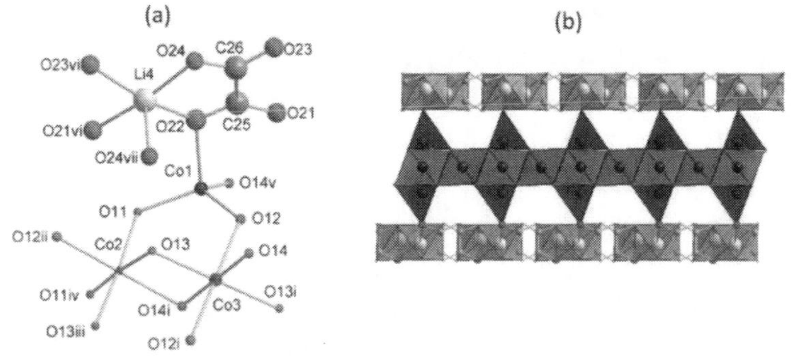

Figure 1 (a) Asymmetric Unit of PM1 (Excluding Hydrogen Atom) (b) Stacking of layers in PM1. $[Li(ox)]^{-1}$ layer is shown in the top and bottom layer and $[Co_3(OH)_8]^{2+}$ is represented in the middle layer.

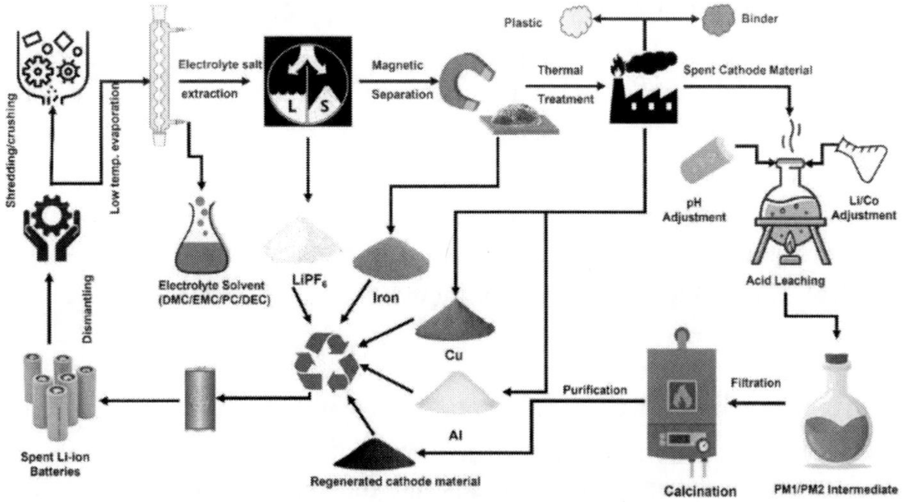

Figure 2 Graphical Abstract of the Proposed Process

195d From Landfill to Cement: Developing a Negative Emission Hydrothermal Pathway to Sequester Waste Carbon

David Kenney, Shuai Wang, Nima Rahbar, Michael T. Timko, Andrew R Teixeira

The 2021 greenhouse emissions for the United States were reported to be 6.4 billion tons of CO_2-eq. Nearly 12% of all emissions are related to two industries: 1) Municipal and Industrial Solid Waste Management and 2) Cement Production. Waste management primarily relies on landfilling to sequester waste, however, once capped landfills become a large emitter of methane (CH_4) which has 30 times the global warming potential of CO_2. On the other hand, ordinary Portland cement (OPC) requires calcium oxide (CaO). To produce CaO at the rate and quantities needed for modern cement demands, calcium carbonate ($CaCO_3$) is calcined to remove CO_2 and produce CaO. This step alone generates 50% of all emissions from cement production. To offset the massive production of greenhouse gases, strategies often focus on direct air capture (DAC) or point source capture (PSC) which look to take dilute or low-pressure CO_2, concentrate it, and then either sequester it or utilize the recovered CO_2 for alternative processes. Focusing these conditions requires significant energy inputs that inherently draw away from the removal capabilities. However, by using a more carbon dense feed such as organic municipal waste and coupling it with a downstream highly CO_2-selective precipitation, the CO_2 removal potential greatly increases.

We took a two-pronged approach to validate this system for commercial deployment. Experimentally, food waste was driven to high purity CO_2 at temperatures and pressures well within the defined supercritical point for water with a stoichiometric excess of hydrogen peroxide to act as an oxidant. Next, a batch pressure reactor was used to demonstrate the enhanced kinetics of mineralization for a high pressure CO_2 feed. The high pressure CO_2 was then mixed with a buffered aqueous solution with free calcium ion (Ca^{2+}). Elevated pressures (>100 psi) of CO_2 were found to achieve 10% conversion to $CaCO_3$ in 10 minutes. Atmospheric CO_2 (P_{CO2} = 0.87 psi) under the same reaction conditions was found to have negligible conversion. Lastly, the resulting mineral has been used as part of the aggregate material to produce alternative cement mortars. Using the experimental results, an extensive lifecycle analysis revealed that when these industries are combined, dry food waste can remove ten times its weight in CO_2 rather than emitting back from a landfill.

195e On the Use of Low Carbon Marine Fuels with Options for on-Board Carbon Capture: A Holistic Evaluation Using Life Cycle Analysis and an Integrated Approach

Nefeli Roufogali, Antonios Kokosis

The shipping industry is under increasing pressure to reduce Greenhouse Gas (GHG) emissions due to increasingly stringent emission regulations worldwide. The International Maritime Organization envisages a 40% reduction of CO_2 emissions per transport work across international shipping by 2030 compared to 2008. In order to comply, technologies involving low-carbon and carbon-free alternative fuels must mature to a point of economic viability and longevity. This study deploys Life Cycle Assessment (LCA) methodologies to evaluate the total environmental impact of carbon capture solutions combined with the use of prominent alternative fuels, namely; Liquified Natural Gas (LNG), methanol, biodiesel and ammonia, while adopting a systems engineering approach for energy optimization and general design considerations, with the objective of supporting the design of low environmental impact ROPAX ships. Stakeholders require such evaluations to maintain an accurate view on the technical performance and limitations of combined technologies focused on GHG emissions reduction.

While currently there is increased interest in decarbonization technologies from the shipping sector, significant research on the technologies themselves and on energy system studies is needed to evaluate possible comprehensive solutions for both the short and long term. Various energy efficiency measures have potential to reduce carbon emissions, however widespread adoption of low- or zero-carbon fuels and energy carriers is critical in order to achieve GHG goals while ensuring longevity of investments. A key advantage of alternative fuels is that they offer energy integration and cogeneration opportunities that are different and often additional to those in conventional fuels. Notably, in the case of LNG there is opportunity for cogeneration and auxiliary firing that one could utilize to support the hotel demands on ships such as ROPAX and ferries with larger service areas on board. Additionally, required conditions of fuel storage are similar to those of CO_2 storage, presenting a unique advantage for energy integration, reduction of methane slip and significant OPEX reductions when carbon capture systems are implemented on board. Considered together, alternative fuels combined with carbon capture technologies have the potential for a near-zero environmental impact ship at a reasonable financial cost. For extensive fuel transition, multiple options in fuel choice must be considered as each presents different opportunities and risks.

LNG presents a promising alternative to Marine Diesel Oil because of its increased availability relatively to other alternative fuels, zero Sulphur emissions and lower carbon emissions. In addition, due to its cryogenic storage it is a viable source of cold energy which can be used in the liquefaction of CO_2 from flue gas, resulting in a lower energy penalty for CO_2 capture and short-term storage on board. Another cost-competitive although potentially hazardous alternative marine fuel is methanol, which features reduced air pollutant and GHG emissions, however on a life-cycle basis the production method of methanol (sources range from coal to biomass) determines the carbon intensity of its use. Combined with carbon capture technologies and green hydrogen, methanol usage has the potential to be part of a circular economy cycle, where its carbon emissions are used to produce more methanol. Ammonia, as a carbon-free alternative fuel, has the potential for significant reduction in GHG emissions, while being a balanced solution in terms of volumetric energy density. The main specific risk of ammonia use as marine fuel, which is included in the impact assessment, is its high toxicity and the potential exposure to humans and to the environment. One of the limitations of alternative fuels in terms of viability is the minimum requirement of retrofitting a preexisting ship. This aspect is eliminated when considering biodiesel fuel as an alternative fuel for marine diesel engines applications. Biodiesel is suitable for wide spread use as it is non-toxic, renewable and biodegradable and while it is associated with an increased production of nitrogen oxide gas, it conversely reduces other toxic gas emissions. In this study, the carbon capture technology chosen was membrane air separation technology, with the final product, after extensive cooling, being liquid CO_2, which is unloaded at the port of destination.

The paper supports a systems approach that considers multiple degrees of freedom, that include the type of alternative fuel, the design and efficiency of the cogeneration process, the engine used, as well as CO_2 capture methods, by making use of data from experts in the maritime and energy industries when available and from literature and computational models when unavailable. Possibly varying or dynamic aspects are included, to illustrate the impact of critical parameters on the available options feasibility and environmental benefit. Design parameters include engine efficiency, air to fuel ratio, combined heat and power generation system use, extend of CO_2 separation, and energy use reduction due to process integration. The proposed approach supports an energy and process integration methodology that considers the ship as a system to integrate in order to minimize CO_2 emissions at different levels of CAPEX consideration. The LCA results function as the main criteria of comparison to evaluate the optimal selection of parameters for minimal environmental impact.

With the objective of maximizing the efficiency of energy resources, methods of heat and power integration in all presented cases of fuels are applied including when CO_2 is captured on board when applicable, resulting in uniquely optimized scenarios for each energy carrier and combined CO_2 capture system, within the same energy network of the ship. The methodology for this analysis combines thermodynamic cascade methods and Pitch Analysis while utilizing the synergy potential for further energy optimization between storage requirements of alternative fuels and carbon capture technology.

The functional unit of the LCA is a sea travel roundtrip in Greece involving a ROPAX vessel, with utilization during the summer months, including maneuvering and at-sea travel which is used as a case study in this paper. A comparative assertion Life Cycle Assessment was conducted between the alternate energy carriers and carbon capture technologies that includes all the sub-processes within the vessel that are identified to relevantly contribute (attributional modelling). All processes throughout the life cycle of fuel use are included in the assessment (cradle to grave) however in the primary data of the analysis the impact of capital goods is considered minimal across the lifespan of the vessel and left out, while the databases used for background secondary data about processes beyond the scope of this study (e.g. fuel production and transport to the ship) include the environmental impact of capital goods. Primary data refer to the storage, integration and use of the proposed alternative fuels within the boundaries of the vessel. The assessment results are the basis for a sensitivity analysis to support identification of the main contributing elementary flows as part of the stepwise improvement of relevant inventory data and uncertainty management, thus allowing for quantification and precision improvement of the study. The tools used for this process were Sima Pro v9.4 and the ecoinvent v3 Life Cycle Inventory database for increased precision in data based on geographical location and in traceability of impact.

Preliminary results indicate that for the use of LNG the adoption of energy recovery and cogeneration methods can lead to an increase in thermodynamic effectiveness of up to 38%, depending on the configuration used. Notably, the application of 2-stroke engines combined with a heat recovery steam generator (the produced steam is used for hotel demands) presents a very efficient scenario with an efficiency improvement of 26%, while also reducing methane slip by 67%. With the additional use of turbines, up to 33,6% of total electricity requirements of the ship can be fulfilled. When considering the capture and liquefaction of CO2 from flue gas by 70% to 80%, the potential for an energy efficient ship with low carbon emissions is evident. Further valorization of energy losses from LNG storage to combustion can lead to even higher overall efficiency of fuel use, provided comprehensive heat integration systems.

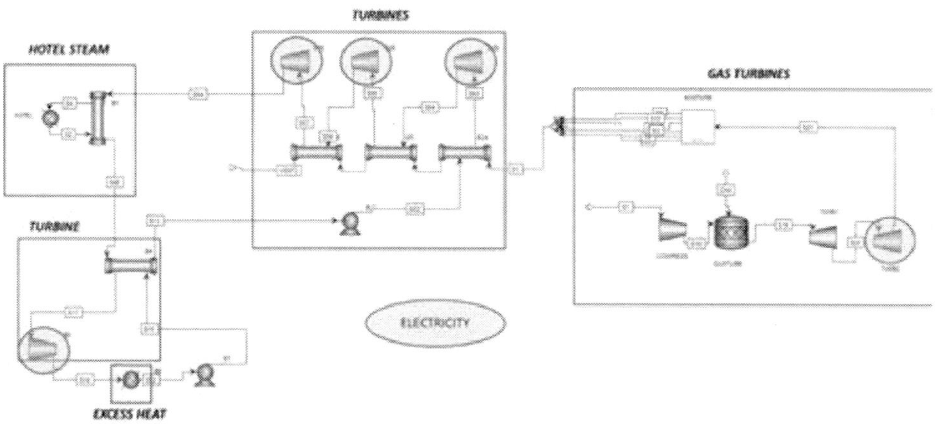

Figure 1. Flowsheet of heat and power cogeneration system using gas turbines with 38% efficiency improvement

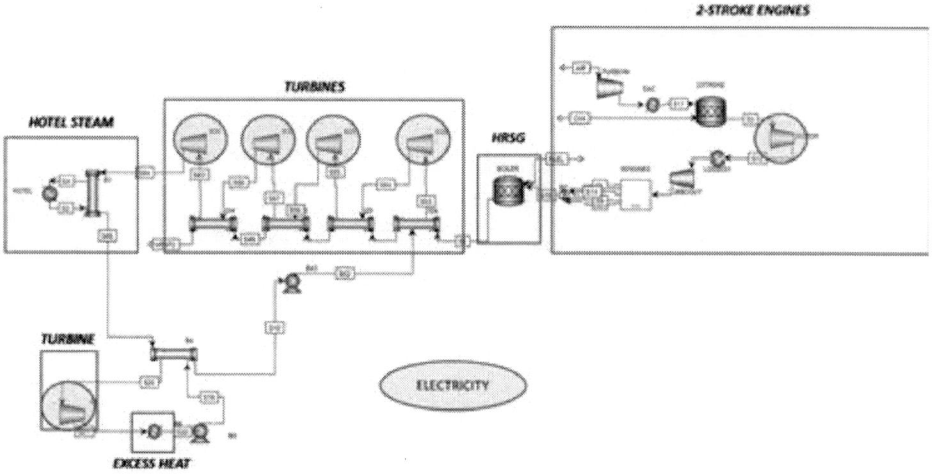

Figure 2. Flowsheet of heat and power cogeneration system using 2-stroke engines and a HRSG with 26% efficiency improvement

195f Sustainable Nitrogen By Design: Mechanochemically Synthesized Co-Crystal Materials for Enhanced Efficiency Fertilizer Products

Mohamed Eisa, Jonas Baltrusaitis

Urea production costs can be attributed, among other things, to the energy cost of atmospheric nitrogen fixation, which is estimated to be from 1% to 2% of global energy consumption. The recent global events, including the global pandemic and Ukraine's invasion by Russia, led to unprecedented market volatility and food supply chain disruptions which resulted in high fertilizer costs. Additionally, the useful yield of N from mineral N fertilizer factory-to-mouth is only ~4% with only 47% of N taken up from the field to crops. We postulate that the overall N fertilizer production and use cycle, not only NH_3 synthesis needs to be evaluated to obtain a more sustainable nitrogen fertilizers but also novel materials have to be engineered that have higher stability in the environment.

In this talk we will focus on the mechanochemically synthesized urea cocrystal materials as both sustainable process and a product with the decreased strain on the environment. In particular, green mechanochemical methods avoid use of aqueous solutions and the resulting evaporation and crystallization processes while urea cocrystals with Ca- and Mg-salts exhibit decreased solubility and lower reactive nitrogen losses. Presentation will focus on the discovery and synthesis of urea cocrystals, their mechanochemical synthesis kinetics, scale up engineering as well as the resulting fundamental surface properties in the presence of water.

98a Increasing Diffusion Selectivity Via Free Volume Manipulation (FVM) and *in Situ* Thermal Oxidative Crosslinks of Amine-Functionalized PIM-1 Membrane

Taigyu Joo, Tae Hoon Lee, Wan-Ni Wu, Sungsool Wi, Zachary P. Smith

Polymers of intrinsic microporosity (PIMs) show promise as membrane materials because of their high free volume and diffusion selectivity from their rigid backbone structures. PIMs can be functionalized with polar moieties like amines to enhance CO_2 sorption; however, the introduction of polar groups that form hydrogen bonds also reduces free volume, making it challenging to increase CO_2 sorption without compromising diffusivity. To overcome the loss of free volume induced by polar moieties, a solid-state deprotection approach called free volume manipulation (FVM) has been developed to reconstruct physical packing structures of polymers without significantly changing the polymer chemistry. FVM protects polar moieties with labile and bulky groups to disrupt chain packing and widen intersegmental distances. When the labile groups are deprotected from the solid-state film using heat treatments, the polymer recovers its original polar moiety, increases free volume, and gives access to new diffusion pathways for gas penetrants. FVM was demonstrated to be highly effective for amine-functionalized PIM-1 (PIM-NH_2) using labile tert-butoxycarbonyl (tBOC) group and resulted in a simultaneous boost in permeability and selectivity, which is a difficult enhancement for polymer membranes due to the trade-off relationship between permeability and selectivity. During FVM, light urea crosslinks form and play a significant role in preserving the increased free volume, narrowing the free volume element distribution, and providing high energetic barriers for the diffusion of larger penetrant molecules, thereby increasing the diffusion selectivity of the membrane.

In this study, the diffusion selectivity of PIM-NH_2 is further increased by systematically dosing in a small amount of oxygen during the deprotection of tBOC groups to induce thermal oxidative crosslinks. The oxidative crosslinks were found to occur starting at approximately 170 °C in the presence of oxygen, and chemical characterizations indicated the formation of isocyanate and new oxidative crosslink species like amides. The pure-gas permeation and sorption results significantly changed with oxygen dose concentrations, indicating that the new oxidative crosslinks impacted the gas transport properties. For example, dosing in 2 mol % O_2 during FVM yielded a 6-fold increase in CO_2/CH_4 selectivity and a 76% increase in CO_2 permeability while dosing in 3 mol % O_2 yielded a 10-fold increase in CO_2/CH_4 selectivity and 30% decrease in CO_2 permeability compared to PIM-NH_2. In addition, the increased density of crosslinks that covalently bond polymer segments together provided a high resistance to CO_2 plasticization and physical aging. This study demonstrates a low-temperature crosslinking strategy using labile pendant groups that tunes the diffusion properties of amine-functionalized polymers and yields stable long-term membrane performance.

98b Selective CO_2 Separation By Facilitated Transport in Ionic Liquid Gel Membranes

Philip Cox, Alex Ferguson, Vincent Storhaug

Membrane-based gas separation represent an attractive low energy separation route to a wide range of gas and liquid based separations providing an improvement to the overall economic and environmental acceptability. High performance membranes offer the potential to provide game-changing process energy advances, provided that sufficiently high levels of selectivity and permeability can be achieved for the target separation. However, selectivity and permeability are typically inversely related and increasing the selectivity typically comes with a penalty of decreased permeability. One critical process is the separation and capture of CO_2 to reduce greenhouse gasses. Numerous membrane-based processes for CO_2 separation exist, but improvements in selectivity will continue to drive down costs. Thin, nonporous membranes or membrane coatings have been demonstrated with processes such as facilitated transport to enhance the selectivity for a range of important target gases. Facilitated transport membranes incorporating targeted redox species have been shown to provide significant benefits in selectivity, however, the permeability penalties persist even with extremely thin, nanometer thick membrane layers.

Mainstream conducted research on an electrochemically-driven facilitated transport membrane to provide both enhanced selectivity for carbon dioxide, while simultaneously driving the permeability of the target gas. To address these selectivity and permeability limitations, targeted biomimetic redox molecules are integrated into an ionic liquid-based gel membrane to provide controlled transport of CO_2 across the membrane and membrane electrode assemblies (MEAs). This presentation will discuss the simultaneous enhancement of the gas separation selectivity and permeability for separation of CO_2 from a range of mixed gas feed compositions using low power, low voltage electrochemically driven separation. The membrane was evaluated in a single 5 cm^2 cell and scaled to a five cell 54 cm^2 stack. Results will be presented on the impact of the configuration, applied field, and CO_2 concentration, achieving an increased permeability of up to 100 times the target CO_2 over the nitrogen and oxygen, significantly increasing its selectivity. We will demonstrate the effect of CO_2 concentration (410 ppm to 5%) in air, with resulting permeate gas concentrations of 54% CO_2 for 410 ppm CO_2 and 90% CO_2 for the 5% CO_2 in room air.

98c CFD Modeling of Flue Gas Separation with Hollow Fiber Modules and Experimental Verification Via 3D Printing

Victor A. Kusuma, Lili Sun, Comfort Oluleke, Lingxiang Zhu, Nathan Diemler, Grigorios Panagakos, Glenn Lipscomb, David Hopkinson

CO_2 removal from flue gas is an important emerging application for membrane technology. A salient challenge is the low concentration of CO_2 in flue gas that coupled with use of highly permeable and selective membranes leads to rapid driving force loss within modules.

CFD is a powerful tool for optimizing module performance. Simulations are reported for a laboratory scale hollow fiber module that are the first to include the full module geometry for comparison to experiment. Geometries were selected to emphasize potential detrimental effects from poor flow distribution (at fiber bundle/case interface and between shell inlet/outlet ports) and axial concentration gradients that arise from permeation through highly selective membranes.

Module cases were 3D printed to control internal dimensions and external port locations. Thin film composite hollow fiber membranes were formed with a Pebax 2533 coating and axially aligned using buttons to control spatial location. Figure 1 illustrates one module. Mixed gas permeation experiments were performed with feed flowrates, pressures, and compositions for CO_2/N_2 mixtures representative of carbon capture applications.

Module performance was evaluated using COMSOL Multiphysics®. Simulations evaluated velocity, pressure, and concentration fields in: 1) shell (external to fibers), 2) lumen (internal to fibers), and 3) membrane domains. Two performance metrics were calculated: 1) retentate recovery (ratio of retentate to feed flowrate) and 2) dimensionless feed flowrate (ratio of feed flowrate to product of CO_2 permeance, membrane area, and feed pressure).

Experiment and simulation were in good agreement (Figure 1). As the external ports are moved away from the module tubesheets, simulations capture the reductions in performance metrics that occur. Moreover, simulations capture the subtle effect of axial concentration gradients that occur due to selective permeation and are detrimental to performance.

Figure 1. Sample experimental module (left) and comparison of simulation to experiment (right).

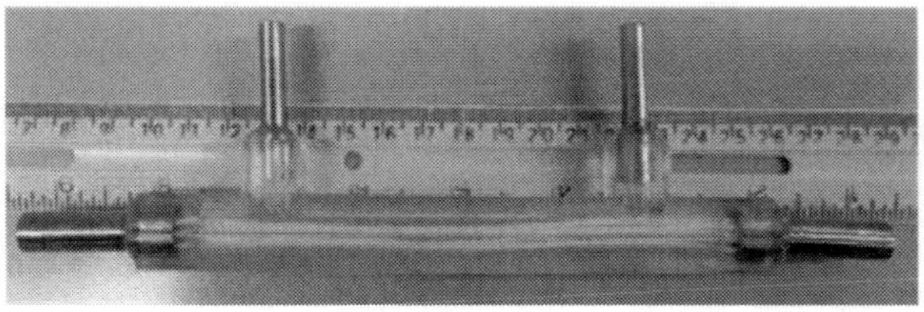

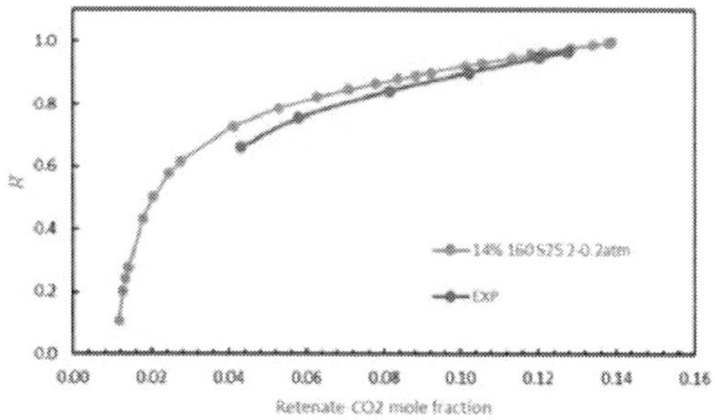

98d CFD Modeling of High-Flux Plate-and-Frame Membrane Modules for Post-Combustion Carbon Capture

Cheick Dosso, Lingxiang Zhu, Victor A. Kusuma, David Hopkinson, Larry Biegler, Grigorios Panagakos

Carbon capture technologies have gained significant attention in recent years due to their potential to mitigate the impact of greenhouse gas emissions on the environment. Membrane-based separation processes offer several advantages over conventional separation processes, such as low energy consumption, compact size, and scalability, making them a preferred option for carbon capture applications. The National Energy Technology Laboratory (NETL) previously developed an innovative rubbery thin film composite (TFC) membrane characterized with high CO_2 permeance, high CO_2/N_2 selectivity, stable performance in the presence of water vapor and non-aging behavior [1]. Upscaling the technology into stacked flat sheet configurations, necessitates insights about optimal operating conditions and geometrical features of the stack that can be elucidated by computational fluid dynamics (CFD) models. High fidelity simulations have been proven a key technology for systematically studying the complex transport phenomena in membrane systems [2]. Module modeling has thus the potential to significantly increase the overall efficiency of the separation process and make the technology more economically competitive.

In this work, we initially focus on setting CFD simulations for a series of plate-and-frame membrane modules that enclose the developed rubbery TFC membranes to investigate how geometries of the membrane modules influence gas transport through the TFC membranes, delineate convective and diffusive mechanisms and determine the effect of various design parameters on its efficiency. The performance of the developed model is assessed using two metrics: the dependence of retentate recovery (fraction of the feed recovered as the retentate product) and the dimensionless feed flow rate. The latter two metrics can be used to assess the process energy requirements (operating costs) and the impact of the membrane area (capital costs), respectively. The simulations aim to drive the optimization and design of the plate and frame membrane stacks, and to create a generic computational framework that could be leveraged universally for membrane technology applications.

References

1. https://netl.doe.gov/sites/default/files/netl-file/22CM_PSC18_Zhu.pdf
2. Rivero J., Panagakos G., Lieber A., Hornbostel K., Hollow Fiber Membrane Contactors for Post-Combustion Carbon Capture: A Review of Modeling Approaches Membranes 2020, 10, 382

*This work is performed in support of the US Department of Energy's Point Source Carbon Capture program.

98e Development and Evaluation of Novel Biphasic Solvents for Post-Combustion Carbon Capture

Yongqi Lu, Hong Lu, Hafiz Salih, Paul Nielsen, Qing Ye

Amine-based chemical absorption is one of the most developed technologies for post-combustion carbon capture (PCC) from point sources. However, despite its maturity, several challenges still hinder its large-scale deployment. The most significant challenges include high energy use incurred during solvent regeneration and CO_2 compression, high capital costs for equipment, and the cost of solvent loss due to degradation and volatile emissions. To overcome some of these challenges, the CO_2 absorption processes that use biphasic solvents are considered attractive. A key feature of biphasic solvents is the formation of dual liquid phases upon the loading of CO_2 or a temperature swing, with the absorbed CO_2 rich in one phase and lean in the other. Because the CO_2-rich phase for thermal regeneration has a reduced mass of solvent and a concentrated CO_2 loading, the required heat duty can be lowered, and the pressure of CO_2 desorption can be elevated, both of which lead to lower energy use and smaller equipment sizes.

In this study, a new group of biphasic solvents have been developed. These solvents are a blend of chemicals that enhance both capacity and rate in solutions with low water content. The solvent exists as a single phase at low to no CO_2 loading, but it undergoes a phase transition to two liquid phases upon CO_2 loading (see Figure 1a). In the laboratory, approximately 80 biphasic solvents were screened based on multiple criteria, including working capacity, absorption rate, thermal and oxidative stabilities, viscosity, and corrosion tendency (see Figure 1b). Two biphasic solvents emerged as superior to all others. Notably, these two biphasic solvents exhibited high resistance to both oxygen-containing and thermal environments. Thermal incubation measurements revealed that the thermal stability of each biphasic solvent for 2 weeks at temperature as high as 150 °C was similar to that of the benchmark 30 wt.% monoethanolamide (MEA) solvent at 120 °C. Oxidative stability measurements showed that compared to MEA, the biphasic solvents exhibited ~8 times higher resistance to oxygen when bubbled in a gas mixture of 96% O_2 and 4% CO_2 at 50 °C for a period of 2 weeks. Furthermore, under simulated absorption and desorption conditions, the corrosion rates of the two biphasic solvents, as measured with a carbon steel coupon (CS1010), were 2-3 times lower than that of MEA.

To evaluate the performance of the developed biphasic solvents, a 40 kW_e bench-scale capture skid was built and installed at the University of Illinois' Abbott power plant (see Figure 2). The skid is an integrated system comprising the flue gas polishing, CO_2 absorption, phase separation, and desorption steps. Parametric tests were conducted for the two biphasic solvents and the 30 wt.% MEA solution, with respect to multiple process or operating variables such as liquid-to-gas ratio, CO_2 concentration, stripping pressure/temperature, and CO_2 loading, to validate and compare the performance and operational flexibility of the skid. The results of the parametric tests showed that both biphasic solvents appeared to be more energy efficient than the reference MEA. Under representative operating conditions, the regeneration heat duty required for the two biphasic solvents was reduced by ~40-43%, and the stripping pressures was elevated by ~3 times compared with MEA. After completing the parametric testing, continuous 24/7 testing was performed with a slipstream of actual coal combustion flue gas from the power plant for one selected biphasic solvent over a course of totally four weeks in two test campaigns. Both test campaigns were operated under steady state at pre-selected conditions. The first campaign aimed for a 90% CO_2 removal rate, while the second one targeted a 95% removal rate with the stripping pressure slightly adjusted. During the slipstream testing, continuous, steady-state operations were validated during the cold wintertime. Test results also revealed that the heat duty required for solvent regeneration ranged between 2,180 and 2,450 MJ/tonne of CO_2 captured, and the stripping pressure ranged between 65 and 60 psia, depending on the CO_2 removal rate target (i.e., 90% or 95%).

We further conducted a techno-economic analysis (TEA) study to evaluate the biphasic solvent-enabled absorption process for large-scale PCC applications. The TEA study followed the methodology and assumptions consistent with those used in the US Department of Energy (USDOE)'s cost baseline study (Version of Revision 4). The results of the TEA study revealed that the parasitic power loss for the biphasic solvent-based process installed in a net 650 MWe power plant could be reduced by ~19% and the cost of CO_2 capture by ~20% compared with the State-of-the-Art technology adopted in USDOE's Base Case B12B. Such technoeconomic advantages resulted from the higher energy efficiency for both CO_2 capture and power generation, smaller power plant for the same net output, and lower capital costs associated with CO_2 absorption, stripping, and compression equipment. This work will continue to investigate and test the performance of the biphasic solvent-based process for CO_2 capture from other industrial point sources such as waste-to-energy facilities.

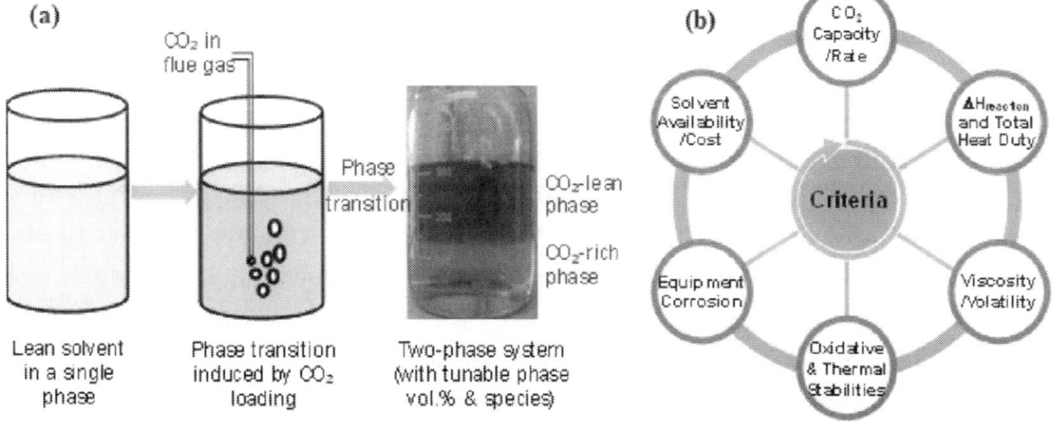

Figure 1. (a) Phase transition of the biphasic solvent with the loading of CO_2; (b) Criteria used in screening biphasic solvents.

Figure 2. Photograph of the 40 kWe biphasic solvent-based CO_2 absorption unit installed at Abbott Power Plant (DCC: direct contact cooler; LLPS: liquid-liquid phase separator; PLC/MCC: programmable logic controller/motor control center).

98f Data-Driven Design and Characterization of Aprotic *N*-Heterocyclic Anion (AHA) Ionic Liquids for CO_2 Capture

Austin N. Keller, Pranav J. Thacker, Louise M. Cañada, Michael Baldea, Mark Stadtherr, Joan Brennecke

Anthropogenic carbon dioxide emissions and resulting climate change represent an existential threat to society. However, deployment of low carbon energy sources may struggle to feasibly match the growing world demand in the near-term. Furthermore, low-carbon substitutes to many high emission industrial processes have yet to reach technological maturity for deployment. For these reasons, efficient point source carbon capture technologies must be developed to decarbonize these sectors. Aqueous amine-based solvents have been proposed as one such technology,[1] but suffer from toxic emissions due to solvent volatility and degradation,[2] as well as energy intensive regeneration due to unfavorable reaction thermodynamics[3] and solvent and water evaporation[4]. Aprotic *N*-heterocyclic anion-based ionic liquids (AHA ILs) are a class of ionic liquids which reversibly react with CO_2 and have been proposed as a substitute for aqueous amines due to their high thermal stability, low corrosivity, negligible vapor pressure, and tunability.

The large variety of possible AHA ILs which may be synthesized presents an opportunity to find an optimal candidate for many diverse CO_2 capture scenarios. This potentially gives us the ability to explore complex tradeoffs between key chemical and physical properties; however, the expansive design space and nonlinear relationships for these compounds are difficult for the designer to navigate. For post-combustion capture from NGCC, a sensitivity analysis of the most influential IL properties for performance has been performed by Seo *et al.*, identifying optimal ranges for solvent CO_2 capacity, density, viscosity, and heat capacity[5]. Machine learning models have been developed by our group to predict the latter three properties of AHA ILs in previous work,[6] but gaps remain in prediction of CO_2 capacity and meaningful implementation of these models toward new solvent discovery.

In this work, we develop a framework to apply first principles calculations toward AHA IL CO_2 capacity prediction, use this framework and that of our previous work to identify potentially high performance ILs, and finally synthesize and test the performance of these species. We have performed over 700 density functional theory calculations to estimate the enthalpy of the CO_2 reaction with 258 unique anions. We implement a linear enthalpy correction and have fit CO_2 uptake predictions to experimental data with an average error 1.5 times the experimental uncertainty. Finally, using these models we have identified and characterized several IL candidates whose properties are better than the economically competitive benchmark IL, [P_{2228}][2CNPyr].

(1) Leung, D. Y. C.; Caramanna, G.; Maroto-Valer, M. M. An Overview of Current Status of Carbon Dioxide Capture and Storage Technologies. *Renewable and Sustainable Energy Reviews* **2014**, *39*, 426–443. https://doi.org/10.1016/j.rser.2014.07.093.

(2) Koornneef, J.; Ramirez, A.; van Harmelen, T.; van Horssen, A.; Turkenburg, W.; Faaij, A. The Impact of CO2 Capture in the Power and Heat Sector on the Emission of SO2, NOx, Particulate Matter, Volatile Organic Compounds and NH3 in the European Union. *Atmospheric Environment* **2010**, *44* (11), 1369–1385. https://doi.org/10.1016/j.atmosenv.2010.01.022.

(3) Shiflett, M. B.; Drew, D. W.; Cantini, R. A.; Yokozeki, A. Carbon Dioxide Capture Using Ionic Liquid 1-Butyl-3-Methylimidazolium Acetate. *Energy Fuels* **2010**, *24* (10), 5781–5789. https://doi.org/10.1021/ef100868a.

(4) Hong, B.; Simoni, L. D.; Bennett, J. E.; Brennecke, J. F.; Stadtherr, M. A. Simultaneous Process and Material Design for Aprotic N-Heterocyclic Anion Ionic Liquids in Postcombustion CO_2 Capture. *Ind. Eng. Chem. Res.* **2016**, *55* (30), 8432–8449. https://doi.org/10.1021/acs.iecr.6b01919.

(5) Seo, K.; Tsay, C.; Hong, B.; Edgar, T. F.; Stadtherr, M. A.; Baldea, M. Rate-Based Process Optimization and Sensitivity Analysis for Ionic-Liquid-Based Post-Combustion Carbon Capture. *ACS Sustainable Chem. Eng.* **2020**, *8* (27), 10242–10258. https://doi.org/10.1021/acssuschemeng.0c03061.

(6) Keller, A. N.; Kelkar, P.; Baldea, M.; Stadtherr, M. A.; Brennecke, J. F. Thermophysical Property Prediction in Anion-Functionalized Ionic Liquids for CO_2 Capture. *Manuscript in Preparation*. **2023**.

98g Simulation and Uncertainty Quantification of a Cryogenic Carbon Capture System for Cement Industry

Brandon Paul, Daison Manuel Yancy Caballero, Miguel A. Zamarripa, Benjamin P. Omell, Michael S. Matuszewski

Cement production is an energy-intensive industry and a major point source of carbon emissions, accounting for around 7% of global CO_2 emissions in 2018 [IEA, 2018]. During 2019, US cement plants produced 40.9 MMT CO_2 equivalent which contributed 24.4% of total industrial emissions that year [EPA, 2021]. Cement is typically produced in rotary kilns where a mixture of aluminate, silicate and ferrite minerals are sintered, particularly the calcination of limestone ($CaCO_3$) in air and natural gas to produce solid clinker (CaO) and CO_2-rich flue gas [Voldsund et al., 2019]. The flue gas typically contains 14-33 mol% CO_2 wet basis as a function of kiln operating conditions such as fuel type, air-to-fuel mass ratio, raw material properties and combustion temperature [Hasan et al, 2012]. This high CO_2 concentration in cement flue gases represent a good opportunity for implementing cryogenic carbon capture technologies, which yield high product purity from low to high CO_2 streams with no required chemical reaction [Nasrifar, 2022].

This work presents simulation of a cryogenic carbon capture system applied to industrial CO_2 emission point sources, particularly from cement plants. A vapor-solid equilibrium model is regressed against experimental flue gas data publicly available in literature at cryogenic conditions, and a 90% capture base case desublimation model is developed. The desublimation model is implemented in Aspen Plus as a 10-stage desublimation column composed of 10 Gibbs reactors in series. Stage outlet temperatures and contact liquid delivery flows to each stage are optimized to maximize the produced slurry temperature at the specified capture level. The cryogenic capture process consists of a direct contact liquid desublimation column, product recovery and heat integration in multi-stream heat exchangers with an external natural gas cooling loop design to take advantage of recirculating refrigerant streams, auto-refrigeration between hot and cold CO_2 streams, and multi-stream heat exchangers to streamline heat integration in the process [Jensen, 2015]. This study investigates the main multi-stream heat exchanger for feed cooling against cold product streams and warm LNG. Additional multi-stream heat exchangers for LNG cooling against a natural gas refrigerant loop to achieve cryogenic temperatures and contact liquid/CF4 refrigerant cooling against cold LNG are studied as well.

This work leverages the open-source Carbon Capture Simulation Initiative's (CCSI) toolset (https://github.com/CCSI-Toolset), regression capabilities within Microsoft Excel, and process models in Aspen Plus. Specifically, the cryogenic capture model is integrated with the Framework for Optimization, Uncertainty Quantification, and Surrogates (FOQUS) to assess model robustness and perform uncertainty quantification [Miller et al., 2016]. Sensitivity analysis is performed on capture efficiency and energy requirements as a function of feed state variables and process conditions, such as inlet compositions (notably carbon and water content) flue gas temperature and pressure, to capture the high variability of produced flue gases from available kiln technologies. Column temperature and flow profiles at 90-99.5% capture, the impact of contact liquid temperature on capture efficiency, and contamination of contact liquid in process streams are quantified. In addition, heat integration uncertainty and co-benefit potential for the removal of SO_x, NO_x, acid gas and mercury are explored. The results show capture efficiency and energy requirements as a function of feed gas state variables including pollutant impurities, and demonstrate contact liquid temperatures required to achieve 90-99.5% capture from 14-33% CO_2 in wet flue gas. Heat integration is assessed for uncertainty in filtration efficiency, heat transfer efficiency and changes in ambient temperature and pressure.

The authors graciously acknowledge funding from the U.S. Department of Energy, Office of Fossil Energy and Carbon Management, through the Carbon Capture Program.

References

EPA. Inventory of U.S. Greenhouse Gas Emissions and Sinks: 1990-2019- Chapter 4. Industrial Processes and Product Use, 2021. https://www.epa.gov/ghgemissions/draft-inventory-us-greenhouse-gas-emissions-and-sinks-1990-2019. [Accessed Mar. 27, 2023]

Hasan, M.F., Baliban, R.C., Elia, J.A. and Floudas, C.A., 2012. Modeling, simulation, and optimization of postcombustion CO_2 capture for variable feed concentration and flow rate. 1. Chemical absorption and membrane processes. Industrial & engineering chemistry research, 51(48), pp.15642-15664.

IEA. Technology Roadmap – Low-Carbon Transition in the Cement Industry: Foldout, 2018, https://webstore.iea.org/technology-roadmap-low-carbon-transition-in-the-cement-industry-foldout. [Accessed Mar. 27, 2023]

Jensen, Mark, 'Energy Process Enabled by Cryogenic Carbon Capture' (2015). Theses and Dissertations.

Miller, D. C., Agarwal D. A., Bhattacharyya D., Boverhof J., Cheah Y-W., Chen Y., et al. (2016). Innovative computational tools and models for the design, optimization, and control of carbon capture processes. 26th European Symposium on Computer Aided Process Engineering – ESCAPE 26. 2391-2396

Nasrifar, K., Moshfeghian, M. (2022). Thermodynamics of carbon dioxide mixtures at cryogenic conditions. Cryogenics 121 103404.

Voldsund, M., Gardarsdottir, S.O., De Lena, E., Pérez-Calvo, J.F., Jamali, A., Berstad, D., Fu, C., Romano, M., Roussanaly, S., Anantharaman, R. and Hoppe, H., 2019. Comparison of technologies for CO2 capture from cement production—Part 1: Technical evaluation. Energies, 12(3), p.559.

Disclaimer: This project was funded by the Department of Energy, National Energy Technology Laboratory an agency of the United States Government, through a support contract. Neither the United States Government nor any agency thereof, nor any of its employees, nor the support contractor, nor any of their employees, makes any warranty, expressor implied, or assumes any legal liability or responsibility for the accuracy, completeness, or usefulness of any information, apparatus, product, or process disclosed, or represents that its use would not infringe privately owned rights. Reference herein to any specific commercial product, process, or service by trade name, trademark, manufacturer, or otherwise does not necessarily constitute or imply its endorsement, recommendation, or favoring by the United States Government or any agency thereof. The views and opinions of authors expressed herein do not necessarily state or reflect those of the United States Government or any agency thereof.

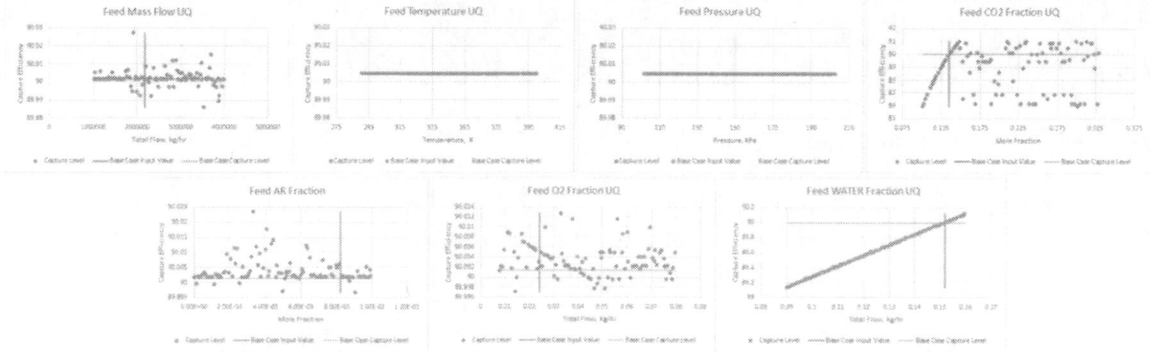
Feed Mass Flow UQ
Feed Temperature UQ
Feed Pressure UQ
Feed CO2 Fraction UQ
Feed AR Fraction
Feed O2 Fraction UQ
Feed WATER Fraction UQ

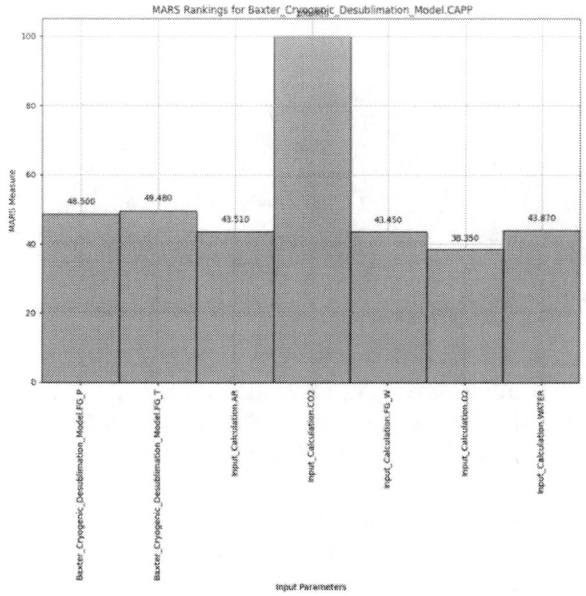
MARS Rankings for Baxter_Cryogenic_Desublimation_Model.CAPP
MARS Measure
48.500
49.480
43.510
43.450
38.350
43.870
Baxter_Cryogenic_Desublimation_Model.FG_P
Baxter_Cryogenic_Desublimation_Model.FG_T
Input_Calculation.AR
Input_Calculation.CO2
Input_Calculation.FG_W
Input_Calculation.O2
Input_Calculation.WATER
Input Parameters

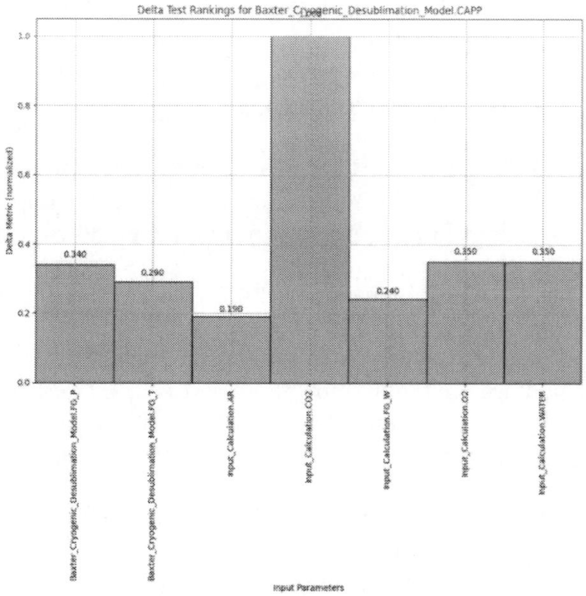
Delta Test Rankings for Baxter_Cryogenic_Desublimation_Model.CAPP
Delta Metric (normalized)
0.340
0.290
0.190
0.240
0.350
0.350
Baxter_Cryogenic_Desublimation_Model.FG_P
Baxter_Cryogenic_Desublimation_Model.FG_T
Input_Calculation.AR
Input_Calculation.CO2
Input_Calculation.FG_W
Input_Calculation.O2
Input_Calculation.WATER
Input Parameters

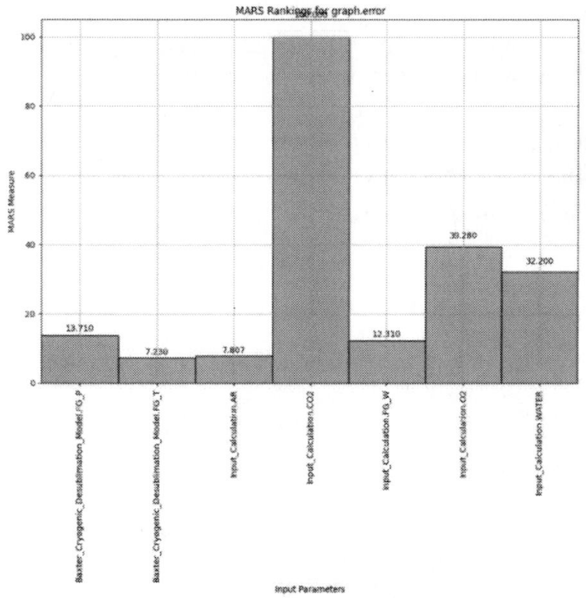
MARS Rankings for graph.error
MARS Measure
13.710
7.230
7.807
12.310
39.280
32.200
Baxter_Cryogenic_Desublimation_Model.FG_P
Baxter_Cryogenic_Desublimation_Model.FG_T
Input_Calculation.AR
Input_Calculation.CO2
Input_Calculation.FG_W
Input_Calculation.O2
Input_Calculation.WATER
Input Parameters

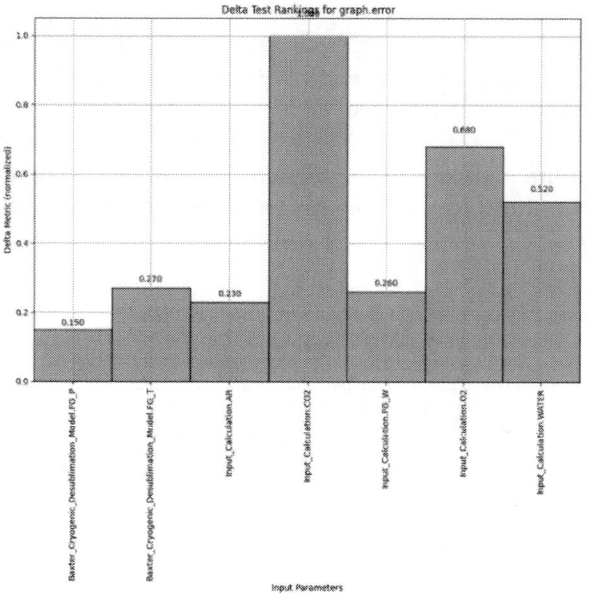
Delta Test Rankings for graph.error
Delta Metric (normalized)
0.150
0.270
0.230
0.260
0.680
0.520
Baxter_Cryogenic_Desublimation_Model.FG_P
Baxter_Cryogenic_Desublimation_Model.FG_T
Input_Calculation.AR
Input_Calculation.CO2
Input_Calculation.FG_W
Input_Calculation.O2
Input_Calculation.WATER
Input Parameters

98h Continuous Carbon Capture Via Oxygen/Water Electrolysis in a Modular Solidelectrolyte Reactor

Peng Zhu, T Alan Hatton, Haotian Wang

Carbon dioxide (CO_2) capture from diffuse sources is essential for migrating global CO_2 emissions and promoting downstream CO_2 utilization and storage. Electrochemical carbon capture technology utilizing redox-active carriers, such as quinones system and pH swing methods has emerged as versatile and more economical alternatives compared to traditional chemical scrubbing. However, the practicality of such systems suffers the requirement of toxic and flammable organic electrolytes, additional downstream septation process, etc., which make it complicated and challenging for decentralized carbon captures. Herein, we demonstrate a different carbon capture design, by coupling the O_2/H_2O electrolysis with our porous solid electrolyte (PSE) reactor, for a continuous and modular CO_2 capture-release from a wide range of CO_2 sources with industrial-relevant capture rate. No chemical inputs were needed, nor side products generated during the whole carbon absorption/release process. Industrially relevant carbon capture rates (440 mA cm^{-2}, 0.14 $mmol_{CO2}$ min^{-1} cm^{-2} or 90 kg_{CO2} day^{-1} m^{-2}), high Faradaic efficiencies (> 90% based on carbonate), and low energy consumptions (starting from ~ 150 kJ/mol_{CO2}) were demonstrated in our carbon capture solid electrolyte reactor, suggesting its promising practical applications. Also, by tuning and optimizing various parameters of the reactor, we achieve continuous carbon capture-release with outstanding capacity, stability, and over 98% carbon removal efficiency under simulated flue gas. The economic applicability is further improved by facilitating different redox-couples to tuning the reaction pathway, and corresponding tech-economic analysis prove the proves its value to meet the needs of actual industrialization.

186a The Important Role of Hydrogen in Solar-Assisted Synthesis of Fuels and Chemicals

Eric Miller, William Gibbons

Harnessing the power of the sun to produce energy-rich fuels and chemicals from abundant resources offers the promise of providing plentiful supplies to meet growing U.S. energy and industrial needs. Opportunities for the solar-assisted production of strategically-important fuels and chemicals include hydrogen from water, hydrocarbon fuels from carbon dioxide and hydrogen/water, and ammonia from di-nitrogen and hydrogen/water. Most of these fuels and chemicals are currently produced from fossil resources using energy-intense high temperature processes, but advanced processes for fuel synthesis utilizing sunlight in conjunction with air and water can provide critical supplements to help meet near- and longer-term demands. These processes rely heavily on thermochemical and/or photochemical manipulation of chemical bonds to facilitate incorporation of hydrogen into molecular structures. Important research, development, and demonstration activities with the U.S. Department of Energy's Hydrogen Program focus on the large-scale production of hydrogen-rich synthetic fuels and chemicals using diverse domestic resources, including solar energy. These activities, coordinated through the Department's Hydrogen and Fuel Cell Technologies Office, will be discussed in context of national priorities related to industrial decarbonization and environmental justice.

186b Investigating Lanthanides and Transition Metal Pair Charged Compensating Doped CeO_2 Materials

Jayni Hashimoto, Alonzo Mendez, Christopher Muhich

Thermochemical gas splitting has been identified as a promising method of harnessing solar energy to produce solar fuels. In this process, a metal oxide first undergoes an endothermic reduction using concentrated solar energy, followed by an exothermic oxidation utilizing CO_2 or H_2O, re-oxidizing the metal oxide and forming CO or H_2 respectively. CeO_2 has emerged as the state-of-the-art material due to high stability and redox kinetics. However, ceria requires extremely high reduction temperatures, ~1500 °C and low oxygen partial pressures, 10^{-6} atm to access these favorable oxygen exchange capabilities. Previous work has identified a paired charged compensating doped (PCCD) ceria strategy as a means of lowering the O-vacancy formation energy. In this strategy, ceria is co-doped with trivalent and pentavalent cations to mimic the effects of successful tetravalent doping, but with a larger, trivalent-dopant dependent, O-vacancy formation energy range. Here, we report on the thermodynamics of additional PCCD materials outside of the IIIA and VA groups previously studied. Fe/Ni/Co/Pr and Nb co-doped ceria is investigated using thermogravimetric analysis and van't Hoff to elucidate the thermodynamic trends of lanthanides and transition metal substituted PCCD materials compared to native ceria.

186c Thermodynamic and Kinetic Characterization of Al- and Ga-Doped La-Sr-Mn Perovskites for Solar Thermochemical Water Splitting

Dylan McCord, Elizabeth Gager, Ximeng Wang, Simon Phillpot, Juan C. Nino, Anthony H. McDaniel, Jonathan Scheffe

Redox cycling of strontium doped lanthanum-manganese perovskite type metal oxides has been shown to be a promising pathway for thermochemically splitting water to produce solar H_2. In this work, we characterized the thermodynamic and kinetic properties of Al and Ga doped LSM perovskites, namely $La_{.60}Sr_{.40}Mn_{.60}Al_{.40}$ (LSMA6464), $La_{.60}Sr_{.40}Mn_{.40}Al_{.60}$ (LSMA6446), and $La_{.60}Sr_{.40}Mn_{.80}Ga_{.20}$ (LSMG6482), for solar thermochemical water splitting. Pelletized samples of each composition were tested in a thermogravimetric analyzer (TGA) with controllable oxygen partial pressure (pO_2), monitored using a ZrO_2 based oxygen sensor. Thermodynamic and kinetic data was gathered via relaxation experiments by stepwise changing pO_2 isothermally and observing the change in mass towards equilibrium. The isothermal experiments were run in the range between 1200°C and 1400°C. Materials were further subjected to testing in a stagnation flow reactor, to compare water splitting capabilities of each material and extend thermodynamic characterization into a wider pO_2 space. The subsequently measured nonstoichiometry, as determined from changes in mass, was used to construct an Ellingham diagram in order to determine partial molar enthalpy and entropy as a function of oxygen non-stoichiometry, for each material. Results from this study lay the groundwork for understanding the role that thermodynamic and kinetic properties of LSM based perovskites has in the solar thermochemical H_2 splitting efficiency, and the conditions and reactor configurations that are most conducive for their utilization.

186d Improving Cycle Performance of Copper Oxide Thermochemical Heat Storage Materials By Doping Cheap Metal Oxides

Gang Xiao

1. **Introduction**

Thermochemical redox couples can satisfy the demand of high-temperature energy storage in the next-generation concentrated solar power plants. Copper oxide owns the advantage of high reaction temperature and high energy storage density, but suffers from poor reversibility and cycling stability due to the sintering problem. The purpose of our research is to inhibit the sintering of copper oxide, so as to improve its reversibility, prolong its cycle life, and increase its application value.

2. **Methods**

2.1 Synthesis of samples

Mixed oxides were synthesized using high-temperature solid-phase synthesis method. They were mixed by ball milling for 30 min via a planetary ball mill. The mixed powders were then calcined at 900 °C for 4 h with a heating rate of 10 °C/min to obtain fresh samples.

- **Redox reactivity of samples**

The redox reactivity tests of the powders were monitored using a Netzsch STA 449 Jupiter F3 simultaneous thermal analyzer in a typical run, a certain amount (8–10 mg) of the mixed powders was placed into an alumina crucible and heated up to 1100 °C and subsequent cooled down to 700 °C. Heating and cooling rates was 20 °C/min to better evaluate the thermal reaction performance of the powders and a constant air (80 vol % N2 + 20 vol % O2) flow of 100 mL/min was used. The redox cycles were carried out in a tubular muffle furnace under a dynamic air flow of 100 mL/min. In the first cycle, the sample was heated from room temperature to 1100 °C and cooled to 700 °C at 20 °C /min. In the following cycles, the sample was heated from 700 °C to 1100 °C and cooled to 700 °C at 20 °C /min. We took about 1 g of sample out every 100 cycles to test the thermogravimetric (TG) and other characterization analyses.

- **Characterization of samples**

Powder X-ray diffraction (XRD) analyses were carried out employing a PANalytical B.V. (Netherlands) X-pert powder diffractometer (Cu Kα radiation) in a diffraction angle (2θ) range of 10–80° with a step of 0.02° and 30 s counting time per angle. The microstructure and morphology of the materials before and after redox cycles were observed by scanning electron microscope (SEM) via a SU8010 field emission scanning electron microscope. Transmission electron microscope combined with energy dispersive spectrometer (TEM-EDS) was identify the elemental distribution of samples after different processes.

3. **Results**

3.1 Reaction characteristics of samples

Inexpensive metal oxides MgO and Al_2O_3 are used to be doped in copper oxide through the high-temperature solid-phase synthesis method. The molar ratio of MgO and Al_2O_3 is 1:1, in order to generate stable $MgAl_2O_4$ spinel. In the TG result, the reoxidation rate of CuO doped with 10 wt% $MgAl_2O_4$ (CuO-10%MgAl) can reach 99.9%. DSC test result shows that the enthalpy change of reduction reaction of CuO-10%MgAl is 507.8 kJ/kg, and the enthalpy change of oxidation reaction is 507.4 kJ/kg. The reversibility of CuO-10%MgAl is excellent up to 250 cycles, remaining above 96%.

3.2 Microstructural characterization of samples

SEM and SEM-EDS images of CuO-10%MgAl at different stages (fresh sample, after reduction and after oxidation) show that $MgAl_2O_4$ is loaded on the surface of CuO as small particles. SEM images of CuO-10%MgAl after different cycles show that the particle size of CuO and $MgAl_2O_4$ particles did not change much compared with that at the first cycle, and $MgAl_2O_4$ particles were still evenly distributed on the surface of CuO. It indicates that $MgAl_2O_4$ prevents the further sintering and growth of CuO particles. Besides, Cu_2O has enough area exposed to air to better absorb oxygen for oxidation reactions.

Conclusion

In summary, we have prepared mixed oxides through solid-phase synthesis, and found an optimal doping ratio of 10wt% of $MgAl_2O_4$ dioxide (CuO-10%MgAl) through thermogravimetric analysis, where the reoxidation rate reaches 99.9%. While keeping a high reduction reaction enthalpy of 507.8 kJ/kg, its oxidation reaction enthalpy reaches 507.4 kJ/kg. The material also shows an excellent performance during the long-term operation, where the oxidation reaction degree maintains 96% after 250 cycles. Through characterizations, we find $MgAl_2O_4$ attaches the surface of CuO particles to prevent sintering and agglomeration of CuO.

186f Thermodynamic and Computational Modeling of Thermochemical Syngas Production Via the CeO_2-CH_4-CO_2 Redox Reforming Cycle in a Tubular Reactor

Mario Zuber, Simon Ackermann, Aldo Steinfeld

Solar thermochemical processing offers an efficient pathway for producing sustainable fuels using concentrated solar energy as the source of high-temperature process heat. The targeted precursor fuel is syngas – a specific mixture of H_2 and CO that can be further processed to drop-in transportation liquid fuels. Within this context, a 2-step dry redox reforming process is considered, using CeO_2 as the redox material and CH_4 and CO_2 as a reducing and oxidizing agent, respectively. The thermodynamics of this process have been extensively studied [1][2] and its on-sun feasibility was demonstrated on a reasonable scale [3]. The dry redox reforming process can operate isothermally in the range 800-1200°C and achieve high non-stoichiometries (δ>0.05), indicating a high proportion of O_2 exchange and therefore a high specific fuel output per reactive mass. The dry redox reforming process is represented by two steps [2]:

Endothermic reduction:

$CeO_2 + \delta CH_4 \leftrightarrow CeO_{2-\delta} + \delta CO + 2\delta H_2$ (eq. 1)

Exothermic oxidation:

$CeO_{2-\delta} + \delta CO_2 \leftrightarrow CeO_2 + \delta CO$ (eq. 2)

We report on the thermodynamic investigation, and its implementation into a computational fluid dynamic model using a customized solver in OpenFOAM. The solver considers all conservation equations in gas and solid phases (species mass, continuity, energy, and momentum) and allows one to model reversible heterogenous reactions of a non-stoichiometric solid whose inherent properties are dependent on δ. Validation of the model was carried out via experiments in a tubular reactor (19 mm-diameter, 300 mm length). The tubular reactor was filled with a packed bed of ceria pellets, and was heated by an electrical furnace. The main operating conditions for purposes of validation were: 1000°C, ambient pressure, mass flowrates of 1 L/min (L: normal liters), and up to 5% reactive gas concentrations diluted in Ar. The outlet product composition was used as the primary means of model validation. Subsequently, the model is used in parametric studies for the purposes of scale-up.

References

[1] B. Bulfin *et al.*, 'Statistical thermodynamics of non-stoichiometric ceria and ceria zirconia solid solutions,' *Phys Chem Chem Phys,* vol. 18, no. 33, pp. 23147-54, Aug 17 2016, doi: 10.1039/c6cp03158g.

[2] Bulfin, B., et al., *Thermodynamic comparison of solar methane reforming via catalytic and redox cycle routes.* Solar Energy, 2021. 215: p. 169-178.

[3] M. Zuber *et al.*, 'Methane dry reforming via a ceria-based redox cycle in a concentrating solar tower,' *Sustainable Energy & Fuels,* 2023, doi: 10.1039/d2se01726a.

398b Designing a Li-Ion Battery Supply Chain Under Uncertainty for Nature-Positive Decisions

Ying Xue, Bhavik Bakshi

The last decade has seen the highest levels of net anthropogenic greenhouse gas emissions, resulting in extreme events such as wildfires, heat waves, and floods. To address this, global goals such as 'Nature-positive by 2030' and 'Net-zero CO2 by 2050' have been proposed [1] . The transportation sector is responsible for the fastest-growing emissions, and electric vehicles (EVs) have emerged as a promising solution [2]. The urgent need to combat global warming and climate change also requires the development of sustainable supply chains that reduce greenhouse gas (GHG) emissions. In this study, from environmental, economic, and social aspects, we designed a Li-ion batteries supply chain which are essential components of EVs, integrating life cycle assessment (LCA), ecosystem services (ESs), and multiobjective stochastic optimization. Minimizing the transgression level of ecological and social thresholds, total cost of the supply chain are three objectives. The environmental objective function is defined by absolute environmental sustainability (AES) metric which compares environmental impacts against ecological thresholds. The environmental impact is quantified by a cradle-to-grave LCA of the Li-ion battery, from raw material extraction to cell assembly to end-of-life. We integrate a multiscale AES assessment method based on ecosystem services (ESs) to assess ecological thresholds for each process along the Li-ion battery supply chain. For social threshold, also referring to socially just space, we calculate a minimum threshold of necessary goods and services to meet basic food, energy, and water needs for each process. A two-stage stochastic model is proposed to account for uncertainties [3-5] in demand of Li-ion batteries and dynamic nature of ecosystem services. Bender's decomposition method is applied to solve this problem. The final pareto-optimal plane reveals the tradeoff between the economic, environmental, and social dimensions.

Previous studies on environmentally sustainable or green supply chain design have primarily focused on minimizing environmental impacts or defining environmental constraints [6]. Sustainability assessment methods, such as LCA and ecological footprint, have been integrated into supply chain design models. However, from an absolute sustainability point of view, this is not sufficient, as nature's capacity to absorb emissions or provide resources is limited and is ignored in most studies [7]. Therefore, sustainable development and global goals require that human impact does not exceed ecological thresholds which require AES metrics. These metrics compare environmental impacts versus ecological thresholds. Our environmental focus is on CO2 emission versus carbon sequestration amount which is defined as the objective function. To quantify the ecological thresholds for each process along the Li-ion battery supply chain, a multiscale method based on ecosystem services (ESs) is implemented. This method quantifies the capacity of ecosystems from different spatial scales as ecological thresholds using biophysical models. It also considers public and private ownership of ESs separately, avoiding issues such as subjectiveness during downscaling and ignoring the heterogeneity of ecological thresholds that are present in current Planetary Boundary-based (PB) methods [8].

The study incorporates the life cycle objective into a country-level supply chain design that includes four major steps in manufacturing Li-ion batteries: cathode production, cell production, non-cell materials production, and battery pack assembly. A total of 13 processes are included in the system boundary. The functional unit is defined as the annual demand of Li-ion batteries for medium-sized EVs, with reference to Tesla. The overall CO2 emission consists of manufacturing emissions and transportation emissions. The objective function is defined by comparing the overall CO2 emission of the supply chain to its ecological threshold. The location of suppliers and transportation modes are the major decision variables, while life cycle flow balance, production capacity, and mass balance are the main constraints. Uncertainties in this supply chain are captured by a two-stage stochastic optimization model. One part of uncertainties comes from the market demand of Li-ion batteries modeled by three scenarios: high, low, and normal. Another major source of uncertainty stems from ecological thresholds which are values of ecosystem services. Ecosystems composed of living organisms are dynamic systems that are constantly changing overtime. In this study, carbon sequestration rates of ecosystems change seasonally, and the carbon sequestration abilities of ecosystems differ across different species. Similarly, the social threshold is defined as the required emissions from food production and electricity generation to meet the energy and caloric intake thresholds for the entire population. A stochastic model is applied to capture the demand variations and ecosystem services availability in different scenarios. Bender's decomposition method is used to solve the problem, decomposing it into master and subproblems, with additional constraints added at each iteration. The multiobjective optimization problem is solved with a ε-constraint method to show tradeoffs among environmental, social, and economic objectives.

The optimal absolutely sustainable and socially safe supply chain is compared with current practical supply chains, and the results are analyzed to find hotspots and identify potential improvements towards sustainable global goals. The contribution of this work includes modeling, methodology, and application. The two-stage stochastic model is incorporated with LCA and multiscale AES framework, designing a supply chain that meets the requirements of transitioning to a 'nature-positive' planet.

Reference

1. Global Goal for nature. https://www.naturepositive.org/. Accessed: 2022-12-13.

2. Parmesan, C., M.D. Morecroft, Y. Trisurat, R. Adrian, G.Z. Anshari, A. Arneth, Q. Gao, P. Gonzalez, R. Harris, J. Price, N. Stevens, and G.H. Talukdar, 2022: Terrestrial and Freshwater Ecosystems and their Services. In: Climate Change 2022: Impacts, Adaptation, and Vulnerability. Contribution of Working Group II to the Sixth Assessment Report of the Intergovernmental Panel on Climate Change [H.-O. Pörtner, D.C. Roberts, M. Tignor, E.S. Poloczanska, K. Mintenbeck, A. Alegría, M. Craig, S. Langsdorf, S. Löschke, V. Möller, A. Okem, B. Rama (eds.)]. Cambridge University Press, Cambridge, UK and New York, NY, USA, pp. 197-377, doi:10.1017/9781009325844.004.

3. Awudu, Iddrisu, and Jun Zhang. 'Stochastic production planning for a biofuel supply chain under demand and price uncertainties.' Applied Energy 103 (2013): 189-196.

4. Aranguren, Maria, Krystel K. Castillo-Villar, and Mario Aboytes-Ojeda. 'A two-stage stochastic model for co-firing biomass supply chain networks.' Journal of Cleaner Production 319 (2021): 128582.

5. Shabani, Nazanin, and Taraneh Sowlati. 'A hybrid multi-stage stochastic programming-robust optimization model for maximizing the supply chain of a forest-based biomass power plant considering uncertainties.' Journal of Cleaner Production 112 (2016): 3285-3293.

6. You, Fengqi, et al. 'Optimal design of sustainable cellulosic biofuel supply chains: multiobjective optimization coupled with life cycle assessment and input–output analysis.' AIChE Journal 58.4 (2012): 1157-1180.

7. Xue, Ying, and Bhavik R. Bakshi. 'Metrics for a nature-positive world: A multiscale approach for

absolute environmental sustainability assessment.' Science of The Total Environment 846 (2022): 157373.

8. Bjørn, Anders, Katherine Richardson, and Michael Zwicky Hauschild. 'A framework for development and communication of absolute environmental sustainability assessment methods.' Journal of Industrial Ecology 23.4 (2019): 838-854.

398c Optimization of Steel Decarbonization Strategies in the United States

Sydney Johnson, Emre GenÃ§er

Hard-to-abate sectors, such as iron and steel, have fossil fuel intensive processes that have been used to materials for centuries[1]. Given the US climate goals for 2030 and beyond, the iron and steel sector will require decarbonization strategies to reduce CO_2 emissions[2]. Currently, majority of the world produces primary (non-recycled) steel with coal-based production methods. Within the US, primary steel is often produced from iron using the blast furnace (BF) In the ironmaking process, coal is used as a reducing agent to transform raw ore (iron oxide) into hot iron metal. Then, the hot metal becomes the desirable steel in the basic oxygen furnace (BOF) where composition of carbon is controlled[1]. Raw material preparation, such as coking, sintering, and pelletization, is also required. The emissions for this process is over 2 tonnes of CO_2 per tonne of crude steel (tCO_2/t steel)[3]. There is also direct reduced iron (DRI) process, which is occurs in three US plants, where iron pellets are reduced in a shaft furnace by natural gas[4]. Afterwards, steel can be produced using an electric arc furnace (EAF). This process generates about 1-1.4 tCO_2/t steel[1,5]. Both of these processes rely on fossil fuels, which is not suitable for US climate goals: 50-52% GHG reduction compared to 2005 levels by 2030 and net zero emission by 2050[2]. As the fourth largest steel producer, the US has to transition its steel sector to more environmentally friendly manufacturing[3].

The cost and emission analysis are developed and evaluated using the Sustainable Energy Systems Analysis Modelling Environment (SESAME). SESAME is a software tool that conducts life cycle, techno-economic, and systems-level analysis for power, transportation, and industrial sectors[6,7]. Specifically, the SESAME Industry Model calculates cradle-to-gate CO2 emissions for the raw material preparation, ironmaking, and steelmaking steps. A combination of literature and industry sources provide material and energy inputs for the plant-level carbon balance. SESAME currently includes the following steel production routes: BF-BOF (and with CCS), HIsarna-BOF (and with CCS), COREX-BOF (and with CCS), DRI-EAF (Coal, Natural Gas, Natural Gas with CCS, Electrolytic Hydrogen), and scrap-based EAF. Molten oxide electrolysis (MOE) and DRI-EAF natural gas with CCS is also to be incorporated. Costs are based on capital and operational expenditures sourced from literature and industry reports and scaled accordingly. Overall, the SESAME Industry Model is a user-facing tool that assists in estimating emissions & costs for decarbonization for the iron and steel industry within the US and beyond. For optimizing the steel sector production routes, a linear program in Python, using Pyomo, is applied to provide greenfield analysis. Constraints are based on Ryan et al, which limits the percentage of primary production routes and restricts the scrap-based production based on available scrap[8].

Current preliminary results indicate (with 2040 electricity grid intensity) a large difference in cost and emissions depending on the objective function. For minimization of gate-to-gate emissions, the cost is about $70.8 billion and 69.5 Mt of CO_2 with a mix of NG-DRI, H2-DRI, EAF, BFBOF with CCS, and COREX with CCS. Meanwhile, the minimization of cost leads to $61.6 Billion and 158 Mt of CO_2 with a mix of BFBOF, COREX, NGDRI, and H2DRI. Comparing the two scenarios yields a nearly $104 cost of CO_2. In addition, there is nearly 20 MT more scrap consumed in the emission minimization case. Further analysis will evaluate the optimum mix to reach US climate goals and consider brownfield (current infrastructure) scenarios.

Sources

(1) IEA. Iron and Steel Technology Roadmap - Towards More Sustainable Steelmaking. **2020**, 190.

(2) *National Climate Task Force*. The White House. https://www.whitehouse.gov/climate/ (accessed 2023-02-26).

(3) Christopher D. Watson. *Domestic Steel Manufacturing: Overview and Prospects*; Summary; Congressional Research Service, 2022. https://crsreports.congress.gov/product/pdf/R/R47107 (accessed 2023-02-21).

(4) Global Energy Monitor. *Global Steel Plant Tracker*. https://globalenergymonitor.org/projects/global-steel-plant-tracker/ (accessed 2022-05-16).

(5) Sarkar, S.; Bhattacharya, R.; Roy, G. G.; Sen, P. K. Modeling MIDREX Based Process Configurations for Energy and Emission Analysis. *Steel Res. Int.* **2018**, *89* (2), 1700248. https://doi.org/10.1002/srin.201700248.

(6) Gençer, E.; Torkamani, S.; Miller, I.; Wu, T. W.; O'Sullivan, F. Sustainable Energy System Analysis Modeling Environment: Analyzing Life Cycle Emissions of the Energy Transition. *Appl. Energy* **2020**, *277*, 115550. https://doi.org/10.1016/j.apenergy.2020.115550.

(7) Gençer, E. *SESAME*. https://sesame.mit.edu/ (accessed 2022-04-11).

(8) Ryan, N. A.; Miller, S. A.; Skerlos, S. J.; Cooper, D. R. Reducing CO2 Emissions from U.S. Steel Consumption by 70% by 2050. *Environ. Sci. Technol.* **2020**, *54* (22), 14598–14608. https://doi.org/10.1021/acs.est.0c04321.

398d Dynamic Systems Analysis of PET and Polyolefin Plastic Supply Chains in Circular Economy

Utkarsh Chaudhari, Tasmin Hossain, David W. Watkins, Damon Hartley, Anne Johnson, Kamand Sedaghatnia, Robert Handler, Barbara Reck, Vicki Thompson, Alejandra Peralta, David Shonnard

Polyethylene terephthalate (PET, resin code #1), and polyolefin plastics such as High Density Polyethylene (HDPE, resin code #2), (Linear) Low Density Polyethylene ((L)LDPE, resin code #4), and Polypropylene (PP, resin code #5) have become an integral part of our day-to-day life owing to their low cost and excellent material properties suitable for multiple applications in several manufacturing industries. High consumption and production of these plastics with low collection rates has led to a global plastic waste crisis. Unfortunately, the plastics supply chains globally, and in the U.S., are highly dependent on fossil resources leading to high greenhouse gas (GHG) emissions and consumption of energy. In the U.S., PET and polyolefin plastic waste accounts for ~80% of the total plastic waste produced, most of which is currently being landfilled (77%), incinerated with energy recovery (16%), and with the remainder collected for sorting and recycling (7%). These mostly linear supply chains in the U.S. released 101 MMT CO_2-eq and consumed 3,248 PJ of energy in 2019, which accounted for 1.5% of the total U.S. GHG emissions and 3.1% of total energy consumed in the U.S.[1]

To break this linearity of plastic supply chains, emerging chemical recycling technologies have been proposed to be a part of this solution to implement a circular economy for plastics. The sustainability of this transition from linear to circular economy is not well understood. In this work, we combine multiple systems analysis tools[2,3] such as material flow analysis, process simulation, life cycle analysis, technoeconomic analysis, transportation logistics and system optimization tools to look at the effect of incorporating a circular economy in the U.S. plastic supply chains for these plastics. In a preliminary application of this systems analysis model to U.S. PET bottles and packaging supply chain, the optimum material flow through mechanical and chemical recycling (including dissolution/purification, glycolysis, methanolysis, and enzymatic depolymerization) resulted in up to a 37% reduction in greenhouse gas emissions and 47% reduction in energy and a circularity metric of 0.8. This presentation extends this preliminary analysis to U.S. HDPE, LDPE, LLDPE, and PP plastics supply chains.

References

1. Chaudhari, U.S., Johnson, A.T., Reck, B.K., Handler, R.M., Thompson, V.S., Hartley, D.S., Young, W., Watkins, D. and Shonnard, D., 2022. Material Flow Analysis and Life Cycle Assessment of Polyethylene Terephthalate and Polyolefin Plastics Supply Chains in the United States. *ACS Sustainable Chemistry & Engineering*, *10*(39), pp.13145-13155.

2. Chaudhari, U.S., Lin, Y., Thompson, V.S., Handler, R.M., Pearce, J.M., Caneba, G., Muhuri, P., Watkins, D. and Shonnard, D.R., 2021. Systems analysis approach to polyethylene terephthalate and olefin plastics supply chains in the circular economy: A review of data sets and models. *ACS Sustainable Chemistry & Engineering*, *9*(22), pp.7403-7421.

3. Shonnard, D., Tipaldo, E., Thompson, V., Pearce, J., Caneba, G. and Handler, R., 2019. Systems analysis for PET and olefin polymers in a circular economy. *Procedia CIRP*, *80*, pp.602-606.

398e Designing Robust Energy Policies for Low-Carbon Technology Adoption

Tom Savage, Antonio del Rio Chanona, Gbemi Oluleye

Increasing the adoption of alternative technologies is vital to ensure a successful transition to net-zero emissions in the manufacturing sector. Yet there is no model to analyse technology adoption and the impact of policy interventions in generating sufficient demand to reduce cost. Such a model is vital for assessing policy-instruments for the implementation of future energy scenarios. The design of successful policies for technology uptake becomes increasingly difficult when associated market forces/factors are uncertain, such as energy prices or technology efficiencies. In this work we formulate a novel robust market potential assessment problem under uncertainty, resulting in policies that are immune to uncertain factors. We demonstrate two case studies: the potential use of carbon capture and storage for iron and steel production across the EU, and the transition to hydrogen from natural gas in steam boilers across the chemicals industry in the UK. Each robust optimisation problem is solved using an iterative cutting planes algorithm which enables existing models to be solved under uncertainty. By taking advantage of parallelisation we are able to solve the nonlinear robust market assessment problem for technology adoption in times within the same order of magnitude as the nominal problem. Policy makers often wish to trade-off certainty with effectiveness of a solution. Therefore, we apply an approximation to chance constraints, varying the amount of uncertainty to locate less certain but more effective solutions. Our results demonstrate the possibility of locating robust policies for the implementation of low-carbon technologies, as well as providing direct insights for policy-makers into the decrease in policy effectiveness that results from increasing robustness. The approach we present is extensible to a large number of policy design and alternative technology adoption problems. Figure 1 provides an overview into the work.

Technology investment cost distributions from numerous engineering studies of nuclear, biomass, solar and wind technologies demonstrate that technology change and learning effects are highly uncertain. The presence of uncertainty makes the availability, adoption, and diffusion of alternative technologies slower and more discontinuous than would be if these decisions were made using nominal models. In practice, stakeholders passively seek to realise unknown uncertainties by waiting for them to instantiate, resulting in slower adoption. Traditional deterministic optimisation and social planning models have been criticised for being optimistic in their handling of uncertainties and technologically too pessimistic by neglecting reduction in technology cost over time from increased demand and spillover effects. These stated downsides have contributed to an increasing number of nonlinear stochastic and robust problem formulations to account for these known uncertainties.

The contributions of this work are three-fold. We first present novel robust market potential assessment model to study and analyse technology adoption in the manufacturing sector considering uncertainty in all the model parameters and considering important effects such as technology learning. The robust market potential assessment model is the first to exploit how market size together with policy interventions can create sufficient demand to trigger and sustain technology cost reduction to accelerate uptake. Secondly, we apply a state-of-the-art iterative robust optimisation algorithm and subsequently investigate robust solutions to the model, which immunise a solution against all aspects of uncertainty. This approach enables us to maintain the original model formulation, including nonlinear effects such as the learning rate of a technology, whilst providing an approximate robust solution.

We investigate the effect of worst-case uncertainty on the model, providing conclusions regarding the adoption of an alternative technology such as Carbon capture and storage (CCS), and transition from natural gas to hydrogen under different scenarios. Finally, by making assumptions as to the form of uncertainty we are able to reduce the conservatism of the robust solution at the expense of increased risk. The insights gained by reducing the conservatism of the problem provides direct insights for policy makers, including the probability a solution will be made infeasible under some assumptions. The approach we present results in an approximate solution to worst-case uncertainty for a given nonlinear model. We believe this approach is more beneficial than making simplifying assumptions to gain an exact solution to worst-case uncertainty, particularly for technology adoption related policy optimisation. Our approach is general enough to be extended to different technologies and fuels.

CCS is crucial for emissions abatement in iron and steel production. CCS is crucial to achieving carbon neutrality targets, however, the high capital, operation, and maintenance costs as well as the large uncertainties of oil or methanol prices hinder its development. Integrated steel making using a blast furnace and basic oxygen furnace are the most common routes to make steel. In this case study we investigate the market effects on the uptake of carbon capture and storage within the iron and steel industry. The policy instrument designated as decision variables is the carbon tax. The objective is to minimise the difference between the existing cost of capture, and the new cost of capture determined from increased demand in CCS due to policy intervention. The final nominal problem contains 66 constraints, 64 variables, and 36 uncertain parameters. The data collected for this research includes the cost of capture for CCS technologies applicable to integrated steel plants, the emissions reduction potential of these technologies, the number of integrated steel plants in the EU, the location of each plant, the steel production capacity of the plants, and the carbon intensity of steel. A post combustion capture from blast furnace with 63% capture rate using 3% MEA solvent has been demonstrated for use in integrated steel plants with TRL 6 – 8, though this capture rate is generally not precisely known. Post-combustion involves removal of carbon dioxide from flue gas generated from a combustion process. It is usually carried out through the chemical or physical absorption of carbon dioxide into a solvent. In this case study we present 32 iron and steel plants across the EU using MEA Post Combustion capture from blast furnace with an existing cost of capture ranging from 59.6-116.3$/t.

Rather than make simplifying assumptions within the model to enable reformulation, we take the view that for this case, maintaining the original model and applying an approximate approach for robust optimisation to immunise solutions against uncertainty provides a greater value. To solve the robust market potential assessment problem, we apply the Blankenship & Faulk algorithm. Algorithm 1 presents this approach.

This method allows us to maintain the original form of the market potential assessment model, important for considering nonlinear terms under uncertainty such as learning rate. Algorithm 1 provably converges to the robust solution of a problem, under the assumption that the lower-level problems are solved globally. This ensures that robust constraints are not violated, regardless of the value of uncertain parameters. As global optimisation solvers do not scale well for large problems, we do not solve lower levels globally, and instead, use a local solver with a multi-start heuristic to solve pessimisation sub-problems. While this provides no guarantees on the optimality of the subproblems, it gives us tractability in solving these real-world formulations. The subproblems solved between lines 5 and 11 in Algorithm 1 may be solved in parallel enhancing the performance of the algorithm. In some large problems, the number of constraints added to the upper-level problem becomes unsustainable, and redundant constraints must be `dropped' from the upper-level problem to ensure its tractability. However, in the current work, we find this unnecessary. For illustration purposes, Figure 2 demonstrates the behaviour of the Blankenship & Faulk algorithm applied to a 2-dimensional robust optimisation problem with a single semi-infinite constraint.

Figure 3 demonstrates how the level of uncertainty of each individual parameter affects the robust solution quality. We subsequently change the nature of uncertain parameters, enabling us to trade-off solution feasibility with quality. Here we provide the ability for policy-makers to choose an optimal policy design entailing their perspective of risk. Figure 4 demonstrates the outcome of trading off solution quality with risk. In this work we provide results and analysis across both case studies, providing recommendations for practitioners, and decision-makers in dealing with nonlinear uncertain energy policy problems.

We present a methodology for reducing the conservatism of robust future energy policy models. We demonstrate how our methodology can result in significantly better solutions that are statistically likely to remain feasible. Our results demonstrate the possibility of locating and communicating robust policies for the implementation of low-carbon technologies, as well as providing direct insights for policy-makers into the decrease in policy effectiveness that results from increasing robustness.

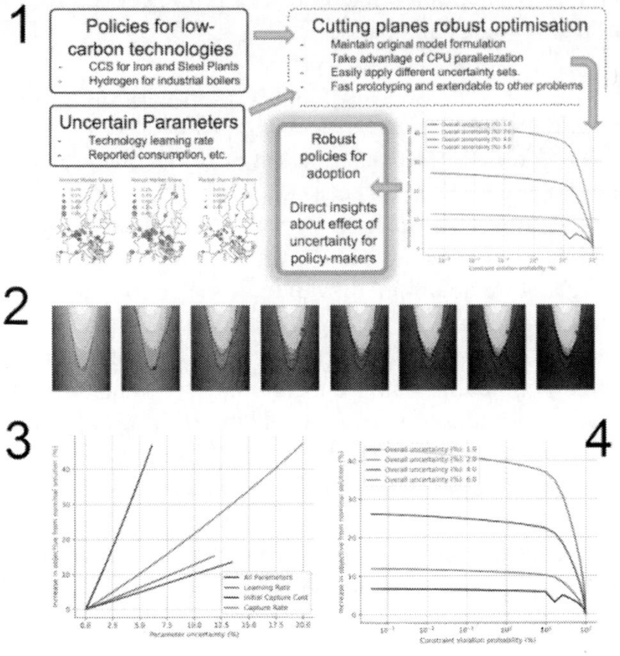

Algorithm 1: Blankenship & Faulk

Input: Objective, $f(\mathbf{x})$, I Semi-infinite constraints $g_0(\mathbf{x},\mathbf{u}),\ldots,g_I(\mathbf{x},\mathbf{u})$, Uncertainty-set $\mathcal{U}$, Nominal parameters $\mathbf{u}_0$, Tolerance ϵ

Output: Robust solution $\mathbf{x}^*$

// Initialise cuts for each constraint
$C_i \leftarrow \{\mathbf{u}_0\}\ i = 1,\ldots,I$
while *True* **do**
    // Flag problem as currently robust
    $R = True$
    // Solve upper level problem with parameter values from current cutting-set
    $\mathbf{x}^* \leftarrow \text{argmin}_{\mathbf{x}\in\mathcal{X}}\{f(\mathbf{x})\,|g_i(\mathbf{x},\mathbf{u}) \leq 0,\ \forall\mathbf{u}\in C_i,\ i = 1,\ldots,I\}$
    for i=$1,\ldots,I$ **do**
        // Maximise constraint violation of current solution
        $\mathbf{u}^* \leftarrow \text{argmax}_{\mathbf{u}\in\mathcal{U}}\, g_i(\mathbf{x}^*,\mathbf{u})$
        // If constraint can be violated by a set of parameters...
        if $g_i(\mathbf{x}^*,\mathbf{u}^*) > \epsilon$ **then**
            // Add worst-case parameters to constraints cutting-set
            $C_i \leftarrow C_i \cup \mathbf{u}^*$
            // Flag solution as not robust
            $R = False$
        end if
    end for
    if $R = True$ **then**
        return $\mathbf{x}^*$
    end if
end while

398f Systems Integration across Multiple Processing Clusters: Design of a Cluster of Clusters

Mohamed Lameh, Dhabia Al-Mohannadi, Patrick Linke

Planning an efficient utilization of material and energy resources should be performed so as the economic development is achieved while considering climate action goals and commitments. This requires an optimal integration of the resources within a system and across multiple systems. The wholistic approach would allow the determination of the best processing pathways that allow resources to be utilized, produced, and exchanged while maximizing profit and meeting global environmental constraints. Previous studies in the field of Process Systems Engineering proposed optimization tools that integrate resources within a processing cluster under different economic and environmental objectives. However, such studies do not consider the supply chain of the exported and imported resources and the interactions through that between multiple systems. This work presents a comprehensive method that allows the integration between different processing clusters by considering the supply chain of the exchanged resources as an additional cluster. The representation of the multiple clusters would allow the optimization of the integration within each cluster and through the supply chain across the different systems to increase the generated profit while meeting allowable levels of emissions and waste generation. The method is demonstrated through a case study in which supply chain of energy carriers considering the processing of resources at an energy importer and exporter is optimized to yield the lowest cost of achieving a sustainable energy transition.

398g Decision Making of Optimal Decarbonization Routes for Industry

Mahdi Ahmed, Joan Cordiner

With the pressing need to address climate change, industries are increasingly seeking low-carbon energy systems to reduce their carbon footprint. Given the wide range of low-carbon energy systems available, the process of selecting the most optimal for a given situation can be a complex task that requires multi-criteria decision-making (MCDM). This involves consideration of a multitude of factors, including economics, technical feasibility, and environmental impact. However, the introduction of new technologies in an ever-evolving environment, coupled with volatile energy markets and the evolution of global emissions trading schemes (ETS), creates a high level of uncertainty. To make secure long-term investments in low-carbon energy systems, it is crucial for companies to consider these temporal variables. We present a review of low-carbon energy systems, with applications to specific industrial cases, analysing the decision making, technical and economic challenges that industries will encounter in achieving net-zero emissions now, and in the future.

457a Assessing the Potential of the St. Peter Sandstone and Everton Formation for CO_2 Storage in Illinois: A Comparison with the Mt. Simon Sandstone

Mansour Khosravi, Kendall Taft, Zohreh Askari

The Illinois Basin (ILB), covering most of Illinois, southwestern Indiana, and western Kentucky, exhibits substantial potential for CO_2 storage. Among the ILB geologic formations, the Mt. Simon Sandstone stands out as a prominent storage resource estimated ranging from 12 billion to 172 billion metric tons. The Illinois Basin-Decatur Project (IBDP) is a noteworthy large-scale injection project targeting the Mt. Simon Sandstone. This injection project, which commenced in November 2011 and was completed in November 2014, entailed the injection of 1 million metric tons of CO_2 into the formation. The successful completion of the IBDP provided localized validation of the capacity, injectivity, and containment of the Mt. Simon Sandstone for CO_2 storage.

While the Mt. Simon Sandstone has been recognized as a viable option for CO_2 storage, other geological formations within the Illinois Basin (ILB), such as the Ordovician St. Peter Sandstone and the Everton Formation, have also emerged as promising candidates for CO_2 storage. Specifically, these formations have been identified as potential options in several counties in southern Illinois, where the thickness of the Mt. Simon Sandstone is limited or absent.

A comprehensive regional study evaluated the storage resource and locations for CO_2 injection into the St. Peter Sandstone and Everton Formation. This study used petrophysical log data and a limited number of core samples and thin sections. Subsequently, a stratigraphic framework was developed by integrating lithological and petrophysical data to assess reservoir characteristics and trends regionally. Furthermore, a regional geocellular model was constructed using this data, with porosity and permeability data populated stochastically within the 3-D grids using regional reservoir property trends. This model enabled the identification of areas where the St. Peter Sandstone and Everton Formation are present and where sandstone intervals display favorable reservoir quality in terms of porosity, permeability, and lateral thickness. Additionally, this study conducted a comparative analysis between the St. Peter Sandstone and Everton Formation, evaluating their reservoir quality and lateral extension in comparison to the Mt. Simon Sandstone.

The evaluation of the petrophysical logs and core intervals indicate that the St. Peter Sandstone is predominantly characterized by its fine to medium, well-sorted pure quartz sandstone composition, which is notably devoid of clay minerals. The St. Peter Sandstone demonstrates an average porosity of 12% and an average permeability of 100 millidarcies in several counties in southwestern and southern Illinois. Furthermore, the Everton Formation locally exists in southwestern and southern Illinois and has two distinct intervals: the upper dolomite and the lower fine to medium grained, well-sorted quartz sandstone intervals. The Everton sandstone interval in southwest Illinois has reservoir properties similar the St. Peter Sandstone. The thickness map and porosity and permeability analysis of the sandstone deposits in southwestern Illinois reveal that the St. Peter Sandstone and the Everton Formation exhibit an average thickness of approximately 180 feet. In contrast, in southern Illinois these formations have a greater average thickness of over 530 feet. However, it is noteworthy that despite the greater thickness of these formations in southeastern and southern tip Illinois, the average porosity measures less than 5%, while the permeability is less than 10 millidarcies.

The thickness map of the Mt. Simon Sandstone in central and northeastern Illinois reveals a thickness variation, ranging from 600 to 2,600 feet. The lower portion of this sandstone is primarily composed of fine to coarse grained sandstone, with an average porosity of 18% and an average permeability of 200 millidarcies. Given its favorable reservoir properties, this lower portion has been identified as a suitable target for CO_2 injection. In designated areas, the thickness of the lower portion generally ranges from 200 to 500 feet.

Comparing these sandstones reveals that the St. Peter Sandstone demonstrates lower reservoir quality and thickness relative to the Mt. Simon Sandstone. However, in several counties in southwestern Illinois and southern Illinois, the St. Peter Sandstone can be considered a viable target for CO_2 storage, particularly in areas where the Mt. Simon Sandstone is absent. The Everton Formation exhibits similar reservoir quality to the St. Peter Sandstone, but its extension is limited to a few counties in southwestern Illinois.

457d Illinois Basin Experiences Estimating the Area-of-Review for Industrial-Scale Carbon Capture and Storage Projects

Roland Okwen, Carl Carman, Scott M. Frailey

One of the key requirements for obtaining a US Environmental Protection Agency's (EPA) Underground Injection Control (UIC) Class VI permit to construct CO_2 injection wells is to estimate the Area-of-Review (AoR), which is "delineated using computational modeling that accounts for the physical and chemical properties of all phases of the injected carbon dioxide stream and displaced fluids, and is based on available site characterization, monitoring, and operational data." AoR delineation is required to ensure the project protects Underground Sources of Drinking Water (USDW) during injection operations. The AoR of a carbon dioxide (CO_2) storage project is either the maximum areal extent of the CO_2 plume, a threshold pressure front during the lifetime of a storage project, or a combination of these. The CO_2 plume extent is estimated based on CO_2 saturation within the injection reservoir using a predefined saturation cutoff. The threshold pressure front is defined by the minimum pressure increase required to initiate fluid flow from the injection zone into the lowermost USDW through a notional conduit. The CO_2 plume extent and threshold pressure front in storage projects are estimated from dynamic reservoir simulation of the injection and post-injection phases, which uses the conceptual and static reservoir models developed during site characterization as inputs.

Determining the threshold pressure front requires the calculation of a threshold differential pressure using measured (1) depths, (2) pressures, and (3) fluid densities at the lowermost USDW and the injection zone. The threshold differential pressure, the difference between the threshold pressure and the initial pressure in the injection zone, can be positive, zero, or negative, corresponding to an injection reservoir that is under-pressurized, at hydrostatic equilibrium, or over-pressurized relative to the USDW. An under-pressurized injection reservoir can accommodate a pressure increase equivalent to the positive threshold differential pressure without the potential migration of fluids to the USDW. For an injection reservoir at hydrostatic equilibrium, migration of fluids to the USDW is likely during CO_2 injection. On the other hand, the migration of fluids to the USDW from an over-pressurized reservoir is likely even without CO_2 injection.

In this study, we use US Department of Energy-sponsored CarbonSAFE project experiences involving the characterization of several storage projects in the Illinois Basin to demonstrate the challenges in determining the AoR. Uncertainty of *in-situ* fluid density propagates to AoR delineation: the accuracy of *in-situ* fluid density measurements can be low due to (1) challenges in obtaining representative fluid samples or (2) the use of a salinity-density correlation from the technical literature when a fluid sample is unavailable. The threshold pressure calculation is highly sensitive to fluid density; therefore, these challenges can have a substantial impact on the calculated threshold pressure. Because the areal extent of the pressure front decreases exponentially with the threshold differential pressure, the pressure front extent is sensitive to changes in threshold differential pressure when threshold differential pressure is low.

457e The Value of 2D Seismic to Identify Regions in a Large Carbon Sequestration Project with Elevated Risk for Long Term CO_2 Storage: A Case Study from the Illinois Basin.

Kendall Taft

One of the key risks associated with long term storage of CO_2 is the integrity and continuity of the confining strata. In a carbon storage project area, the presence of faults that transect both the storage and confining strata can be a significant risk that could allow leakage of injected CO_2 into shallow groundwater aquifers. Two-dimensional (2D) seismic is an effective source of data to evaluate the subsurface for the presence of faults.

The Illinois Storage Corridor project is a United States Department of Energy supported Carbon Storage Assurance Facility Enterprise (CarbonSAFE) project that consists of the subsurface characterization of two different sites in Illinois: The One Earth Energy site in Ford and McClean counties, and the Prairie State Generating Company (PSGC) site in Washington County, which is the subject of this study. As part of the subsurface characterization of the PSGC site, a characterization well, Lively Grove #1, was drilled in 2021. The storage complex penetrated by the well consists of two sandstone reservoirs separated by a dolomite, with an overlying ductile shale as the primary containment interval. After the well was drilled, a 2D seismic survey was also acquired in 2021 that consisted of three lines totaling 25 linear miles. The storage complex intervals display a robust seismic response due to the strong impedance contrasts between adjacent strata.

Interpretation of the 2021 2D seismic data reveals the presence of a fault complex that transects both the storage and confining intervals. The fault complex is interpreted as a strike-slip fault with a component of reverse offset indicated by a small positive flower structure, with a later phase of minor extension. Because it was observed on only one 2021 2D seismic line, the orientation and extent of the complex was unknown. In order to better characterize this fault complex, as well as to characterize a larger area for carbon storage around the project site, additional 2D seismic was acquired and interpreted in 2022. Eight additional lines were acquired totaling 76 linear miles.

The 2022 2D seismic data was able to constrain the extent and likely orientation of the previously identified fault complex. The survey also identified two more faults that transect both the storage and containment intervals: a small normal fault and another strike-slip fault. Taken together, the 2021 and 2022 2D seismic surveys have identified three specific locations within the project area that may have an elevated risk of CO_2 containment due to faulting. Further technical work will be needed to quantify this risk, and determine how and if this risk will impact the placement of CO_2 injection wells across the project area.

457f Supporting and Developing Commercial Carbon Storage Projects in the Western United States: The Carbon Utilization and Storage Partnership

Robert Balch, Brian McPherson, George El-Kaseeh

Introduction

The Carbon Utilization and Storage Partnership of the Western United States (CUSP) is one of four regional Initiatives created in 2019 and sponsored by the US Department of Energy (DOE). The 2019 initiatives were created to support and create commercial opportunities, working directly with companies developing Carbon Capture Utilization and Storage (CCUS) projects. Each of the new initiatives were created from the original seven Regional Carbon Storage Partnerships (RCSP's), first created by the DOE in 2003. The CUSP is built from three of those original initiatives: The Southwest Regional Partnership on Carbon Sequestration (SWP); WestCarb; and Big Sky, all located in the western US. The CUSP region Includes 14 western states, and has 20 founding organizations in the region including Universities, Research Centers, State Geological Surveys, and US National labs (see figure) including key managing partners from the original RCSP's. With recent changes to section 45Q tax credits in the 2021 Build Back Better Bill, and the 2022 Inflation Reduction Act, the US government has accelerated and incentivized an already intense interest in industry to pursue commercial de-carbonization projects. The CUSP team represents over 100 scientists and engineers in more than 40 organizations that are actively working to accelerate and assist in development of 23 projects in the western US.

Description of the CUSP

Through 2022 the CUSP has received $15 million in federal funds, and has generated $3 million in cost-share from its original partners and from companies working on CUSP sponsored projects in the region. Additional non-federal funds have come from direct work by CUSP partners with industry. While the federal funds are not designed, nor able, to add significant infrastructure to most projects, projects benefit from the experience gained in the region over two decades of previous carbon storage research at the three original RCSP's and assistance is most often given in geologic and geophysical characterization of potential storage sites and in preparing for, and assisting with, permits, monitoring plans, and in development of EPA Class VI, and state Class II injection permits. In addition, the CUSP helps projects by developing economic analyses and in pairing regional opportunities that fit sources, mid-stream, and storage components for industries in the region. In 2020 the CUSP funded studies to benefit three projects in the region: conversion of a mature EOR project at Farnsworth Texas from a primarily EOR focus to a primarily storage focus by assisting with a monitoring, reporting and verification (MRV) plan that was approved by the US EPA in May of 2021. The CUSP also assisted in the conversion of a midstream facility from an acid-gas disposal focus to a regional carbon storage hub focus in the Permian basin of New Mexico, with assistance in state permitting a new disposal well, and submission of an MRV plan which was approved by the US EPA in December 2021. The third project involved writing the MRV plan and providing aid in submission of Class VI permits for a storage site near Farmington, New Mexico, this site will have as its first customer a coal plant conversion to a hydrogen burning power source. These three projects alone will store nearly 10 million tonnes of CO_2 per year with injection already occurring at Farnsworth and in the Permian basin, and by 2025 at the Farmington site. In 2021, the CUSP identified 23 additional projects in the region that could be operational within 1-3 years, and funded work to support 12 of those projects. The CUSP is doing unique work in understanding storage in basalt formations, and in the interplay of hydrogen and carbon economies. Large scale projects include engagement with mid-stream gas processors, and iron smelter that is converting to capture and storage, a power plant that is converting to hydrogen burning, and in stacked storage throughout brownfields in the region. The new projects give CUSP a portfolio of 15 federally funded projects and CUSP members are supporting eight additional projects in the region funded directly by industry.

The US Department of Energy (DOE) support and regular funding to US Regional Carbon Sequestration Partnerships such as the CUSP helps foster partnerships between stakeholders including industry, academic research, and regulatory agencies, all of which helps speed up commercialization of CCS/CCUS projects. In addition, expertise developed within the region by CUSP partners provides local expertise to address regional and state issues, public outreach, and to form project teams for industry sponsored work. The CUSP partnership has allowed all participating states to further characterize their CO_2 sources, sinks and potential transportation pathways in the western US, and to then connect those networks to adjacent states and regions. Each state in the CUSP region focuses on local geological knowledge, permitting and regulatory environments, understanding public sentiment, local industries and storage potentials, and the appetite for CCUS projects. Each state benefits from regional and topical expertise provided by partners outside of their states, that may have expertise or tools to assist them in areas they have previously not had resources to delve into. This feedback process has helped CUSP members identify potential regional CCUS hubs and to begin to develop those opportunities. In the CUSP region the CCUS project development process has been optimized through refinement of more traditional static atlas-style data and converting older static databases into self-updating data sources. The CUSP is working towards integrating machine learning tools to manage its data. It has already allowed project teams to rapidly perform base-level surveys of the region for CCUS opportunities and will also help pinpoint specific issues related to those opportunities such as infrastructure challenges, areas with varying regulatory environments, and very importantly, public opinions. CUSP has also been working to improve data regarding social considerations and impacts on communities surrounding CCUS projects, with focuses on environmental justice and socio-economic considerations.

Summary

The teamwork, expertise, and continuity developed through the CUSP has elevated the western region's capability to develop MRV plans, fulfill the complex and comprehensive process of preparing for state applications for Class II applications, and the federal requirements for permitting and drilling of Class VI well for CO_2 storage. Importantly, the regional expertise in preparing and needed documentation for the MRV applications to the EPA, which allow access to 45Q tax credits. As a whole the regional network has created an environment which has improved the ability to create commercialized CCUS projects. This effort has enabled the development of collaborations with industry partners and created trust in the CUSP team capabilities to develop CCUS projects with high readiness levels.

The interest industry has in in CUSP and our focus projects demonstrates that with the right set of incentives CCUS projects can be developed and become commercially viable within the region and across the US. The Inflation Reduction Act (IRA) recently signed into law has given a significant boost to the carbon management industry. The IRA will substantially increase the availability of the federal income tax credits available for domestic CCUS projects (45Q credits). The new rates for qualified carbon oxide are $60/MT for utilization of the CO_2 (instead of $35/MT) and $85/MT for safe geological sequestration (instead of $50/MT). Those rates are even higher for Direct Air Capture projects: respectively $130/MT-$180/MT. In addition, the IRA has significantly lowered volume thresholds for CCUS projects to qualify for 45Q, and has provided new options for monetizing 45Q credits.

The CUSP is actively seeking opportunities to help companies access 45Q credits. CUSP members have proven experience in generating CO2 storage models, MRV applications, and in stakeholder engagement. The databases created for regional use, enables partners to share useful information necessary to create robust geologic models and CCUS business models. This paper will present an overview of CUSP goals, summaries of direct and indirect projects, plans for furthering the program, and describe processes developed for optimizing and commercializing projects in the Western USA using examples of past, current, and potential projects.

PNNL
MONTANA
STATE UNIVERSITY
University
DRI
Stanford
University
MINES
Los Alamos
Sandia
National
Laboratories
AZGS
UT
CARBON
SOLUTIONS
LLC

457g The Investigations of Reservoir Petrophysical Heterogeneity on the Storage Capacity and Injectivity: A Case Study from Dry Fork Carbonsafe Project, Powder River Basin, WY

John Jiao, Jonathan 'Fred' McLaughlin, Matthew Johnson, Ying Yu, Scott Quillinan

ABSTRACT

To achieve the ambitious goal of net-zero carbon emissions in the middle of this century, one of must have approaches is commercial-scale implementation of geological storage of carbon dioxide (CO_2). Storing CO_2 in the subsurface requires understanding the movement (injectivity, sweep efficiency, and pore space utilization) of the injected CO_2 and ensuring that it remains in the target reservoir. Therefore, one of the critical tasks for a commercial-scale storage project is to characterize the reservoir property heterogeneity, and build and maintain static and dynamic models that will predict the reservoir performance throughout the lifecycle of the project. This case study shows that the reservoir petrophysical heterogeneity has a significant effects on the CO_2 injectivity and storage capacity of the targeted reservoirs.

The WY Dry Fork CarbonSAFE projects Phases I, II and III conducted by researchers from the UW and partners have documented that the Dry Fork CCS site is a promise CO_2 storage location characterized by its favorable geologic conditions (i.e., thick stacked saline aquifers and simple structure) and in close proximity to the CO_2 sources. However, Characterizing geological heterogeneity in three dimensions for CO_2 injection, storage and reservoir pressure management represents the greatest uncertainty. This uncertainty can be reduced significantly by integrating geophysical well logs interpretations, core observations and measurements, 3D seismic attribute analysis, facies and property modeling, and dynamic simulation.

The Dry Fork CCS site is located within the coal-rich Powder River Basin and about six miles north of Gillette, Wyoming. The reservoir intervals with the greatest potential for CO_2 storage are the Permian-Pennsylvanian Upper Minnelusa Formation, the Jurassic Hulett Sandstone, and the Lower Cretaceous Lakota Sandstone. Legacy well data, two stratigraphic test wells, more than 625 feet of core from UW PRB 1, 75 sidewall cores from UW PRB 2, complete log suite retrieved from both wells, a 9 square miles of 3-D seismic survey, inter well tomography, and field injection test are the basis of this construction.

To substantially reduce risk in modeling CO_2 storage in these reservoirs, researchers in Center for Economic Geology Research have integrated 3-D seismic attribute interpretations with results from well log analysis and core observations to construct rigorous geologic structural, facies, and property models that delineate reservoir geometries, lithofacies, porosity and permeability distributions, and fracture feature. Within the property models, we can now isolate individual reservoir horizons and construct maps of the distribution of petrophysical properties, and characterize the heterogeneous feature of petrophysical properties. Results of these modeling efforts are used to evaluate the reservoir petrophysical heterogeneity effects on injection feasibility, injected CO_2 migration and plume development, storage capacity, maximum injection pressure, reservoir pressure propagation, confining layer integrity, and determining the Area of Review for Class VI well application in a 576 mi^2 area around the Dry Fork Station.

The detailed static modeling provides a solid foundation for the CO_2 injection simulation. Various injection and storage scenarios run for the DF commercial scale CO_2 geological storage site. The simulation results show the heterogeneity of reservoir petrophysical property has prominent effects on the reservoir injectivity and storage capacity. This project results have significant utility in reducing risk for promoting commercial scale CO_2 storage in stacked deep saline aquifers in the Laramide Basins, Wyoming.

457h A Case Study from Wyoming Dry Fork Carbonsafe Project: The Effects of Wettability Variation on CO_2 Storage Capacity Assessment

Ying Yu, Sumaiya Farzana, Charles Nye, Davin Bagdonas, John Jiao, Jonathan 'Fred' McLaughlin, Scott Quillinan

Meeting global net-zero carbon emission targets will require geologic carbon disposal. The Wyoming Dry Fork Station (DFS) CarbonSAFE project in the Powder River Basin includes subsurface characterization to determine how 50 MT of CO_2 could be permanently stored in less than 30 years. This presentation describes how storage capacity changes with wettability variations in the targeted storage reservoirs.

Contact angle (CA) measurements of CO_2 droplets on three potential storage reservoirs' (the Lower Cretaceous Lakota Sandstone, Jurassic Hulett Sandstone, and Pennsylvanian Minnelusa Formation at the DFS site) rock samples in synthetic formation water (SFW) and interfacial tension (IFT) tests between SFW and CO_2 were conducted. The representative rock samples of each formation were carefully selected. In addition to nine samples of reservoir rock, two sealing formation Fuson and Opeche shale samples were also tested. A high pressure and temperature (HPHT) IFT and CA measurement system was deployed for IFT tests and CA tests at samples' respective reservoir pressures and two additional increased pressures to mimic the pressure change after the CO_2 injection. These tests demonstrated that the average IFT increased with increasing pressure for each sample. These rock samples at the studied reservoir conditions all showed water-wet behavior and became more water-wet as pressure increased further above reservoir pressure, indicating that as CO_2 injection continues, the rock could become more water-wet, making residual trapping more effective. In addition, the Opeche formation cap rock showed the maximum height of the CO_2 column above the free water level at which CO_2 can be trapped by the cap rock is 4002.5 ft, which is double what the Fuson is capable of and suggests the underlying Minnelusa formations could be favored as the main CO_2 storage reservoir.

In addition to the micro-wettability tests, macro-wettability and CO_2 storage potential were assessed via core-flooding tests for the samples from Lakota, Hulett, and Minnelusa Formations. If injectivity could be guaranteed regardless of low permeability, then the lower permeabilities and less water-wet formations promote higher CO_2 storage potential. Nevertheless, permeability appears to be less of a factor in the CO_2 storage potential compared to the wettability in the same reservoir, which dominates the CO_2 storage potential performance and sets the baseline for long-term CO_2 residual trapping.

36a Biological Degradation and Moisture Effects on Corn Stover Comminution

Tiasha Bhattacharjee, Jordan Klinger, Noah Berglund, Bradley Wahlen, Vicki Thompson, Yidong Xia, Neal Yancey

Biomass comminution is the most energy intensive unit operation during preprocessing behind drying. This size reduction stage is simple in concept, but the resulting particle fragments' size and morphology depend on complex interactions of biomass properties, mill operations, and their natural variabilities. These resulting particle sizes and morphologies, contribute to feeding and handling performance as well as convertibility in terms of biological access or thermal transport, etc. In previous developmental work, a predictive population balance model (PBM) was developed for knife milling of corn stover stalks and scaled up from bench to pilot scale. Based on a breakage probability and breakage function, the milled particle size distribution was predicted as a function of variation in mill tip speed, moisture content, initial particle size, and retention screen size. This work presents further developments to the predictive model to probe more complex factors impacting stover stalk strength properties and breakage behavior in a knife mill, including non-linear effects of moisture and material biological degradation. Linkages between pectin and moisture in the middle lamella of the stalks, are parametrically explored for their impact on strength properties and comminution behavior. This work focuses on changing pectin and moisture levels of the corn stalks to further probe these correlations and their effect on breakage behavior. Particle size and distribution shape metrics are matched between experimental measurements and model predictions. Particle size metrics studied include passing diameters and distribution shape metrics include relative span, kurtosis and skewness.

36b Feed System Design Considerations for Biomass and MSW Gasification

Dan Dupuis

A major challenge for biorefineries to adopt new and variable feedstocks is understanding the impacts on the ability to convey those new feedstocks to the reactor. Feeding a consistent feedstock requires ample design considerations that perform well for a single feedstock and single set of specifications. However, when the feedstock changes, the existing feed system design may not function as intended, requiring equipment or feed processing modifications. In this talk, we highlight approaches to test the feedability of different solids from biomass or municipal solid waste streams, like isolated feed testing under relevant conditions. Since biomass and landfill-bound feedstocks have considerably differing characteristics, we made and tested a variety of feed screw prototypes to understand the most important design features for effective material conveyance.

36c Experiment-Informed Virtual Physics for Granular Feedstock Preprocessing and Handling Operations: Overview of Recent Progress

Yidong Xia, Wencheng Jin, Ahmed Hamed, Jordan Klinger, Nepu Saha, Tiasha Bhattacharjee, Mark Small, Neal Yancey, C. Luke Williams, John E. Aston, David Thompson, Lynn Wendt, Vicki Thompson

Virtual physics is an emergent technology in the field of biorefinery design and optimization to assist and expand the quality-by-design approach through development of experiment informed and validated multiphysics models and predictive simulations for de-risking the scale-up of granular feedstock preprocessing and handling unit operations (size reduction, screening, feeding and transport, etc.). This presentation will overview the recent progress and challenges in the development of physics-based computational models for granular feedstock preprocessing and handling operations. A variety of models for simulating particle size reduction, fractionation, and granular flow & transport from the laboratory benchmark scale to the pilot operation unit scale will be introduced. The best practice of using those models regarding their suitability and computational costs will be discussed as well.

36d Assessment of the Abrasiveness of Biomass and MSW Feedstock Particulate Materials

Oyelayo Ajayi, George Fenske, Jun Qu, James Keiser, Jeffrey A. Lacey, Vicki Thompson, Miranda Kuns

Various components in equipment for handling and preprocessing of biomass and MSW feedstock materials are susceptible to excessive wear. This will adversely affect the quality and throughput of different unit operations producing feedstock particles for different downstream conversation processes. Mitigation of wear issues in these various components is the primary focus of FCIC task 1 team. Abrasive and erosive wear modes have been identified as the predominant wear mechanisms in these components. Because of the complex nature of biomass and MSW feedstock in terms of variabilities in compositions and levels of contamination, the wear produced by interaction between preprocessing equipment and feedstock particulate materials is also complex. In order to effectively address the wear issues in these preprocessing equipment, there is a need for an effective method or test protocol to evaluate the performance of different materials and surface treatments being developed to address the wear in materials of construction of the equipment. When possible, the use of a standard test method is always preferable for several reasons: test methods have been vigorously evaluated; repeatability of results tends to be better documented and understood; and occasionally, previous data exists allowing comparison of results. For 3 body abrasive wear testing by particulate materials, the ASTM G 65 –Dry Sand Rubber Wheel Abrasion test is the most commonly used standard test. This test is commonly used to evaluate and rank the abrasive wear resistance of metallic materials subjected to low stress abrasion in service such as agricultural tools, construction and mining equipment. As the name indicates, the standard abradant in ASTM G 65 is 50/70 mesh silica sand. However, the test method can be appropriated to evaluate the abrasiveness of different particulate materials, such as biomass and MSW particles, instead of standard sand. This paper presents our adaptation of the ASTM G 65 –Dry Sand Rubber Wheel Abrasion test protocol to assess the abrasiveness of pine particle biomass and MSW particulate materials. Tests were conducted with three types of abradants namely (1) the standard AFS 50/70 silica sand with an average mesh size of 212-300 μm; (2) loblolly pine biomass particles with sieve of 2 mm; and (3) paper-rich MSW with an average of 2 mm particle size. 1045 steel plate was used as the test material. All the three particulate produced measureable amount of wear. Although there were significant differences in the amount of wear by different type of particles, optical microscopy of the worn surfaces confirmed the mechanism of wear in tests with all type of particles was abrasion. The result of this study demonstrates the applicability of a standard ASTM test method for the evaluation and assessment of the abrasiveness of biomass and MSW particulate materials.

36e Effect of Particle Size, Moisture Content, and Tissue Fraction on Mechanical and Feeding Behavior of Milled Corn Stover and Loblolly Pine Residues

Nepu Saha, Jordan Klinger, Tiasha Bhattacharjee, Noah Berglund, Wencheng Jin, Yidong Xia

Biomass has enormous potential as a promising and sustainable feedstock for the production of bioenergy. However, this potential has been hampered due to handling and feeding issues resulting in the inability to operate the biorefineries at full capacity. To improve operational reliability, it is essential to accurately characterize the mechanical and physical properties, and flow performance of these diverse resources. Hence, this work overviews the impact of material types (woody, herbaceous) and attributes (particle size, moisture, anatomical tissue distribution) on rotational shear, compressibility, elastic recovery, and flow performance. Shear tests are conducted on a rotary shear tester (Schulze-style tester), while compressibility and elastic recovery were characterized from a uniaxial consolidation test. Flow performance was evaluated in a customized wedge hopper with variable wall inclination angle and an Acrison® screw feeder. Results showed that the mass flow rate decreased with the increase of particle size while increasing with the increase of moisture content. Individual tissue fractions are contributed towards the mass flow. The power consumption followed the same trend as the mass flow rate for most materials. These findings are discussed in the context of feedstock variability, their potential mitigation in material preprocessing, and how these variable properties translate to flow handling challenges that are facing in the growing biorefinery operations.

36f Failure Modes and Effects Analysis of Biorefinery Pathways

Rachel Emerson, Pralhad Burli, Lorenzo Vega-Montoto, Tiasha Bhattacharjee

This talk provides an overview of failure modes and effects analysis (FMEA) development and implementation as a systematic criticality and risk assessment tool for biorefinery pathways within the FCIC. This supports a quality by design (QbD) approach, and this talk provides a high-level overview of the results for the FMEA evaluation focused on the generation of pine residue materials for high-temperature pyrolysis conversion. the FMEA interviews included two approaches. The first approach was based around the entire system of unit operations giving a wholistic system level view. The second approach used detailed interviews from individual unit operation within the system allowing for specific failures for individual system components. These two approaches provide different resolutions of information about the reliability and risk. The FMEA results focused on failures associated with meeting critical quality attributes (CQAs) identified for the high temperature conversion of loblolly pine residues and were supplemented with experimental data supporting process upsets and reliability also collected withing the consortium. Estimations of risk scores for meeting each given CQA specification, identification of the impacts for not meeting a CQA specification, capturing causes associated with material attributes and process parameters for each failure, identification of current detection methods, and speculation of potential mitigation strategies for decreasing a failure's risk score were gathered through the FMEA interviews, and were combined to understand the overall process risk metrics and where technology, process, and knowledge improvements are needed in order to de-risk emerging biorefineries.

332a Variability of Sulfur Content in Loblolly Pine Residues across Age, Soil Type, and Harvest Method

Jian Shi, Gary Lopez

The variability of sulfur content in biomass feedstocks can negatively impact a biorefinery using thermochemical conversion technologies due to deactivation of catalysts and equipment wear. It also poses a challenge for designing and sizing sulfur removal systems. A collection of loblolly pine residues
from South Carolina was sorted into anatomical fractions and analyzed to find correlations between sulfur content and soil type, age, and harvest method. Logging residues collected from similarly aged trees grown on coastal plain soil type showed on average higher sulfur content as compared to the sulfur content of the samples collected from sandhill and piedmont soil regions. While looking at the age
factor, logging residues collected from the sandhill soil regions displayed decreasing sulfur contents for
trees of 16 to 41 years old. Furthermore, the bark and needles anatomical fractions had proportionally higher sulfur-percentage-to-mass-percentage ratios as compared to the limbs and particulates fractions.
This study suggests that soil type, age and harvest method are important factors when selecting loblolly
pine residues for thermochemical conversion.

332b Production and Characterization of Biochar from Individual Anatomical Fractions As Biorefinery Co-Products.

Bradley Wahlen, William A. Smith, Wenqi Li, Asif Hasan Rony

Fractionation of biomass enables tissue-specific processing of feedstocks at a biorefinery. Fractions meeting specifications can be processed to fuels and chemicals directly and fractions out of specification can be considered for additional processing to achieve a quality standard or to develop as a co-product. Biochar is a co-product of interest as it can be used to improve soil quality while also sequestering carbon. Little is known of the properties of biochar produced from individual fractions or how storage might affect biochar properties produced from individual fractions. The storage operation within a biorefinery offers the opportunity to condition feedstock using little energy or chemical input, taking advantage of the long residence time. Previous work showed that treatment of whole corn stover with formic acid before storage improved preservation and reduced pyrolysis activation energy. Pyrolysis-GC-GC/MS analysis also revealed improved lignol pyrolysis efficiency indicating formic acid-initiated lignin degradation in storage. Formic acid treatment in storage could therefore impact the properties of biochar produced in pyrolysis. To understand this potential, individual fractions of corn stover (cob, stalk, leaves, husk) were stored either without treatment or after treatment with formic acid. The stored material was then used as feedstock for biochar production, which was then characterized for yield, ash content, elemental composition, thermal recalcitrance, surface area, pH and electrical conductivity. Based on the outcome and techno-economic analysis the fractions best suited for biochar production were identified and preferred uses for the remaining fractions were suggested.

332c The 1st- and nth-Plant Cost Analyses for Incorporating Advanced Biomass Fractionation into Feedstock Processing Systems

Yingqian Lin, Damon Hartley, David Thompson, Pralhad Burli

The U.S. Department of Energy (DOE) Bioenergy Technologies Office (BETO) funds fundamental and applied research to advance the production of advanced liquid transportation fuels from lignocellulosic biomass. Corn stover is an important biofuel feedstock because it is a readily available, low-cost agricultural residue that does not compete with food production and has the potential to reduce greenhouse gas (GHG) emissions and promote advanced liquid transportation fuels development. Challenges exist in terms of logistics and variability in feedstock quality, and research efforts are underway to improve the efficiency of corn stover utilization as a sustainable and cost-effective biofuel feedstock. This study evaluates the 1st-plant and nth-plant costs for an advanced biomass fractionation and processing system to produce tissue fraction-enriched pellets from three-pass corn stover. The advanced biomass preprocessing system was developed by incorporating the results of BETO-funded research and development efforts conducted by INL. The resulting tissue fraction-enriched pellets could be converted in campaigns, taking advantage of their different degrees of recalcitrance, or blended based on their differing chemical compositions to meet the conversion critical material attributes (CMAs). Unused fractions could be sent to a value-added pathway to offset the system's preprocessing cost. The nth-plant analysis predicted modeled delivered costs of $78.80/dry ton (2016$), $79.17/dry ton, and $76.92/dry ton for the enriched husk, stalk, and cob pellets, and $56.15/dry ton for the enriched leaf fraction which was collected for use in the feed market and was not pelleted. The modeled weighted cost for the pellets delivered as the biorefinery feedstock was $78.64/dry ton. While the nth-plant analysis assumed a constant quality feedstock, the 1st-plant analysis utilized variability in the material to assess the impact. The average cost per ton of biomass (using the blending strategy) in the 1st-plant analysis was $78.97/dry ton (2016$), which was $0.33/dry ton higher than the nth-plant estimate. The 1st-plant analysis also revealed that the average throughput was 1,823 dry tons/day or 82.69% of the name plate capacity, due to operability impacts that were dependent on the stochastic variability of the feedstock properties. After a year of operation, 638,152 tons (88.02% of the annual nameplate capacity) of preprocessed feedstock were produced; the variability in throughput was primarily caused by the interaction in flow rates amongst the different biomass fractions and the draw rates of the comminution line and densification line. Equipment failures in the system also contributed to the variability of throughput. The resulting Overall Operating Effectiveness (OOE) for the system, from the simulation, was found to be 74.8%. From this simulation, 84.98% of the preprocessed material met or exceeded the minimum quality specification, requiring the disposal of 15.2% of the biomass. When accounting for quality the average delivered cost of the material increased to $104.47/dry ton, ranging from $88.40 to $8,767.78. A feedstock performance factor of 84.98% and a fully burdened cost that is $25.83/dry ton higher than the estimated nth-plant cost suggest that additional technology development is needed to further address the variability of quality within the feedstock to attain an OOE of 90% and cost parity with the nth-plant analysis.

332d Using Municipal Wastes with Conventional Feedstocks for Biofuel Production

Pralhad Burli, Yingqian Lin, Damon Hartley, David Thompson

The U.S. Department of Energy promotes production of advanced liquid transportation fuels from a variety of biomass feedstocks by funding fundamental and applied research. Supply system designs depend on the properties of the materials and variability in feedstock quality. Meanwhile, increasing generation of municipal solid wastes (MSW) presents a problem of waste accumulation in landfills and oceans accompanied by a wide range of negative environmental consequences. MSW also exhibits some of the same challenges associated with terrestrial biomass feedstocks including material heterogeneity, inconsistency in feedstock quality, and issues around handling during preprocessing. In this study, we evaluated the potential to use "cost advantaged" MSW (grass clippings and mixed paper) along with a conventional biomass feedstock (corn stover) to evaluate the potential to meet specific quality targets for downstream conversion (≥ 59 wt% total carbohydrates and ≤ 4.93 wt% ash). Feedstocks such as grass clippings and mixed paper have relatively unfavorable quality characteristics such as high ash content that necessitate an alternation of the biomass supply logistics system. Grass clippings have average carbohydrate and ash contents of 28.1% and 13.1%, respectively, whereas mixed paper contains on average 61.4% carbohydrate and 10.1% ash. Our analysis indicates that it is possible to meet the quality requirements for conversion by delivering to the biorefinery a blend of pellets comprising 88.9% two-pass corn stover pellets, 7.7 % grass clipping pellets, and 3.3% mixed paper pellets (with a starch binder 2 wt%) at a delivered cost of $82.57/dry ton (2016$) for the blend including the binder. Moreover, it is also possible to deliver a pellet blend that exceeds the quality requirements by blending 86.7% two-pass corn stover pellets and 13.3% mixed paper pellets with a resulting carbohydrate content of 61% and an ash content same as the previous scenario at 4.93%. The blend, which excludes grass clippings, also results in a small cost saving as the per ton cost of the feedstock blend was $82.18/dry ton (2016$). Our analysis indicates that it is possible to use MSW to support future fuel production pathways, including the production of sustainable aviation fuels.

332e Advancing Forest Biorefineries Towards Commercial Applications through Preprocessing of Biomass Wastes

Clayton Wheeler, Sampath Gunukula, C. Luke Williams, Sampath Karunarathne, Peter van Walsum

The Forest Bioproduct Research Institute (FBRI) at the University of Maine has been demonstrated the acid hydration and dehydration (AHDH) of lignocellulosic biomass to fuel intermediates at a one-ton-per-day pre-commercial pilot plant. In the AHDH process, cellulose is converted to levulinic, and formic acids and hemicelluloses are converted to furfural. The integrated AHDH process consists of sawdust preprocessing, conversion of forest residues to organic acids (levulinic and formic acids), biochar separation, and separation of organic acids. For the last few years, the FBRI has experienced a wide number of challenges while demonstrating the unit processes in each process area. The FBRI has collaborated with the Idaho National Laboratory to address these challenges through preprocessing and establishing the critical material attributes of forest residues. A combination of bench and pilot scale experiments coupled with computational modeling are employed to study the effect of feedstock composition, particle size, bulk density, moisture content, and particle size distribution on the feedstock flow flowability and the feedstock conversion to organic acids. The preliminary results have indicated that particle size has little importance on the effectiveness of a strong acid pretreatment step and the organic acid yields. Further it is found that the loss of mannan and galactan of forest biomass must be reduced in the chemical preconditioning of forest residues. Finally, the initial cold flow tests at pilot scale have indicated that the presence of fine particles aid flowability of forest biomass slurries in progressive cavity pumps.

496a Analysis of Economic Impacts and Emissions across Energy Exporters, Importers and the Shipping Industry for Alternative Low Carbon Energy Supply Chain Scenarios

Mohamed Lameh, Dhabia Al-Mohannadi, Gareth Burton, Joseph Rousseau, Patrick Linke

A low-emissions supply chain of energy carriers at a feasible cost is necessary to ensure a sustainable energy transition. The transition to the future energy system results from many inter-connected decisions taken by the different stakeholders involved in the energy supply chain. This work considers an energy carrier supply chain consisting of an exporter, an importer, and the shipping. Previous studies in literature which analyze the environmental and economic performance of energy carrier processing focus on one section in the supply chain; either production or shipping. Life cycle assessment of the potential energy carriers focus only on the environmental impact of the supply chain. This work proposes a systematic analysis of the entire supply chain to evaluate the environmental and economic impacts of the decisions made by the importer on shipping and production, and on the system overall. Starting from the base case in which Japan imports liquefied natural gas (LNG) from Qatar to satisfy a defined demand for energy, the study considers the transition in Japan towards low-emissions energy system through either capturing CO_2 and shipping it back to Qatar for storage, or through shifting to alternative energy carriers such as H_2 and H_2-based fuels. The analysis determines the required sizes of production, shipping, and energy extraction to satisfy the energy demand. The costs on the different stakeholders are determined based on the capital and operating costs of production, shipping, and energy extraction, and considering the depreciation cost of the existing assets that will become invaluable. The different cases are assessed based on the corresponding CO_2 reduction cost of the entire system and the impact of the decisions on the stakeholders.

496b Dynamic Modeling and Techno-Economic Optimization of a Low-Pressure Microwave Assisted Ammonia Synthesis Process

Opeyemi Ogunniyan, Md Emdadul Haque, Yuxin Wang, Debangsu Bhattacharyya, Jianli Hu

Ammonia is an essential chemical being used as a fertilizer[1], specialty chemical, hydrogen carrier, and recently being investigated more as an energy fuel[2] due to its high energy intensity. Industrially ammonia is produced by the Haber-Bosch (HB) process that traditionally operates at very high pressure and temperature which results in costly energy demands, high capital investments, high CO_2 footprint[3], as well as safety concerns. Microwave (MW)-assisted ammonia synthesis can occur close to ambient pressure and moderate temperature [45]. MW-assisted ammonia reactors can be modular with considerably smaller footprint than the traditional HB process thus enabling the opportunity for generation of ammonia in remote and isolated locations. MW-assisted reactors also offer fast startup and shutdown, thus opening great opportunities for using intermittent renewable energy sources. However, as the HB reaction is equilibrium-limited, single-pass conversion is often limited to less than 30% and therefore unreacted reactants needs to be separated from the product. While this separation is efficiently achieved under cryogenic condition for the traditional high-pressure HB processes, this separation becomes challenging under the low-pressure condition. One potential technology for this separation under low pressure condition is to use a solid sorbent[6]. However, modeling, synthesis and optimization of low-pressure ammonia synthesis processes using solid sorbent-based separations is largely lacking in the literature. One difficulty for economic analysis and optimization of this process is due to the transient nature of the adsorption-desorption process.

A kinetic model of the MW-assisted reactor is developed by using the experimental data from an in-house reactor. A plant-wide model of the ammonia synthesis process is developed with multi-stage reactors with interstage cooling and heat recovery and recycle of unreacted reactants. Two potential sorbents are considered- an alkali metal sorbent and a zeolite-based adsorbent[7,8]. Isotherm models are developed for these sorbents. A dynamic model of a fixed bed contactor is developed, and pressure temperature swing adsorption (PTSA) cycles are simulated. An economic model of the process is developed. However, as the separation process is dynamic, it is difficult to use it directly for techno-economic optimization. Surrogate models are developed for the key performance measures of interest for techno-economic optimization as a function of the decision variables (both design and operating variables) of the separation process. Using design of experiments [9], input-output data are generated from the dynamic contactor model. A number of candidate basis functions are considered, and optimal selection of the surrogate model and parameter is done by using an information-theoretic criterion. The surrogate models are then used to solve a mixed-integer nonlinear optimization problem for techno-economic optimization. Among several decision variables that are optimized, one of the key decision variables is the optimal operating pressure of the reaction-separation system. It was observed that operation under atmospheric pressure may not be optimal under certain circumstances.

REFERENCES

1. Humphreys J, Lan R, Tao S. Development and Recent Progress on Ammonia Synthesis Catalysts for Haber–Bosch Process. *Adv Energy Sustain Res*. 2021;2(1):2000043. doi:10.1002/aesr.202000043
2. Smith C, Hill AK, Torrente-Murciano L. Current and future role of Haber-Bosch ammonia in a carbon-free energy landscape. *Energy Environ Sci*. 2020;13(2):331-344.

doi:10.1039/c9ee02873k

3. Rouwenhorst KHR, Krzywda PM, Benes NE, Mul G, Lefferts L. Ammonia, 4. Green Ammonia Production. *Ullmann's Encycl Ind Chem*. Published online 2020:1-20. doi:10.1002/14356007.w02_w02
4. Wildfire C, Abdelsayed V, Shekhawat D, Spencer MJ. Ambient pressure synthesis of ammonia using a microwave reactor. *Catal Commun*. 2018;115(May):64-67. doi:10.1016/j.catcom.2018.07.010
5. Wildfire C, Abdelsayed V, Shekhawat D, Dagle RA, Davidson SD, Hu J. Microwave-assisted ammonia synthesis over Ru/MgO catalysts at ambient pressure. *Catal Today*. 2020;(January 2020). doi:10.1016/j.cattod.2020.06.013
6. Malmali M, Le G, Hendrickson J, Prince J, McCormick A V., Cussler EL. Better Absorbents for Ammonia Separation. *ACS Sustain Chem Eng*. 2018;6(5):6536-6546. doi:10.1021/acssuschemeng.7b04684
7. Helminen J, Helenius J, Paatero E, Turunen I. Adsorption equilibria of ammonia gas on inorganic and organic sorbents at 298.15 K. *J Chem Eng Data*. 2001;46(2):391-399. doi:10.1021/je000273+
8. Helminen J, Helenius J, Paatero E, Turunen I. Comparison of sorbents and isotherm models for NH3-gas separation by adsorption. *AIChE J*. 2000;46(8):1541-1555. doi:10.1002/aic.690460807
9. Farooq S, Ruthven DM. Heat Effects in Adsorption Column Dynamics. 2. Experimental Validation of the One-Dimensional Model. *Ind Eng Chem Res*. 1990;29(6):1084-1090. doi:10.1021/ie00102a020

496c Optimizing the Sustainable Energy Transition for Small Island Developing States Using Multiple Decision Criteria

Sherard Sadeek, Dhurjati Prasad Chakrabarti, Maria Papathanasiou, Keeran Ward

Abstract

Trinidad and Tobago is one of the largest per capita emitters of CO_2 globally. A significant amount of these emissions are attributed to the power and petrochemical sectors. The country is rare amongst small island developing states as it is abundant in oil and gas and thus 98% of the country's power generation is natural gas based. With these abundant resources, the country's power costs are amongst the lowest in the region, with these low prices linked to subsidies. Trinidad and Tobago has signed agreements like Paris Agreement, which require a reduction in greenhouse gas emissions of 15% by 2030. The country requires a holistic approach to aid its sustainable energy transition. Our study implements mixed integer linear programming (MILP) to model 5 scenarios related to the national power generation system via the General Algebraic Modeling System (GAMS). MILP is commonly used in supply chain optimisation problems across various areas and this framework is tailored to resource availability, local infrastructure, and energy system capability of a SIDS. This is a key novelty of the study as the majority of published work in this field relates to developed nations. The basis of our modelling framework is to examine the efficiency of the current national power generation system using MILP. In accordance with current decision-making, the 2019 power system outlook used a single objective optimization (SOO), whereby levelized cost of electricity (LCOE) was minimized, in line with current national perspective on power generation. To further investigate the effects of process and feedstock constraints on both economic (LCOE) and environmental impacts, based on life-cycle greenhouse gas emissions (cradle-to-power) (GHG_{LC}) in 2019 and 2030, a multi-objective optimization approach (MOO) was used. The model involves the use of binary integer variables to represent each power plant, thus allowing the model the opportunity to select which ones can be used. Each plant was constrained to a minimum capacity factor. The key variables used included the energy required by each plant for both peak and normal times (assumed to be 12 hours each), along with the natural gas utilized. These variables link directly with costs and emissions. The system was constrained to meet the national electricity demand of 8.87 TWh (2019), with a daily peak load of 1350 MW.

The scenarios examined include a Business-as-Usual case (BAU), which is split into two parts; one including the present power purchase agreement (PPA) implemented, while the second examines the system without the restriction of this agreement, thus allowing the model to choose which power plants can be operated. Scenarios 2 – 5 also allows the model to choose which power plants are used. Scenario 2 examines the upgrade of a large single cycle (SC) power plant to combined cycle power plant, while Scenarios 3 – 5 considers the inclusion of polymer electrolytic membrane produced (PEM) Hydrogen (H_2) and downstream products in the form of green methanol (MeOH) and ammonia (NH_3) from the electrolytic H_2.

Scenario 1B results in an approximate LCOE reduction of $\$_{USD}$ 18 per MWh (29% decrease) compared to Scenario 1A ($\$_{USD}$ 61.72 per MWh). Elimination of the PPA allows the grid demand to be satisfied by only three power plants. The use of more efficient power plants also resulted in a reduction of GHG_{LC} from 756 $kgCO_{2eq}$ to 743 $kgCO_{2eq}$.

Benefits in both economic and environmental sustainability can be noted after upgrading Plant A (Scenarios 2 – 5) from single cycle to combined cycle. Scenario 2 resulted in a reduction of LCOE of approximately $\$_{USD}$ 25 per MWh (40% decrease) and GHG_{LC} decreased to 572 $kgCO_{2eq}$ (24% reduction) compared to Scenario 1A. These improved results were as a result of the improved efficiency due to technological improvements on the power grid.

Scenarios 3-5 demonstrated a similar LCOE reduction as Scenario 2, ranging from $\$_{USD}$ 36 – 37 per MWh. However, these cases required improved commodity prices (H_2, MeOH and NH_3) for the production of downstream products to be considered feasible in this system. Also, GHG_{LC} increased by 8 – 12% when compared to Scenario 2. Thus, it can be noted that Scenario 2 is the best performing case for both economic and environmental in assessment despite the use of single objective optimization (SOO). Thus, our results illustrate that efficient power generation processes takes precedence over downstream diversification in effectively reducing both energy costs and carbon footprint.

The MOO analysis for 2030 allows for insight into the current power system's outlook and assessment with respect to the national goals. Due to the projected increase in grid demand and peak loads, four power plants were online across all scenarios. The results for 2030 followed a similar trend to 2019. However, there was an approximate increase in LCOE of $\$_{USD}$ (8.5 – 14) per MWh compared to the equivalent scenarios in 2019 SOO. The GHG_{LC} increased by approximately (8 – 27) $kgCO_{2eq}$, when compared to 2019 equivalent, with Scenario 3 being an exception. Scenario 3 demonstrated a reduction in the GHG_{LC} of approximately 39 $kgCO_{2eq}$ when comparing 2030 to 2019. This was due to a reduction in electrolyser capacity, as a result of grid constraints. Overall, our results indicate an estimated increase of 29% and 5% for LCOE and GHG_{LC}, respectively compared to the equivalent scenarios in 2019. These results suggest Scenario 2 as the most sustainable pathway to clean and affordable energy. Despite the increase in GHG_{LC} impact from 2019 to 2030, the upgrade of Plant A to combined cycle (CC) operations provides a 21% improvement in GHG emissions when compared to BAU operations (Scenario 1A). Thus, the MOO 2030 outlook underlines the importance of increasing efficiency across the grid and utilizing accurate price forecasts to guide decision making.

The system's sensitivity to natural gas prices, commodity (H_2, MeOH and NH_3) prices and PEM efficiency was examined. All cases are impacted by adjustments in natural gas price as Trinidad and Tobago's power generation system is exclusively linked to natural gas utilization. The sensitivity to natural gas price is particularly volatile given the global shift away from coal as an energy source. When natural gas prices are set to $\$_{USD}$ 4.00 per MMBtu (57% higher than current prices), LCOE increases over the range of $\$_{USD}$ 11 – $\$_{USD}$ 23 per MWh across Scenarios 1A, 1B and 2. The highest impact was noted in Scenario 1A, which was constrained by the PPA. An increase in natural gas price would increase the country's reliance on its subsidies to maintain its low electricity rates, which is unsustainable.

Our results also highlight potential economic benefits among commodity-based scenarios, specifically when natural gas prices are reduced. For example, for natural gas prices ranging $\$_{USD}$ 1 – 1.5 per MMBtu, there was a 30-43% reduction in LCOE across Scenarios 3, 4 and 5. However, at natural gas prices above $\$_{USD}$ 2.60 per MMBtu, production of H_2, MeOH and NH_3 were deemed infeasible, as the model chooses to neglect electrolysis- mainly due to high capital charges. Regardless of the natural gas price forecast, increased costs only impact the LCOE, thus leading to potential conflicts within our MOO model. Hence, our results illustrate the importance of using multiple decision criteria when proposing future energy policies that align to both economic and environmental sustainability. It must also be noted that the viability of utilising commodity-based scenarios to reduce LCOE is dependent on an increase in selling price of the products (H_2, MeOH and NH_3).

Improved efficiency of PEM in the system demonstrated minimal benefits to both LCOE and GHG_{LC}, thus highlighting that the downstream production of electrolytic H_2, MeOH and NH_3 does not necessarily support national GHG reductions targets.

Overall, our results support the need for an upgrade of the current power system, driving down susceptibility to increasing natural gas prices and promoting higher energy and resource efficiencies through technological advancement. This must be coupled with new policies to reduce GHG emissions and incentivize sustainable energy sources. A reconsideration of the present PPA must take place, as well as introduction of carbon taxes and emissions trading. The use of carbon capture and sequestration processes (CCS) should be explored. The introduction of feed-in tariffs to promote alternative energy sources in the power generation system can also be beneficial.

This work highlights the use of multiple decision criteria, through multi-objective optimization, and its relevance in promoting clean and affordable energy among SIDS, using Trinidad and Tobago as a case study. This work demonstrates the ability to link energy supply chains for less developed countries in the Caribbean and globally, thus linking to sustainable development. Our results illustrate that the most optimal scenario would be the improvement of gas turbine technology from SC to CC for the nation's largest power plant. Our results underpin viable solutions-driving impact and distributing guidance on the sustainable energy transition across the Caribbean Region, through the deployment of clean and affordable energy.

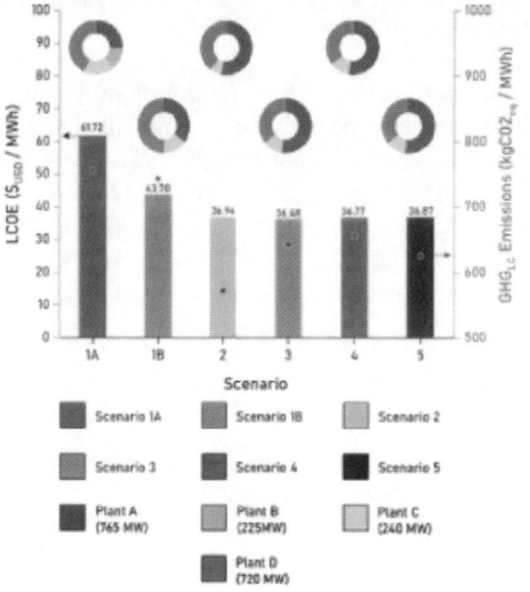

Figure 1: SOO 2019 highlighting LCOE, GHG_{LC} and Grid share attributed to each Power Plant. Bars show LCOE prices while • illustrates GHG_{LC}.

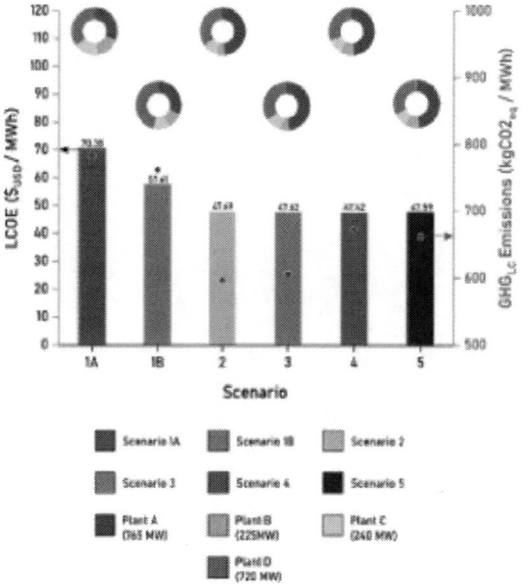

Figure 2: MOO 2030 highlighting LCOE, GHG_{LC} and Grid share attributed to each Power Plant. Bars show LCOE prices while • illustrates GHG_{LC}.

496d Long-Term Planning for Energy Transition Under Endogenous Uncertainty

Molly McDonald, Christos Maravelias

Growing concern for the effects of greenhouse gas emissions has led to the creation of emission-reducing policies by governments around the world. Stricter emission regulations have created an opportunity to advance underdeveloped and underutilized technologies, and to further develop new technologies that share the goal of reducing emissions. Not only is development of new technologies important in achieving sustainability goals, but the way in which new technologies are deployed can have significant impacts on the choice of systems to develop, the cost of implementation, and the ability to meet emission regulations. The decisions about which technologies should be invested in can be difficult to make since they are based on information about the future, which is uncertain. One way to study the effect of uncertainty on energy system models is through sensitivity analysis, as was done in the Net Zero America study [1]. Another method is multistage stochastic programming (MSSP), which accounts for uncertainty through scenarios that represent different outcomes of uncertain parameters [2].

When considering emerging technologies, a source of uncertainty to consider is how the costs will develop over time. Learning curves are used to model the decrease in cost as the total installed capacity of a technology increases. However, the extent to which the cost decreases is uncertain. Uncertainty is classified into two types, exogenous and endogenous. If the uncertainty is exogenous, it is known when the uncertainty will be revealed, and decisions can be made based on when this observation will occur. In the case of endogenous uncertainty, it is not known when uncertainty will be observed and, thus, in MSSP setting when scenarios will become distinguishable; rather, the timing depends on decisions. For example, a cost reduction for a technology will only be observed if that technology is chosen to be invested in; otherwise, it will remain unknown. Previous work in this area has led to an MSSP model for capacity expansion that considers uncertain learning curves and was applied to a small case study of a network of energy generation technologies [3].

In this work, we develop a linear mixed-integer MSSP model and apply it to a large-scale system, using real-world data, with the goal of reaching net-zero carbon emissions in the United States by 2050 [4]. The model includes nine technologies in the transportation sector and twelve technologies in the energy sector for the entire United States over a 30-year time horizon broken into six 5-year periods. Endogenous uncertainty is included in the learning curve of biomass technologies. Additional exogenous uncertainty in the demand of both sectors and policy related to carbon emissions is also included. The model is used to analyze how uncertainty in the cost reduction of emerging technologies will affect the investment decisions into fossil fuel-based processes and their low-carbon alternatives.

Furthermore, given that an energy transition will involve a sequential decision-making process, a rolling horizon approach is applied to both the deterministic (meaning that the problem has perfect foresight when it comes to parameters) and stochastic models, which allows for the models to incorporate feedback. We present a number of auxiliary methods that allow us to simulate this system that has a number of challenging real-world characteristics (e.g., large time delays, lagged capacity installation, impact of social acceptance). This allows for feedback to occur in the next iteration, considering any realizations of random parameters that might have occurred. After all iterations have been completed, a new energy transition (investment timeline), in the presence of feedback, is generated.

When the models are employed to generate a single (predictive) solution, it is seen that in scenarios when costs of emerging technologies decrease to competitive prices, decisions to invest in these technologies should be made earlier to allow for the decrease in costs to be taken advantage of in the future. It is also seen that a wider variety of energy and biomass technologies are expanded when uncertainty is included. Interestingly, we also see lower carbon emissions when uncertainty is considered, because the stochastic model includes scenarios of high energy demand, and in order to meet those demands, more investment must be made into different, including renewable, energy technologies. When a rolling horizon approach is used the cost of the new investment timeline of the stochastic model is slightly lower than the solution obtained using, iteratively, the deterministic model. Importantly, the transition plan generated using the stochastic model still has a wider variety of technologies installed.

We close with a short but broader discussion of energy transitions. Specifically, we discuss how the proposed model allows us to understand how real-world constraints and technology readiness will impact the implementation of different energy transition pathways and how these technologies can affect the probability of meeting certain emission goals at lower cost.

References

1. Larson, E., Greig, C., Jenkins, J., et al. (2021). Net-Zero America: Potential Pathways, Infrastructure, and Impacts, Final Report Summary. *Princeton University.*
2. Birge, J.R., Louveaux, F. (1997). Uncertainty and modeling issues. *Introduction to Stochastic Programming*, 54-61.
3. Rathi, T., Zhang, Q. (2022). Capacity planning with uncertain endogenous technology learning. *Computers and Chemical Engineering*, 164, 107868.
4. "U.S. Energy Information Administration - EIA - Independent Statistics and Analysis," https://www.eia.gov/state/search/#?5=126&r=false.

496f Decarbonization Solutions

Michael Hartmann

The global need for effective technology to capture, use, and store CO2 has intensified due to increasing public and governmental awareness of the importance of meeting climate goals while sustaining energy demands. This presents a lucrative opportunity for the chemical and process industries to take advantage of. With the advent of CO2 capture technologies in the 1970s, existing facilities around the world have the potential to capture over 40 MtCO2 annually. There has been a growing impetus in this field, supported by investment incentives and the civic responsibility of achieving climate targets. Chemical absorption and physical separation are the two most frequently utilized technologies, which have proved successful for decades.

At SwRI, we have established teams specializing in all aspects of Carbon Capture, Utilization, and Storage (CCUS), including related infrastructure designed for real-world CO2 producers. Our wide-ranging expertise encompasses chemisorption, membrane and oxy-fuel separations, chemical and calcium looping, direct air capture, infrastructure and logistics, and supercritical CO2 power cycles. Our organization is significantly invested in developing technologies for the impending net-zero carbon economy by 2050 and achieving a 100% clean electrical grid by 2035.

In addition, we have substantial knowledge and experience in syngas production and upgrading, CO2 conversion to fuels and chemicals, fermentation, and electrolysis. We strive to advance CCUS technologies through fundamental research, process development, and industrial application. Our years of diverse and numerous projects make us a leading promoter of achieving sustainability and curbing global warming. Our presentation will provide insight into various CCUS approaches.

359a Scalable and Sustainable Valorization of Lignin Using Single-Step Aerosol Method

Sujit Modi, Pratim Biswas

Growing demand for energy and materials has led to increased greenhouse gas emissions from the use of fossil resources. Lignocellulose biomass is one of the few renewable resources of carbon that has the abundance and geographical distribution to displace fossil resources. However, the under-utilization of by-products (i.e., lignin fraction) remains one of the challenges to the rapid growth of biorefinery. Thus, developing technologies that can valorize lignin to high-value products is crucial. Lignin could be used as a precursor for the synthesis of a diverse range of high-value nanomaterials, including lignin nanoparticles (NPs), carbon NPs, and functionalized carbons [1, 2]. However, conventional synthesis methods of these nanomaterials involve multistep and batch processes or large volumes of solvents/activating agents that are not cost-effective and environmentally benign, limiting the potential for their large-scale synthesis.

To overcome these challenges of conventional synthesis methods, the present work explores the synthesis of high-value nanomaterials from lignin using a novel and simple furnace aerosol reactor (FuAR) technique. The as-synthesized nanomaterials are tested for their performance in various modern applications (e.g., efficient energy storage and carbon capture). Key features of FuAR are (i) ultra-fast processing (order of seconds); (ii) continuous and single-step operation; (iii) minimal use of solvents; and (iv) simple and precise control over material properties; making it scalable and environmentally benign for lignin valorization [3, 4].

First, the FuAR is demonstrated for controlled synthesis of lignin NPs of mean sizes between 50 and 68 nm [5]. The as-synthesized lignin NPs are analyzed for size and functional groups. The mean size of lignin NPs showed an increasing trend with lignin solution concentration. Furthermore, the bulk and as-synthesized LNPs were tested for UV protection applications. The observed improvement in UV protection with a decrease in lignin particle size is systematically investigated using the optical absorption parameter [5].

Secondly, with the systematic understanding of the role of temperature and residence time in the FuAR on the pyrolysis product of lignin, high surface area carbon nanoparticles (up to 925 m^2/g) are synthesized without the use of activating/templating chemicals [6]. Furthermore, the as-obtained carbon nanoparticles are tested for specific capacitance which showed a linear trend with surface area. The best-performing material with a surface area of 925 m^2/g exhibited a specific capacitance of 247 F/g at 0.5 A/g with excellent capacity retainment of over 98 % after 10,000 cycles [6]. This is a clear demonstration of their superior performance compared to supercapacitors synthesized earlier from lignin. Moreover, the size and morphology of carbon particles, which are impacted by interparticle collision and sintering, are known to play a crucial role in their stability as well as energy storage performance. Consequently, the kinetics of reaction, sintering, and collisions in the FuAR is systematically studied using a novel and generalizable numerical model based on a geometric modeling approach [7].

Finally, the FuAR is used for in-situ nitrogen functionalization to synthesize nitrogen-functionalized porous carbons. Urea is added to the lignin solution as a precursor for nitrogen functional groups. The effects of temperature, residence time, and the urea-to-lignin ratio on carbon nanoparticle functional groups, pore size distribution, surface area, and morphology are systematically investigated. Our findings indicate that surface area increases with temperature and residence time in the FuAR.

Furthermore, the as-obtained carbon nanoparticles are tested for CO_2 adsorption and the material with a maximum surface area of 1051 m^2/g exhibited a CO_2 adsorption capacity of 62 mg/(g of carbon) and excellent cyclability.

Overall, these studies demonstrate the unexplored potential of FuAR for the scalable valorization of lignin as well as the excellent performance of obtained nanomaterials in respective applications, which will help advance displacing the fossils with renewable resources.

References:

[1] W. Zhang, X. Qiu, C. Wang, L. Zhong, F. Fu, J. Zhu, Z. Zhang, Y. Qin, D. Yang, C.C. Xu, Lignin derived carbon materials: current status and future trends, Carbon Research 1(1) (2022) 1-39.

[2] M. Kienberger, Potential Applications of Lignin, in: Y. Krozer, M. Narodoslawsky (Eds.), Economics of Bioresources: Concepts, Tools, Experiences, Springer International Publishing, Cham, 2019, pp. 183-193.

[3] M. Ago, S. Huan, M. Borghei, J. Raula, E.I. Kauppinen, O.J. Rojas, High-throughput synthesis of lignin particles (~ 30 nm to~ 2 μm) via aerosol flow reactor: size fractionation and utilization in pickering emulsions, ACS applied materials & interfaces 8(35) (2016) 23302-23310.

[4] H. Zhou, M. Kouhnavard, S. Jung, R. Mishra, P. Biswas, One-step aerosol synthesis of a double perovskite oxide (KBaTeBiO 6) as a potential catalyst for CO 2 photoreduction, Nanoscale 13(27) (2021) 11963-11975.

[5] S. Modi, M.B. Foston, P. Biswas, Controlled Synthesis of Smaller than 100 nm Lignin Nanoparticles in a Furnace Aerosol Reactor, ACS ES&T Engineering (2023).

[6] S. Modi, O. Okonkwo, S. Saha, M. Foston, P. Biswas, Reuse of Lignin to Synthesize High Surface Area Carbon Nanoparticles Using a Continuous and Single-step Aerosol Method (under review).

[7] S. Modi, O. Okonkwo, H. Zhou, S. Kavadiya, M. Foston, P. Biswas, Geometric Model for Predicting the Size and Morphology Evolution of Multiparticle Aggregates during Simultaneous Reaction and Sintering, Chem. Eng. J. (2023) 141423.

359b Carrier-Free Enzyme-Assembled Hydrogel Based on Synergistic Weak Interactions to Boost Enzyme-Immobilized Microflow Biocatalysis

Qiang Chen, Yujun Wang, Guangsheng Luo

Immobilized enzymes can be potentially used as biocatalysts; however, their industrial applications are limited by the low specific activity and poor stability. We proposed a feasible strategy utilizing the synergistic bridging of triazoles and metal ions to induce the formation of porous enzyme-assembled hydrogels (EAGs) with increased activity. The catalytic efficiency of the prepared EAG toward acetophenone reduction was 6.3 times higher than that of the free enzyme, and the activity retention of 90.3% after 12 use cycles confirmed its reusability. A near-atomic resolution (2.1 Å) structure of the EAG enzyme was successfully analyzed via cryogenic electron microscopy, which indicated a structure–property relationship for the enhanced performance. In addition, the possible mechanism of EAG formation was elucidated, revealing the indispensability of triazoles and metal ions, which guided the use of two other enzymes to prepare EAGs capable of good reusability. The described strategy can pave the way for the development of new catalytic biomaterials and immobilized biocatalysts.

Furthermore, we demonstrated the microfluidic induction of alcohol dehydrogenase to form EAG in a microreactor, with controllable mechanical properties and porous architecture. Six chiral aryl alcohols were successfully synthesized by using this approach in flow asymmetric reduction. Moreover, the support-free EAG was capable of in situ flow immobilization and reversible de-immobilization without reactor disassembly, while exhibiting sufficient retention against cofactors NADP(H). The EAG monolithic microreactor presented an excellent space-time yield of 88.05 g L^{-1} h^{-1} and a high total turnover number of 62600 after 100 h of operation (over 1300 reactor volumes), highlighting its high activity and stability.

359c Routes to Renewable Production of Paracetamol: A Green Chemistry Analysis of Industrial Viability

Jimin Park, Caria Evans, Jacob Maier, Marta Hatzell, Stefan France, Carsten Sievers, Andy Bommarius

The active pharmaceutical ingredient (API) paracetamol, an important ingredient in many over-the-counter (OTC) drugs such as Tylenol®, is one of the most widely produced pharmaceuticals worldwide. Supply chain shortages of paracetamol due to recent world events such as COVID-19 have increased interest in developing new methods to synthesize paracetamol that are green enough to facilitate permitting the establishment of new production plants.

Current methods to produce paracetamol are characterized by chemistry involving petrochemical feedstocks as well as dated and environmentally harmful reaction conditions, including batch processing and harsh reactants. To introduce new paracetamol supply chains, a new process is necessary that is based on renewables, environmentally friendly, and economically viable on a large scale. Lignocellulosic material is a promising alternative source of aromatic unit structures owing to its regional abundance and renewability, and may be the starting point for developing viable new routes to paracetamol.

A key bottleneck in the synthesis of paracetamol is access to the important intermediate 4-aminophenol, which is unstable and difficult to obtain selectively. Traditional syntheses obtain 4-aminophenol through the nitration of petrochemically derived benzene. We have examined the viability of three novel, alternate routes to 4-aminophenol and paracetamol, all starting from phenol, a monomer product of lignin depolymerization, as the feedstock. These three routes are the acetamidation of hydroquinone, the imination and reduction of p-benzoquinone, and the hydrogenation of 4-nitrophenol. We have examined the performance of each of these routes with the goal of determining their industrial viability. High reaction rate (indicating viability for continuous synthesis) and high Green Chemistry metrics (such as yield, RME, and E-factor) were considered.

Through our analysis, we identified that the 4-nitrophenol route is the most promising in terms of Green Chemistry metrics and continuous viability. Important problems holding back other reaction routes include slow reaction speed, low selectivity/yield from side reactions, and poor Green Chemistry metrics stemming from too many required reactants or reaction steps.

359d Green Solvent Exploration in Li Ion Battery Electrode Fabrication: A Sustainable Approach

Amrita Sarkar

State-of-the-art Li ion battery manufacturing primarily focuses on the electrode fabrication process with minimum environmental concern. Current method "*slurry casting*" process involves a toxic organic solvent N-methyl pyrrolidone (NMP). Though successful, NMP is declared as a priority substance in USA under the Toxic Substance Control Act (TSCA), due to its slow biodegradability and high carcinogenic, mutagenic and reprotoxic risks to the human health. Thus, a greener alternative of toxic NMP is urgently required for the sustainable fabrication of electrode. Recently our group developed an advanced cathode fabrication process using two greener alternatives of NMP, methyl-5-(dimethylamino)-2-methyl-5-oxopentanoate (PolarClean) and dimethyl isosorbide. I will elaborate the finding here

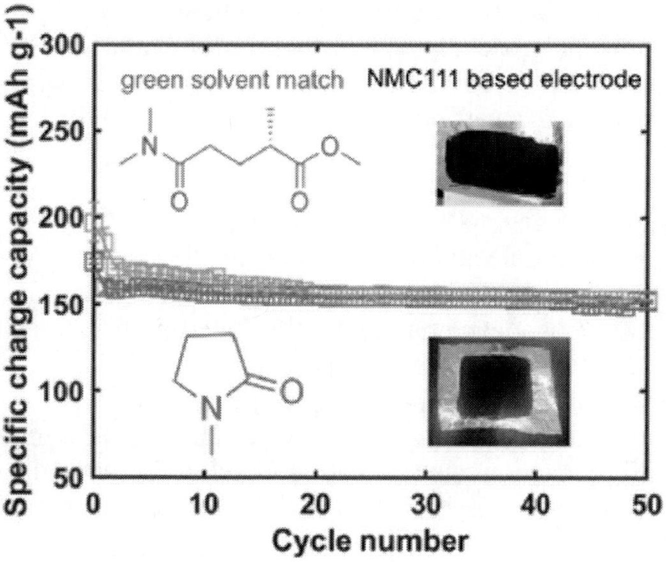

359e Potential Utilization of Nitrogen Oxide from Exhausted Gas for Enhancement of Nitrogen Cycle

Isshin Ino, Yuka Sakai, Yuki Sato, Yasuki Kansha

Under the stress of environmental hazards and resource scarcity, our society is forced to transform from resource-consuming to sustainable. To achieve the sustainability goal, energies and materials that used to be discarded should be transformed into valuable products and returned to society again. Therefore, this research focuses on the re-utilization mechanism for the Nitrogen component contained in industrial flue gas as valuable materials.

In general, the Nitrogen component in waste gas often exists as toxic Nitrogen Oxide (NOx) form. Because of its toxicity, NOx is neutralized before releasing into the atmosphere. Currently, industries usually apply Selective Catalytic Reaction (SCR) method on waste gas detoxification, which neutralizes NOx into harmless Nitrogen gas and water by reacting with ammonia and catalysts. However, using the SCR method requires purchasing external ammonia as the reductant, and even though the product, Nitrogen gas, is harmless, it is not valuable materials to return to society.

In contrast, this research examines the NOx resource transformation methods for the NOx-contained flue gas. According to the literature review, reacting the NOx component with alkaline solution was selected as one of the possible pathway because nitrate salts were produced, which often utilized as fertilizer and gunpowder's raw material. In addition, the reaction was analysed for scale-up of this method.

Unlike the ammonia-consuming SCR method, this method produces valuable by-products through different approaches, enabling further resource utilization and construction of the Nitrogen cycle for a sustainable society. Furthermore, along with applying this approach to various cases, the resource utilization cycles may blossom in different areas.

359f Sequential Extraction of Triglycerides and Fucoxanthin with Supercritical Carbon Dioxide for the Valorization of the Integrated Biorefinery

Elise Gilcher, Mary Kate Lane, Paul T. Anastas, Julie B. Zimmerman

The costs associated with producing renewable biofuels from algae have been prohibitive to their successful realization. However, pursuing an integrated biorefinery model through the valorization of additional high value and high purity products from algae can favorably shift the economics of the system. Of importance to realizing renewable fuels and chemicals sustainably, our work aligns with the Principles of Green Chemistry by replacing organic solvents, typically used for extraction from biomass, with nontoxic, green solvents like carbon dioxide (CO_2) through supercritical fluid extraction (SFE). In the supercritical phase, the solvation properties of CO_2, such as solvent polarity, hydrogen donating and accepting capacity, can be tuned by adjusting the temperature and pressure to enable the selective extraction of different biomass fractions. In this work, carotenoids ("super-antioxidants" like fucoxanthin) are selectively extracted from the microalgae *Phaeodactylum tricornutum* in a supercritical CO_2 flow through system to yield a solvent-free, high value enriched extract with commercial applications in pharmaceutical and nutraceutical products. We then examine a series of sequential, selective extractions targeting triacylglycerides or fucoxanthin and other carotenoids by systematically changing the supercritical solvent properties as well as the order of extractions targeting the different high value products. Solvent polarity were tuned by modulating pressure and volumes of ethanol co-solvent to achieve desired solvent properties, and the optimal order of extractions is determined to maximize product yields and purity. This work advances a greater initiative towards a large scale, algae-based bioeconomy via an environmentally benign, integrated biorefinery approach.

359g Preparation of Silica By Sol-Gel Methods: How Sustainable Are Current Research Methods?

Ethan Errington, Jerry Heng, Miao Guo

Silica-based nano-materials (SBNMs) have been adopted in a wide range of engineering applications including separation technologies, filler materials, and novel biomedical applications (Singh *et al.*, 2014). This is primarily due to the maturity of methods available for controlling the size, porosity, and surface chemistry of SBNMs. Importantly, a significant portion of SBNMs are manufactured according to a sol-gel process.

The sol-gel process is a "wet" production method which relies on the hydrolysis and condensation of an orthosilicate reagents in aqueous solution (Singh *et al.*, 2014; Minju *et al.*, 2021). For example, this includes the use of the sodium silicates in commercial settings due to their low reagent cost, and alkoxide-silicates such as tetraethyl-orthosilicate (TEOS) in academia - due to the greater control of SBNM properties that they afford. However, our previous work (Errington *et al.*, 2023) and other research (Baccile *et al.*, 2009) suggested that the use of SBNMs derived from alkoxide-silicates may carry environmental burdens that are significantly larger than sodium silicate-derived SBNMs.

This work investigates the extent to which use of alkoxide-silicate reagents may be affecting environmental burdens associated with SBNMs being produced in literature. By performing Life Cycle Assessment, the current study aims to:

- address the knowledge gaps and create Life Cycle Inventory (LCI) models for SBNMs derived from alkoxide precursors using TEOS-derived SBNMs as an example,
- estimate the environmental impacts associated with TEOS-derived SBNMs based on the newly produced LCI models,
- compare the environmental impact of TEOS-derived and TMOS-derived silica with that of sodium silicate-derived silica - as detailed modelled previously by the authors (Errington *et al.*, 2023).

The functional unit used is one kilogram of SBNM product and impact assessment is carried out according to the ReCiPe 2016 Hierarchical method (Huijbregts *et al.*, 2016).

Following a review of patent and academic literature, the LCI model for TEOS-derived SBNMs is developed based on consideration of a total of total of 70 unique implementations reported across 9 separate sources. For fairness of comparison, this is done according to an inventory averaging approach in which special consideration is made for the material efficiency of each implementation would make it a fair representation of any commercial process.

Our LCA results indicate that the environmental impact of TEOS-derived SBNMs may be significantly larger than that of a sodium silicate-derived equivalent SBNM (depending on the impact category). Global warming potential of TEOS-derived silica is at least 4 times greater than that of the sodium-silicate alternative considered (see Figure 1). Importantly, the source of the higher impact is discussed from the perspective of the low atom economy of the TEOS reagent when compared to sodium silicate (29% vs. 42–59% (Errington *et al.*, 2023)) and additional insight is provided via a contribution analysis which identifies the need extent of organic solvents use as also being significant.

Our presented case study shows that alkoxide-derived SBNMs may be significantly more impactful than sodium silicate-derived alternates. These results form a new contribution to the literature by providing the first quantitative assessment of the environmental impact of alkoxide-derived SBNMs. A discussion is opened surrounding the social responsibility of investigating alkoxide-derived SBNMs given their high environmental impacts (shown here).

References

Errington, E., Guo, M. and Heng, J.Y., 2023. Synthetic amorphous silica: environmental impacts of current industry and the benefit of biomass-derived silica. Green Chemistry.

The functional unit used is one kilogram of SBNM product and impact assessment is carried out according to the ReCiPe 2016 Hierarchical method (Huijbregts et al., 2016).

Minju, N., Nair, B.N. and Savithri, S., 2021. Sodium silicate-derived aerogels: effect of processing parameters on their applications. RSC advances, 11(25), pp.15301-15322.

Singh, L.P., Bhattacharyya, S.K., Kumar, R., Mishra, G., Sharma, U., Singh, G. and Ahalawat, S., 2014. Sol-Gel processing of silica nanoparticles and their applications. Advances in colloid and interface science, 214, pp.17-37.

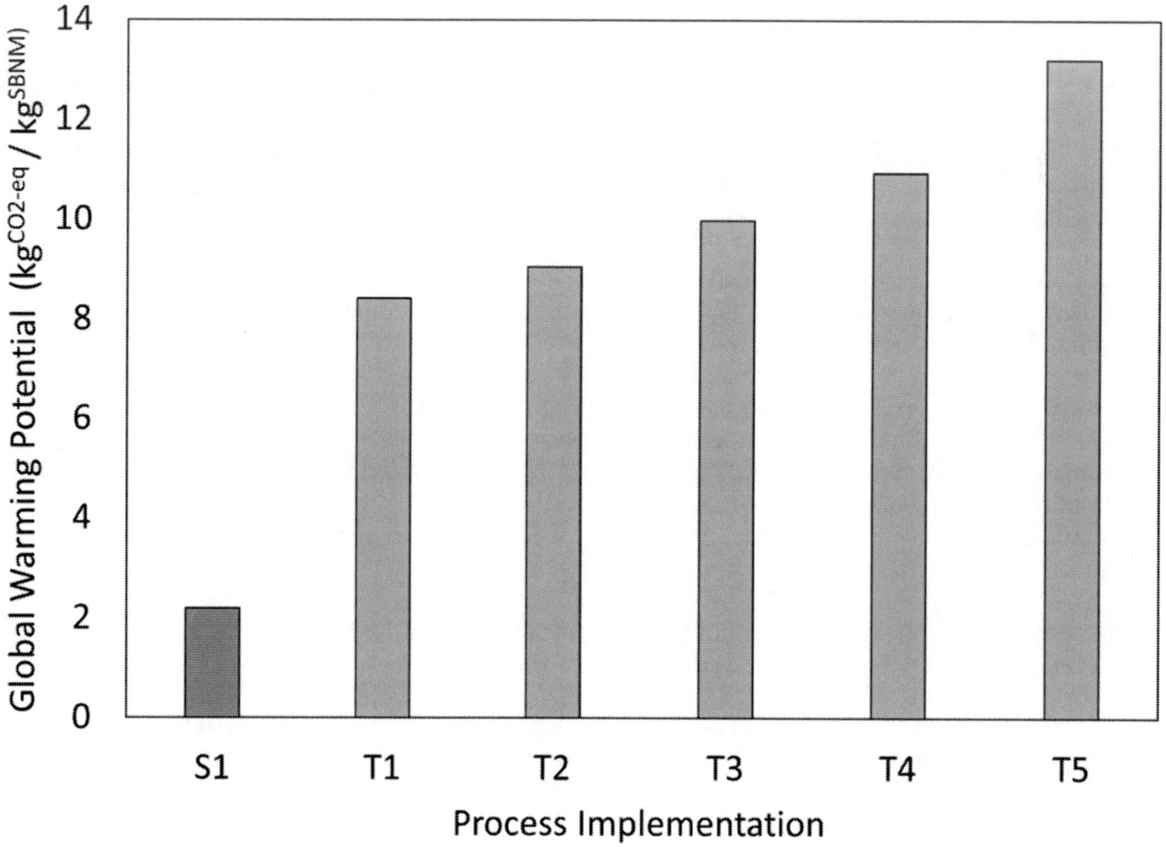

Figure 1: Global Warming Potential of sodium silicate-derived SBNM (S1) compared to that of the five lowest global warming potentials predicted for TEOS-derived SBNMs (T1-T5). Impact of sodium silicate-derived SBNM taken from previous work of the authors (Errington *et al.*, 2023).

390a LCA Comparison of Char Impregnation and Direct Chemical Activation for Synthesizing Ultraporous Superactivated Hydrochars from Food Waste

Swarna Saha, Al Ibtida Sultana, Samira Hossain, Kawnish Kirtania, Toufiq Reza

Food waste management creates an enormous concern, which has made it urgently necessary to upcycle it into usable materials, such as ultraporous adsorbent. Chemical activation of food waste hydrochar with potassium hydroxide (KOH) has developed as one of the approaches for increasing the porosity of such waste-derived porous sorbents. Direct chemical activation (DCA) and char impregnation (CI) are the two separate processes used to produce ultraporous adsorbent. The purpose of this study is to compare the environmental impacts of DCA and CI utilizing the life cycle assessment method (LCA). The results demonstrate that CI is a more environmentally friendly method of producing ultraporous carbons from food waste, with freshwater ecotoxicity (57.2%) being the category with the greatest environmental impact. This is primarily due to acid neutralization in the mixer unit of the CI technique of activation, whereas the dryer unit, which is powered by natural gas combustion, was the category with the greatest environmental impact. Hence, as a substitute, the use of renewable solar energy (from solar thermal power plants) was also explored, and the findings emphasized the prospect of achieving lower climate change and acidification potential.

390c Life Cycle Assessment of Hydrogen Generation from Biogenic Waste Via Anaerobic Co-Digestion Integrated with Reforming

Samira Hossain, Mst. Lucky Khatun, Chayan Kumer Saha, Md. Monjurul Alam, Mohammad Ismail, Toufiq Reza, Kazi Bayzid, Kawnish Kirtania

Hydrogen generation from biogenic waste or intermediates is a promising route for many countries lacking in typical renewable resources such as solar or wind. The characteristics of such waste, however, varies widely and require specific integration of technologies. For example, in Bangladesh, over 80% biogenic waste consists of agricultural residue containing over 70% moisture content. Majority of the waste is handled via landfilling while a portion of is anaerobically mono-digested, thus yielding high environmental implications. For high moisture biogenic waste, hydrogen production can be achieved via coupling anaerobic co-digestion (ACoD) technology with catalytic reforming. However, the environmental benefits of such integration, especially for highly populous developing countries, has been largely unexplored. In this work, we consider an ACoD-Reforming system for upgrading biogas derived from co-digestion to hydrogen rich syngas for simultaneous handling of waste and energy production-considering regional context of Bangladesh as a case study. A thorough attributional life-cycle assessment (LCA) is carried out for the considered system from cradle-to-grave for determining five environmental impacts such as global warming, acidification potential, eutrophication and eco-toxicity using primary life cycle inventory data. In addition, a comparison with other waste-based hydrogen production technologies is also made for the same feedstock conditions. The results of the life cycle impact assessment can serve as an important decision-making tool for assessing hydrogen production potential in developing countries to kick-start its path towards assimilating hydrogen economy.

390d Regionalized Dynamic Life Cycle Assessment of Waste Biomass Utilization

Yuan Yao

Agriculture and forest residues are waste biomass currently underutilized, e.g., they are burned or left on-site, both of which generate greenhouse gas (GHG) emissions. Valorizing waste biomass is a promising solution for enhancing resource utilization efficiency, reducing GHG emissions, and producing bio-based materials/products to replace fossil-based counterparts. Many valorization pathways, such as biofuels, biochar, and bio-based chemicals, have been explored. Previous studies have used life cycle assessment (LCA) to quantify the potential environmental impacts of different valorization pathway, but most of them have not considered the regional and temporal differences in biomass quality and quantity and how biomass removal affects ecosystems and land-related GHG emissions. Understanding the net carbon consequence and potential environmental benefits/risks of different biomass utilization pathways across diverse temporal and spatial scales is critical to support timely decision-making related to engineering, supply chain, and policies.

This talk will present a life cycle modeling framework that integrates LCA with spatial and dynamic modeling for both biomass conversion and ecosystems. The framework allows for regionalized LCA to understand region-specific environmental impacts of biomass utilization within short- or long-term timeframes. This talk will present two case studies related to forest-residue-derived biofuels for demonstration. The first case study will demonstrate how the temporal effects of GHG emissions from both biorefineries and forests affect the life cycle GHG emissions of biofuels and how dynamic LCA can be used to support short- and long-term decision-making related to engineering design and policies. The second case study will discuss how the spatial and temporal factors of forest ecosystems and biorefineries (e.g., regional environmental conditions and biorefinery locations) affect the GHG emissions from both industrial systems and ecosystems (e.g., soil) that ultimately determines the life cycle GHG emissions of biofuels.

390e A Life Cycle Analysis of the Conversion of Landfill Gas to Fischer-Tropsch Diesel By Tri-Reforming and Fischer-Tropsch Synthesis (TriFTS)

Tuhin Poddar, George G. Zaimes, Saurajyoti Kar, Devin M. Walker, Troy Hawkins

The waste industry remains a hard-to-decarbonize sector with conventional waste management methods being carbon-intensive. A promising alternative to these methods could be the production of renewable liquid fuels from biomass-derived landfill gas. These renewable liquid fuels have the potential to play a significant role in the transition to low-carbon fuel pathways. In this paper, we take a closer look at the production of Fischer-Tropsch diesel from landfill gas using the catalytic conversion Tri-Reforming and Fischer-Tropsch Synthesis process or TriFTS® process. We investigated the life cycle greenhouse gas (GHG) emissions of the TriFTS diesel production process and compared it with conventional petroleum diesel produced from fossil sources. A life cycle-based comparison was also made between TriFTS diesel and other landfill waste management pathways - the conversion of landfill gas to power and landfill gas to compressed renewable natural gas - on a per kilogram of feedstock basis as well as a per MJ of energy basis. The study assumed the conventional waste management method of flaring landfill gas as the underlying counterfactual scenario for the investigated waste-to-energy product pathways. We estimated the life cycle GHG emissions for TriFTS diesel to be -36.4 carbon dioxide equivalent (grams CO_2e)/MJ which is much lower than its fossil fuel counterpart which was estimated to be 90.5 grams CO_2e/MJ on a cradle-to-grave basis. The life cycle emission results from both functional unit perspectives show that TriFTS diesel is a viable alternative energy pathway from LFG when compared to other pathways, primarily due to the main product being a renewable fuel that can serve as a drop-in fuel for diesel-based uses, both within the waste industry as well as the larger transportation fuels market. Additional sensitivity analysis was performed based on the production of TriFTS diesel with counterfactual waste management scenarios of flaring and electricity production from landfill gas. The sensitivity of the TriFTS diesel carbon intensity to flaring efficiency, thermal power combustion efficiency as well as the carbon intensity of the electricity grid were also investigated. An extension of this work to study the life cycle impacts of the production of TriFTS diesel from other biomass-based feedstock including sludge deposits from wastewater treatment plants and swine manure is also considered. The paper highlights the potential for the TriFTS energy conversion technology to contribute to the waste industry's decarbonization initiatives, and to provide low carbon fuel for transportation.

390f Material Flow Analysis of Bioplastics Manufacturing Using Agricultural Residues in US for Pharmaceutical Packaging

Apoorva Bademi, Shweta Singh

Using biomass from waste resources is a low-cost option for bio-based material production without further agricultural inputs such as land, water, and fertilizer. The Billion Ton report of 2016 reported that under the base case scenario, the US was projected to have around 105 and 125 million dry tons of agricultural residue available at 60/dry ton in 2017 and 2022, respectively (Langholtz et al., 2016) . A majority of this residue is left in the field after harvest, and used for fodder, tilled back into the soil for increased soil fertility or incinerated. However, in recent years, there has been an increased growth in bio-based product manufacturing. There are various studies that report the manufacture of bio-based plastics from agricultural residue. Since most of these are biodegradable, it further increases waste utilization efficiency and can reduce plastic pollution. Packaging is one of the largest biobased plastic market segments. Currently, 70% of the bio-based plastics are used in short-life packaging options. This has led to a growing interest in studying the end-of-life options and the potential for circular economy (Spierling et al., 2018) . Pharmaceutical packaging is the packing and packaging processes for pharmaceutical preparations. Optimal drug packaging maximizes the physical and chemical stability of the medication, which may be affected by environmental factors such as light, temperature, and humidity. Various types of packaging options exist in the market today based on the type and form of drug, dosage, quantity, and other specific physical properties of the drug. Since the packaging options are so varied, it is challenging to have a one-size-fits-all solution for a more sustainable end-of-life option. This study focuses on the potential impact of replacing traditional pharmaceutical packaging with bio-based and biodegradable plastics to increase resource efficiency and provide more sustainable end-of-life options, specifically with reference to acetaminophen packaging. Facilitation of environmental and economic sustainability requires an understanding of the structure of the material economy and the material flows/waste generation in industrial networks to minimize the environmental consequences of these industrial systems. Physical Input-Output Tables (PIOTs) are uniquely suited for this purpose by mapping physical flows between industrial sectors. PIOT-Hub, a cloud-based tool, automates the generation of Physical Supply Tables (PSTs) and Physical Use Tables (PUTs), in addition to their conversion into PIOTs using mechanistic models and macroeconomic Input-output framework (Vunnava et al., 2022 . The goal of this study is to study the impact of bio-based, biodegradable pharmaceutical packaging manufacturing on material and elemental flows in regional industrial networks. We develop process flow models for waste-to-plastics manufacturing and employ PIOT-Hub for studying the industrial network flows. We utilize this tool to identify new material circularization opportunities. We further conduct an elemental flow analysis to track the flow of Carbon and Nitrogen within the network.

References

Langholtz, M. H., Stokes, B. J., & Eaton, L. M. (2016). *2016 Billion-Ton Report: Advancing Domestic Resources for a Thriving Bioeconomy*. https://doi.org/10.2172/1271651

Spierling, S., Röttger, C., Venkatachalam, V., Mudersbach, M., Herrmann, C., & Endres, H. J. (2018). Bio-based Plastics - A Building Block for the Circular Economy? *Procedia CIRP*, *69*, 573–578. https://doi.org/10.1016/j.procir.2017.11.017

Vunnava, V. S. G., Shin, J., Zhao, L., & Singh, S. (2022). PIOT-Hub - A collaborative cloud tool for generation of physical input–output tables using mechanistic engineering models. *Journal of*

Industrial Ecology, *26*(1), 107–120. https://doi.org/10.1111/jiec.13204

390g Technoeconomic Analysis and Life Cycle Assessment of Bio-Based Non-Isocyanate Polyhydroxyurethane and Polythiourethane Production with Monomer Recovery

Yasheen Jadidi

Background: In 2016, 2900 kt (kilotonnes) of polyurethane (PU) were produced in the United States at the expense of 1100 kt of crude oil and 1100 kt of natural gas.[1] With a 5.5% of end-of-life recycling rate, polyurethanes are recycled at less than the average recycling rate of U.S. plastics in 2015.[1,2] The use of fossil fuel and natural gas reserves depletes nonrenewable resources while releasing emissions that directly contribute to global warming. To mitigate issues with sustainability, it is necessary to advance technologies for bio-based PU synthesis and to expand methods for recycling PU monomers. To alleviate issues with safety, the U.S. Environmental Protection Agency (EPA) has begun to phase out the use of isocyanates in the synthesis of PU.[3] Polyhydroxyurethane (PHU) and polythiourethane (PTU) can be synthesized and reprocessed from biomass using isocyanate-free methods. Catalytic coupling of epoxide residues with either carbon dioxide or carbon disulfide affords 5-membered cyclic carbonates or 5-membered cyclic dithiocarbonates, respectively. Subsequent ring opening polyaddition with a diamine can afford PHU from the cyclic carbonates or PTU from the cyclic dithiocarbonates. Chemicals that can be derived from biomass, including 1,3-butadiene or 1,4-butanediol,[4] are readily epoxidized via a reaction with hydrogen peroxide or an S_N2 substitution with epichlorohydrin. Cyclic carbonate and dithiocarbonate moieties have also previously been derived from epoxidized plant seed oils for PU biosynthesis.[5,6]

Results: We examine case studies for phasing out non-renewables and shifting toward a more circular economy. To boost the bio-based content of PU consumption in the United States, it is proposed to substitute with non-isocyanate polyurethanes (NIPU), PHU and PTU, synthesized using bio-based monomers. Production processes with monomer recovery for PHU and PTU, modeled using AspenPlus, inform the results of a life cycle assessment (LCA) and technoeconomic analysis (TEA). The LCA component of the study discusses the environmental impacts of PHU and PTU synthesis in terms of energy usage, emissions, and water usage. The TEA component of the study discusses the profitability of selling PHU and PTU and breaks down the major components of their production costs. Preliminary results indicate raw materials drive cost and GHG emissions of PHU and PTU production from bio-based intermediates. Monomer recovery costs are predominantly affected by the depolymerization yield and the cost of solvent.

[1]Liang, C., Gracida-Alvarez, U.R., Gallant, E.T., Gillis, P.A., Marques, Y.A., Abramo, G.P., Hawkins, T.R. and Dunn, J.B. 'Material flows of polyurethane in the United States.' *Environmental Science & Technology* 55.20 (2021): 14215-14224.

[2]Di, J., Reck, B.K., Miatto, A. and Graedel, T.E. 'United States plastics: Large flows, short lifetimes, and negligible recycling.' *Resources, Conservation and Recycling* 167 (2021): 105440.

[3]US Environmental Protection Agency. 'Methylene Diphenyl Diisocyanate (MDI) And Related Compounds Action Plan.' (2011).

[4]Biddy, M.J., Scarlata, C. and Kinchin, C. "Chemicals from biomass: a market assessment of bioproducts with near-term potential." No. NREL/TP-5100-65509. National Renewable Energy Lab.

(NREL), Golden, CO (United States), 2016.

[5]Hu, S., Chen, X. and Torkelson, J.M. 'Biobased reprocessable polyhydroxyurethane networks: Full recovery of crosslink density with three concurrent dynamic chemistries.' *ACS Sustainable Chemistry & Engineering* 7.11 (2019): 10025-10034.

[6]Vanbiervliet, E., Fouquay, S., Michaud, G., Simon, F., Carpentier, J.F. and Guillaume, S.M. 'Non-isocyanate polythiourethanes (NIPTUs) from cyclodithiocarbonate telechelic polyethers.' *Macromolecules* 52.15 (2019): 5838-5849.

514b Highly Efficient System for Direct Air Capture and Conversion of CO2 to Value-Added Chemicals

Codruta Loebick, Christian Junaedi, Subir Roychoudhury

Carbon dioxide capture and utilization is becoming the key concern for investment in the energy and industrial sector due to ever increasing carbon emissions and growing concerns about climate change among public and government agencies around the world. Direct Air Capture and Conversion has emerged as a leading technology for mitigating elevated CO_2 atmospheric concentrations.

The primary barrier to minimizing the cost of direct air capture of CO_2 is the cost of energy. We are developing a high-performance solid sorbent collector technology and materials that operate at low-temperature and low energy utilization for Direct Air Capture (DAC) of CO_2. The recovered CO_2 is then utilized to produce methane via the Sabatier reaction. This process produces methane and water by reacting CO_2 with hydrogen. The methane generated can be used directly as a fuel or it can be converted into higher molecular weight hydrocarbons which can be transported or stored in liquid form.

Precision Combustion, Inc. (PCI), with support from DOE and NASA Small Business Innovation Research (SBIR) awards, has developed a novel efficient and compact Direct Air Capture System that operates with low energy input, coupled with Sabatier (i.e., CO_2 methanation) reactor based on its Microlith® catalytic technology to convert the captured CO_2 to produce methane.

Our approach demonstrated the capability to achieve efficient CO2 capture from atmospheric air, and high CO_2 conversion and CH_4 selectivity (i.e., ≥90% of the thermodynamic equilibrium values) at high space velocities, low operating temperatures, and low H_2 requirements. The ability to both capture CO_2 and achieve high CH_4 yield was made possible using high-heat-transfer and high-surface-area Microlith catalytic substrates and novel nanosorbent and catalyst development.

For the high space velocity DAC sorbent structure, PCI has developed and patented a short contact time mesh-based substrate, coated with a densified nanostructured sorbent. The combination enables higher surface area per unit volume and decreased bed volume with equivalent effectiveness to other types of monolithic or loose packing, without significant pressure drop penalty leading to substantial energy savings in DAC operation. Additionally, up to twenty times higher mass and heat transfer coefficients are obtainable as compared to other sorbent systems, such as monoliths and pellets, due primarily to boundary layer minimization and break-up, boosting CO_2 removal rates with greater sorbent bed utilization and less bypass inherent to packed beds or monoliths. Our sorbent manufacturing technology allows for adherent and durable sorbent coatings on the Microlith substrate. PCI developed and demonstrated in operation carbon capture systems capable of supplying CO2 to the Sabatier reactor with low energy cost.

PCI's approach to Sabatier permits reactor operation under exothermic conditions with efficient heat recuperation for high catalyst performance at high space velocities and extended catalyst lifetime. A key advantage is the ability of the reactor to efficiently sustain its temperature during the CO_2 methanation reaction via an efficient heat recuperator and heat exchanger, thus eliminating the need for an external heat source, resulting in minimal power usage during steady state operation.

Using this Sabatier reactor, PCI designed, developed, and demonstrated a stand-alone CO_2 methanation test system for demonstration and performance validation. The versatility of the test system prototype was demonstrated by operating it under H_2-rich (H_2/CO_2 of >4), stoichiometric (ratio of 4), and CO_2-rich conditions (ratio of <4) without affecting its performance and meeting the equilibrium-predicted CH_4 yield and throughput.

In this presentation, the development of the Microlith-based Direct Air Capture Unit and the and CO_2 methanation reactor and the test system assembly for demonstration and performance validation will be discussed.

Catalytic Conversion of CO_2 over Cu and Ni Metals Supported on CeO_2 and La_2O_3 Via Reverse Water Gas Shift Reaction

Authors: Parisa Ebrahimi, Anand Kumar* and Majeda Khraisheh

Department of Chemical Engineering, Qatar University, Doha, P O Box 2713, Qatar

Corresponding author: Anand Kumar, Department of Chemical Engineering, Qatar University, Doha, P O Box 2713, Qatar,

Email: akumar@qu.edu.qa

Phone: (974) 4403-4164

Fax: (974) 4403-4131

Catalytic Conversion of CO_2 over Cu and Ni Metals Supported on CeO_2 and La_2O_3 Via Reverse Water Gas Shift Reaction

Authors: Parisa Ebrahimi, Anand Kumar* and Majeda Khraisheh

Abstract

The reverse water gas shift (RWGS) process is considered a feasible method for lowering greenhouse gas emissions by utilizing the CO_2 and converting it to CO. Herein, we intend to evaluated the catalytic conversion of CO_2 through RWGS reaction over transition metals nanoparticles supported on lanthanum. Catalysts of selected active metals are investigated in a packed bed tubular reactor within a temperature range of 100–600°C to assess their catalytic activity and selectivity towards CO. The results of catalyst's activity and stability experiments will be presented during the conference. In this report, we provide the details of synthesis method for targeted compounds. Furthermore, the morphological properties of the catalysts, as well as the effect of the reaction conditions on the catalysts surface, have studied using a variety of techniques, including XRD, TEM, SEM-EDX and TPR; and presented during the conference. At this stage, the proceedings is more focused on general synthesis of compounds and readers are referred to the published journal articles to gain more information. As the relevant technical outcomes are already published, to avoid reproducing the same texts here, concerning articles are cited and readers are encouraged to read them to get further insights on the results.

Keywords: Reverse water gas shift reaction, combustion synthesis, catalysts, nanomaterials.

Introduction

Solution combustion synthesis (SCS) has been used as the primary technique for synthesizing nanomaterials in this work [1-12]. The underlying principle in SCS utilizes the exothermic energy generated during redox reaction of its constituent components to synthesize high purity crystalline materials in a single step without requiring further post-synthesis treatments such as separation, purification or calcination. The material produced, based on their desired crystalinity, can be directly for catalytic application. The redox reaction involving metal nitrate precursor and glycine as fuel is believed to be based on following between stoichiometric reaction [1, 5]:

$$M^{v}(NO_3)_v + (\frac{5}{9} v\varphi)\, NH_2CH_2COOH + v\frac{5}{4}(\varphi-1)O_2$$

$$\downarrow$$

$$M^{v}O_{v/2}(s) + (\frac{10}{9}v\varphi)CO_2(g) + \frac{25}{18}v\varphi H_2O(g) + v(\frac{5\varphi+9}{18})N_2(g) \qquad (1)$$

Where M is the metal with valency υ. The value φ known as fuel/oxidizer ratio determine the oxidation state of the metal in the output $\varphi > 1$ is fuel rich condition, $\varphi < 1$ is fuel lean condition and $\varphi = 1$ is the stoichiometric condition with no need of atmospheric oxygen for fuel oxidation.

The SCS synthesized materials have been used for diverse applications including hydrogen generation, CO_2 conversion, RWGS, solar fuel production, solar oxidation, wastewater treatment etc. using various reactions is our lab [4-8, 10, 11, 13-21], along with being used for validation of theoretical predictions [22]. Our focus has also been towards the development of theoretical understanding of the combustion process [23], and more recently towards the utilization of this technique to develop electrodes for fuel cells and other electrochemical systems [24-26], which resulted in promising catalytic performances:

Experimental

Nanoparticles of copper and nickel, along with the precursors of the support (CeO_2 and La_2O_3) were synthesized using their nitrate precursors (e.g. $Cu(NO_3)_2$, $Ni(NO_3)_2$) and glycine as fuel.

These precursors were mixed together and dissolved in deionized water and kept these for 1 hr to obtain a uniform mixture of solution. In selected experiments, the metallic precursors were mixed with the already synthesized oxides. An appropriate ratio of fuel is used to combust the reactive solution generating suitable conditions to synthesize supported catalysts. More details of this particular set can be found in our recent articles on RWGS and methane dry reforming [27-31].

Results and Discussion

The specific results and more detailed analysis of the reaction can be found in our previous publications associated with RWGS on CeO_2 and La_2O_3 based catalysts, in addition to the design of experiments (DOE) analysis of combustion synthesis of ceria supported metals [27,28,31]. Only a summarized version is presented here to avoid repeating the same results. The DOE analysis does indicate optimum conditions for synthesizing copper supported on cerium oxide surface. The optimization analysis was conducted on the basis of fuel ratio, availability of oxygen for combustion as well as the metal content [31]. Results clearly indicate the possibility of optimization with respect to combustion temperature or with respect to volume of gases evolved during the combustion reaction, which in turn can be associated with the porosity of the final compound synthesized. A highly porous material will result in large surface area where reactants can adsorb and react to generate the final products. The RWGS results [27,28], highlight the importance of oxygen mobility in the support for CO_2 adsorption and subsequent conversion to CO in presence of Cu and Ni metals loaded on CeO_2 and La_2O_3 supports. It should be noted that in all the cases, metal loading was limited to 1wt.%, which allows one to systematically study the role of metals of oxygen vacancy of reducible oxides, and subsequently study the role of oxygen mobility on catalytic performance [27,28]. These results indicated that

CeO_2 was active when synthesized by combustion technique, and addition of Cu further enhanced the activity of the catalysts. Lanthanum based catalysts also showed similar results, however, slightly lower activity for Cu/La_2O_3 compared to Cu/CeO_2 was observed. Incorporation of nickel showed a gradual shift towards methanation reaction rather than RWGS reaction, nonetheless, a limited loading of 1wt.% allowed considerable amount of CO formation, possibly via reducible supports, along with methane.

Conclusion

Catalysts containing 1wt% of Cu and Ni nanoparticles on CeO_2 and La_2O_3 supports were synthesized using solution combustion synthesis technique with a predetermined fuel ratio. The related journal articles containing the synthesis and reaction parameters are referred in the texts and the results were summarized. These results do indicate the role of oxygen mobility in RWGS catalysis, where CeO_2 showed a reasonable activity at low temperature. La_2O_3 and CeO_2 loaded with Cu and Ni resulted in improved catalytic activity with high selectivity of CO, nonetheless, addition of Ni showed some tendency towards methane formation.

Acknowledgement:

This publication was made possible by the NPRP grant (NPRP14S-0302-210011) from the Qatar National Research Fund (a member of Qatar Foundation). The statements made herein are solely the responsibility of author(s).

The authors also wish to acknowledge the technical services granted by the Central Laboratory Unit (CLU) and the Gas Processing Centre (GPC) team in Qatar University.

References:

1. A. Kumar, E.E. Wolf, A.S. Mukasyan, Solution Combustion Synthesis of Metal Nanopowders: Nickel-Reaction Pathways , AIChE Journal 57 (2010) 2207-2214.

2. A. Kumar, A.S. Mukasyan, E.E. Wolf, Impregnated layer combustion synthesis method for preparation of multicomponent catalysts for the production of hydrogen from oxidative reforming of methanol, Applied catalysis.A, General 372 (2010) 175-183.

3. A. Kumar A.S. Mukasyan, E.E. Wolf, Modeling Impregnated Layer Combustion Synthesis of Catalyts for Hydrogen Generation from Oxidative Reforming of Methanol, Industrial engineering chemistry research 49 (2010) 11001-11008.

4. A. Kumar, Combustion synthesis of novel catalysts for hydrogen production from light alcohols, University of Notre Dame, 2011.

5. A. Kumar, E.E. Wolf, A.S. Mukasyan, Solution combustion synthesis of metal nanopowders: Copper and copper/nickel alloys, AIChE Journal 57 (2011) 3473-3479.

6. A. Kumar, A. Mukasyan, E. Wolf, Combustion synthesis of Ni, Fe and Cu multi-component catalysts for hydrogen production from ethanol reforming, Applied Catalysis A: General 401 (2011) 20-28.

7. A. Cross, A. Kumar, E.E. Wolf, A.S. Mukasyan, Combustion Synthesis of a Nickel Supported Catalyst: Effect of Metal Distribution on the Activity during Ethanol Decomposition 2012, *51* (37), pp 12004–12008, Ind Eng Chem Res 51 (2012) 12004-12008.

8. A. Kumar, J.T. Miller, A.S. Mukasyan, E.E. Wolf, In situ XAS and FTIR studies of a multi-component Ni/Fe/Cu catalyst for hydrogen production from ethanol, Applied Catalysis A: General 467 (2013) 593-603.

9. G. Carotenuto, A. Kumar, J. Miller, A. Mukasyan, E. Santacesaria, E. Wolf, Hydrogen production by ethanol decomposition and partial oxidation over copper/copper-chromite based catalysts prepared by combustion synthesis, Catalysis Today 203 (2013) 163-173.

10. R.R. Bhosale, I. Alxneit, L.L.P. van den Broeke, A. Kumar, M. Jilani, S.S. Gharbia, J. Folady, D.Z. Dardor, Sol-Gel Synthesis of Nanocrystalline Ni-Ferrite and Co-Ferrite Redox Materials for Thermochemical Production of Solar Fuels, MRS Online Proceedings Library 1675 (2014) 203-208.

11. A. Kumar, A. Cross, K. Manukyan, R.R. Bhosale, L.J.P. van den Broeke, J.T. Miller, A.S. Mukasyan, E.E. Wolf, Combustion synthesis of copper–nickel catalysts for hydrogen production from ethanol, Chem. Eng. J. 278 (2015) 46-54.

12. A. Ashok, A. Kumar, R.R. Bhosale, M.A.H. Saleh, van den Broeke, Leo JP, Cellulose assisted combustion synthesis of porous Cu–Ni nanopowders, RSC Advances 5 (2015) 28703-28712.

13. A. Kumar, A. Ashok, R.R. Bhosale, M.A.H. Saleh, F.A. Almomani, M. Al-Marri, M.M. Khader, F. Tarlochan, In situ DRIFTS Studies on Cu, Ni and CuNi catalysts for Ethanol Decomposition Reaction, Catal. Lett. 146 (2016) 778-787.

14. R.R. Bhosale, A. Kumar, L.J.P. van den Broeke, S. Gharbia, D. Dardor, M. Jilani, J. Folady, M.S. Al-Fakih, M.A. Tarsad, Solar hydrogen production via thermochemical iron oxide–iron sulfate water splitting cycle, Int J Hydrogen Energy 40 (2015) 1639-1650.

15. D. Dardor, R.R. Bhosale, S. Gharbia, A. Kumar, F. Al Momani, Solar carbon production via thermochemical ZnO/Zn carbon dioxide splitting cycle, Journal of Emerging Trends in Engineering and Applied Sciences 6 (2015) 129-135.

16. F.A. Almomani, R.R. Bhosale, A. Kumar, C. Kennes, Removal of volatile sulfur compounds by solar advanced oxidation technologies and bioprocesses, Solar Energy 135 (2016) 348-358.

17. R.R. Bhosale, A. Kumar, F. AlMomani, U. Ghosh, S. Al-Muhtaseb, R. Gupta, I. Alxneit, Assessment of CexZryHfzO2 based oxides as potential solar thermochemical CO2 splitting materials, Ceram. Int. 42 (2016) 9354-9362.

18. R.R. Bhosale, A. Kumar, F. AlMomani, I. Alxneit, Sol–gel derived CeO2–Fe2O3 nanoparticles: Synthesis, characterization and solar thermochemical application, Ceram. Int. 42 (2016) 6728-6737.

19. R.R. Bhosale, A. Kumar, F. AlMomani, U. Ghosh, D. Dardor, Z. Bouabidi, M. Ali, S. Yousefi, A. AlNouss, M.S. Anis, M.H. Usmani, M.H. Ali, R.S. Azzam, A. Banu, Solar co-production of samarium and syngas via methanothermal reduction of samarium sesquioxide, Energy Conversion and Management 112 (2016) 413-422.

20. R.R. Bhosale, A. Kumar, F. AlMomani, I. Alxneit, Propylene oxide assisted sol–gel synthesis of zinc ferrite nanoparticles for solar fuel production, Ceram. Int. 42 (2016) 2431-2438.

21. F.A. Almomani, M. Shawaqfah, R.R. Bhosale, A. Kumar, Removal of emerging pharmaceuticals from wastewater by ozone-based advanced oxidation processes, Environmental Progress & Sustainable Energy 35 (2016) 982-995.

22. A. Kumar, R.R. Bhosale, S.S. Malik, A.E. Abusrafa, M.A.H. Saleh, U.K. Ghosh, M.J. Al-Marri, F.A. Almomani, M.M. Khader, I.M. Abu-Reesh, Thermodynamic investigation of hydrogen enrichment and carbon suppression using chemical additives in ethanol dry reforming, Int J Hydrogen Energy 41 (2016) 15149-15157.

23. A. Kumar, E.E. Wolf, A.S. Mukasyan, Modeling Impregnated Layer Combustion Synthesis of Catalyts for Hydrogen Generation from Oxidative Reforming of Methanol, Industrial & Engineering Chemistry Research 49 (2010) 21, 11001–11008.

24. A. Ashok, A. Kumar, J Ponraj, SA Mansour, F Tarlochan, Single step synthesis of porous $NiCoO_2$ for effective electrooxidation of glycerol in alkaline medium, Journal of The Electrochemical Society 165 (2018) J3301.

25. A. Ashok, A. Kumar, R.R. Bhosale, F. Almomani, M.A.H. Saleh, S. Suslov, F Tarlochan, Influence of fuel ratio on the performance of combustion synthesized bifunctional cobalt oxide catalysts for fuel cell application, International Journal of Hydrogen Energy 44 (2019) 436-445.

26. A. Ashok, A. Kumar, J Ponraj, SA Mansour, Development of Co/Co_9S_8 metallic nanowire anchored on N-doped CNTs through the pyrolysis of melamine for overall water splitting, Electrochimica Acta. 368 (2021) 137642.

27. P Ebrahimi, A Kumar, M Khraisheh, Combustion synthesis of lanthanum oxide supported Cu, Ni, and CuNi nanoparticles for CO_2 conversion reaction, International Journal of Hydrogen Energy 48 (2023) 64, 24580-2459344.

28. P Ebrahimi, A Kumar, M Khraisheh, Combustion synthesis of copper ceria solid solution for CO2 conversion to CO via reverse water gas shift reaction, International Journal of Hydrogen Energy 47 (2022), 97, 41259-41267.

29. YH Ahmad, AT Mohamed, HA El-Sayed, A Kumar, SY Al-Qaradawi, Design of Ni/La_2O_3 catalysts for dry reforming of methane: Understanding the impact of synthesis methods, International Journal of Hydrogen Energy 47 (2022), 97, 41294-41309.

30. YH Ahmad, AT Mohamed, A Kumar, SY Al-Qaradawi, Solution combustion synthesis of Ni/La2O3 for dry reforming of methane: tuning the basicity via alkali and alkaline earth metal oxide promoters, RSC Advances 11 (2021), 53, 33734-33743.

31. P Ebrahimi, A Kumar, M Khraisheh, Analysis of combustion synthesis method for Cu/CeO2 synthesis by integrating thermodynamics and design of experiments approach, Results in Engineering 15, (2022), 100574.

514d Life Cycle Analysis Tools to Evaluate Environmental Footprint of Carbon Conversion Projects

Sheikh Moni, Michelle Krynock, Joseph Chou

Conversion of captured carbon dioxide (CO_2) emitted from power and industrial sources into valuable products for various applications can reduce greenhouse gas (GHG) emissions. Carbon dioxide conversion or carbon dioxide utilization (CO2U) is an emerging research area that encompasses many possible pathways, including biological uptake, catalytic conversion, and mineralization. Life cycle assessment/analysis (LCA) can help determine the potential environmental impacts and net life cycle GHG reductions of conversion technologies from a consistent and unbiased viewpoint, which is crucial for commercialization and acceptance of CO_2-derived products in the marketplace. The U.S. Department of Energy (DOE) and the federal government are increasingly requiring LCA as a condition for funding primary research, grant programs such as Utilization Procurement Grants (UPGrants) for eligible entities to procure and use carbon conversion products, and tax incentives such as "45Q" for CO_2 utilization projects (Treasury Regulations Section 1.45Q-4 (26 C.F.R. § 1.45Q-4), 2021).

To support the creation of useful and consistent LCAs of carbon conversion projects, the DOE National Energy Technology Laboratory (NETL) has developed a CO2U LCA Toolkit and released version 2.0 in 2022. The NETL CO2U LCA Guidance Toolkit is intended to help the project principal investigators meet LCA requirements under the DOE Carbon Conversion Program and support program managers in decision-making by ensuring consistency and transparency across Carbon Conversion Program projects. NETL also released specific addenda to support the development of LCAs for UPGrants and 45Q. The toolkit has also been used to guide a wider audience outside the Carbon Conversion Program in conducting LCAs on carbon conversion systems and advance the global discussion on carbon conversion LCA.

The guidance document should be considered a companion to the requirements established in ISO 14040/14044, providing specific guidance on methodological issues for scoping and completing the LCA. The guidance, data, and tools are publicly available and free. This presentation will provide an overview of each part of this updated LCA toolkit, its utility to the LCA community, and its value in assessing emerging and existing carbon conversion technologies.

References:

Treasury Regulations Section 1.45Q-4 (26 C.F.R. § 1.45Q-4) (2021). Retrieved from https://www.irs.gov/pub/irs-drop/td-9944.pdf on November 18, 2021

Disclaimer: This project was funded by the Department of Energy, National Energy Technology Laboratory an agency of the United States Government, through a support contract. Neither the United States Government nor any agency thereof, nor any of its employees, nor the support contractor, nor any of their employees, makes any warranty, expressor implied, or assumes any legal liability or responsibility for the accuracy, completeness, or usefulness of any information, apparatus, product, or process disclosed, or represents that its use would not infringe privately owned rights. Reference herein to any specific commercial product, process, or service by trade name, trademark, manufacturer, or otherwise does not necessarily constitute or imply its endorsement, recommendation, or favoring by the United States Government or any agency thereof. The views and opinions of authors expressed herein do not necessarily state or reflect those of the United States Government or any agency thereof.

514e Catalytic Conversion of NAS-CO2 to Chemical Feedstock

Moumita Bhattacharya, Marty Lail, Daniel Mogollon

RTI's current generation non-aqueous amine solvent (NAS) technology uses steam to strip CO_2 off the CO_2-rich solvent, followed by an activated carbon bed to capture the amine. As the capture side of the technology is currently undergoing commercial maturation, RTI has started to expand its research efforts in the utilization of the captured CO_2 to important chemical feedstock. Such conversion route will alleviate the cost associated with the energy intensive amine regeneration step as well as establish novel pathways towards carbon utilization. The work presented here will summarize the results of homogeneous and heterogeneous catalytic hydrogenation of CO_2, captured as carbamate to various feedstock. The effect of solvent, temperature and metal/ligand identity on the product selectivity and conversion efficiency will be discussed. This work presented here will showcase RTI's capability of utilizing the CO_2 captured within the novel non aqueous solvent to further bring down the cost of the existing capture technology.

514f CO_2 Utilization for Production of Fuels and Chemicals Using Ereact™

Peter M. Mortensen, Marené Rautenbach, Martin Oestberg, Kim Aasberg-Petersen

Utilizing CO_2 for production of fuels and / or chemicals requires not only hydrogen but also an effective sustainable production of the synthesis gas that enables the synthesis of the desired products. In this presentation, focus is on production routes for fuels and chemicals from CO_2 using sustainable electrical energy. Topsoe's eREACT™ technology allows for the first-of-its-kind electrification of syngas production for reverse water gas shift (RWGS) as well as the traditional steam methane reforming (SMR) process. The reaction heat for eREACT™ is generated directly from (renewable) electricity, thereby eliminating the production of flue gas traditionally seen from heated reactors altogether. The technology has gone through scale-up from bench scale to industrially relevant pilot scale showing that the technology is now ready for industrial application. This includes production of synthesis gas with a H_2/CO ratio of 2 ideal for Fischer-Tropsch synthesis of fuels.

Utilizing CO_2 for catalytic conversion put high demand to the purity of the CO_2 feed. Many CO_2 sources are releant including chemical produced CO_2 as by-product e.g. from hydrogen or ammonia; CO_2 extracted from combustion processes; biological produced CO_2; and CO_2 from direct air capture., Many different chemical species can be encountered in the feed that must be handled to allow an acceptable lifetime of the catalytic reactions. This cleaning process is important if successful utilization of CO_2 for fuels and chemicals should be obtained and possible solutions will be presented.

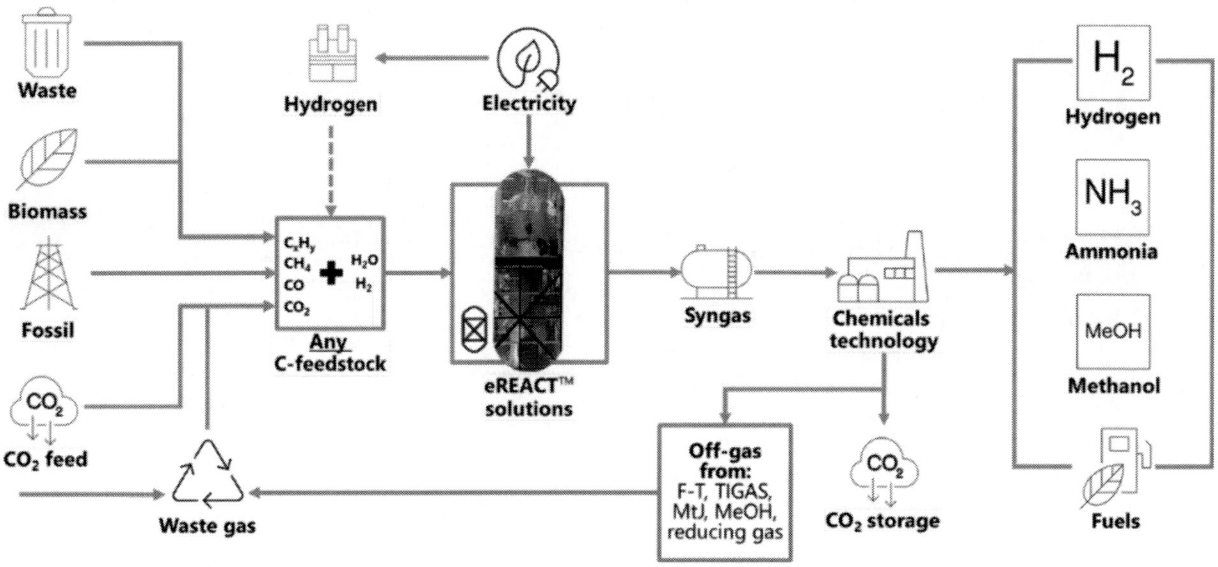

CO_2 utilization through power-to-methanol with high temperature heat pump

Nick Badger, Rahim Boylu, Valentine Ilojianya, Mustafa Erguvan, Shahriar Amini*

Department of Mechanical Engineering, University of Alabama, Tuscaloosa, Alabama, 35401

University of Alabama, Department of Mechanical Engineering, Tuscaloosa, AL, USA

Email: Samini3@ua.edu

Keywords: CO_2 utilization, CO_2 removal (CDR), CCS and hydrogen combinations

ABSTRACT

Carbon Capture Utilization (CCU) provides an attractive opportunity to create value for decarbonization of industrial sectors. DAC is recognized as a crucial technology for atmospheric CO2 removal. Its uniqueness lies in capturing CO2 from air and mixtures of air and CO2-rich industrial emission sources independent of origin and location. The utilization of CO2 through a power-to-methanol with CO2 from Direct Air Capture is studied in this work. The CO2 from Direct Air Capture (DAC) is paired with electrolytically produced hydrogen to synthesize liquid hydrocarbons (e-fuels).

ASPEN Plus software is used to model the energy requirements for the methanol and hydrogen production while the results from experimental work for DAC are integrated for the system analysis. In this research, four distinct configurations of a low-temperature (LT) solid sorbent DAC are explored, differing in their regeneration heat sources: conventional steam, waste heat steam, a high temperature heat pump (HTHP), and microwave-based heat. Various scenarios are analyzed for the integration of renewable energy to operate the systems, offering a comprehensive perspective on sustainable e-fuel generation pathways.

The mass and energy flows are connected using Life Cycle Assessment (LCA) methodology to investigate the overall environmental impacts of various routes to produce green methanol. Preliminary results highlight the pronounced benefits of waste heat in curtailing the carbon footprint of the DAC processes as well as the promise of future electric regeneration heat sources. Moreover, the analysis showcases the pronounced efficacy of renewables, with photovoltaics emerging as particularly advantageous.

2023 AIChE ANNUAL MEETING

November 5 — November 10, 2023
Hyatt Regency Orlando
Orlando, FL

Solar Fuel Production via Germanium Oxide-Based Thermochemical Conversion of CO_2

(Principal Author)
Rahul Bhosale
Department of Chemical Engineering,
University of Tennessee, Chattanooga, TN, 37421, USA

(Co-Authors)
Shelby Adams, Zachary Allen, Gabrielle Bennett, Edvinas Berezniovas, Taylor Bishop, Michael Bonnema, Sequoia Clutter, Ryan Fagan, Jordan Halabrin, Mason Hobbs, Daniel Hunt, Miguel Ivarra, Mattigan Jordan, Pooja Karunanithi, Julianna Mcreynolds, Valerie Ring, Sam Smith, Jonathan West
Department of Chemical Engineering,
University of Tennessee, Chattanooga, TN, 37421, USA

Prepared for Presentation at
American Institute of Chemical Engineers
2023 AIChE Annual Meeting
Orlando, FL, USA
November 5 – 10, 2023

Solar Fuel Production via Germanium Oxide-Based Thermochemical Conversion of CO_2

Rahul R. Bhosale*, Shelby Adams, Zachary Allen, Gabrielle Bennett, Edvinas Berezniovas, Taylor Bishop, Michael Bonnema, Sequoia Clutter, Ryan Fagan, Jordan Halabrin, Mason Hobbs, Daniel Hunt, Miguel Ivarra, Mattigan Jordan, Pooja Karunanithi, Julianna Mcreynolds, Valerie Ring, Sam Smith, Jonathan West

Department of Chemical Engineering, University of Tennessee, Chattanooga, TN, 37421, USA

Keywords: Solar Fuel, Germanium Oxide, CO2 Conversion

Abstract

A thermodynamic evaluation of the feasibility of the GeO_2/GeO redox CO_2 splitting cycle is carried out in this investigation. The assumptions and the process flow configuration are developed in the first part of the paper. In the second step, the thermodynamic modeling equations are drafted to estimate the influence of various process variables. The details associated with the process and thermodynamic model are presented in detail.

1. Introduction

Many investigators worldwide work on air and water pollution treatment technologies [1–14]. The most crucial air pollution issue is the release of CO_2 (a greenhouse gas) into the environment causing global warming. Likewise, the dumping of wastes and chemicals into the water is responsible for water pollution. Both air and water pollution are somehow the products of the incessant application of fossil fuels and tremendous industrialization. One of the options to reduce air and water pollution is to use alternative fuels produced from renewable sources instead of fossil fuels [15–31]. For example, alternative fuels such as methanol and others can be produced from carbon dioxide [32–36].

A metal oxide-based solar thermochemical water and carbon dioxide splitting process can produce alternative fuels such as H_2 and the precursor, syngas, required to produce liquid transportation fuels such as gasoline, diesel, and kerosene. Various pure and mixed metal oxides such as ZnO, SnO_2, Fe_3O_4, ferrites, doped ceria, perovskites, and others have been extensively tested for water and carbon dioxide splitting reactions [37–77]. In addition to the above-mentioned metal oxides, exciting findings have been reported in the case of GeO_2/GeO redox cycles in recent years. However, all the published studies were focused on the water-splitting application. To produce syngas, in addition to the metal oxide's water-splitting ability, it is essential to find out its aptitude for carbon dioxide conversion as well. Therefore, in this investigation, all 18 undergraduate students were divided into six teams to explore the thermodynamic feasibility of the GeO_2/GeO-based carbo dioxide splitting cycle. In this paper, the process and the thermodynamic model developed by the students are explained.

2. Process Model

The two-step GeO_2/GeO-driven solar CO production via thermochemical CO_2 splitting is carried out per the following reactions.

$$GeO_2 \rightarrow GeO + \frac{1}{2}O_2 \quad (1)$$

$$GeO + CO_2 \rightarrow GeO_2 + CO \quad (2)$$

A process model (shown in Fig. 1) is developed to investigate the thermodynamic feasibility of the cycle. The following assumptions were considered for the model development:

(a) ideal gas behavior and steady-state operation
(b) isothermal operation of reduction and oxidation cells
(c) attainment of equilibrium between the oxides and gaseous species
(d) exemplary CO/O_2 fuel cell with 100% operating efficiency
(e) 99.9% separation of the gaseous species in separator-1 and separator-2
(f) 100% conversion of reduction as well as oxidation reactions
(g) 20% thermal energy loss from the reduction cell
(h) negligible kinetic/potential energy and viscous losses

Two separate cells are installed for the thermal reduction (reduction cell) and CO_2 splitting (oxidation cell) steps. Thermal reduction is carried out at the reduction temperature (T_{red}), whereas the CO_2 splitting reaction is performed at the oxidation temperature (T_{oxd}). The thermal reduction of GeO_2 is carried out in the presence of an inert gas environment at T_{red}. The overall analysis is carried out by assuming a continuous flow of GeO_2 (1 mol/s) to the solar reactor.

3. Thermodynamic Model

The following equations can be used to estimate various parameters associated with the GeO_2/GeO-driven solar CO production process.

The energy required to attain complete dissociation of GeO_2:

$$\dot{Q}_{GeO_2-red} = \dot{n}_{GeO_2} \Delta H|_{GeO_2 \rightarrow GeO + \frac{1}{2}O_2(g)} \quad (3)$$

The exothermic energy released from the oxidation cell:

$$\dot{Q}_{GeO-oxd} = -\dot{n}_{GeO_2} \Delta H|_{GeO + 10CO_2(g) \rightarrow GeO_2 + 9CO_2(g) + CO(g)} \quad (4)$$

The thermal energy required to heat GeO_2:

$$\dot{Q}_{heater-1} = \dot{n}_{GeO_2} \Delta H|_{GeO_2 @ T_{oxd} \rightarrow GeO_2 @ T_{red}} \quad (5)$$

The thermal energy required to use the separator-2:

$$\dot{Q}_{sep-2} = \dot{n}_{GeO_2} \left[\frac{T_{sep-2}}{\eta_{sep-2}} (\Delta S_{mix,O_4} - \Delta S_{mix,O_3}) \right] \quad (6)$$

$$\Delta S_{mix,O_3} = -R\{n_{CO_2} \ln(1 - y_{CO,O_3}) + n_{CO,O_3} \ln y_{CO,O_3}\} \quad (7)$$

$$\Delta S_{mix,O_4} = -R\{n_{CO_2} \ln(1 - y_{CO,O_4}) + n_{CO,O_4} \ln y_{CO,O_4}\} \quad (8)$$

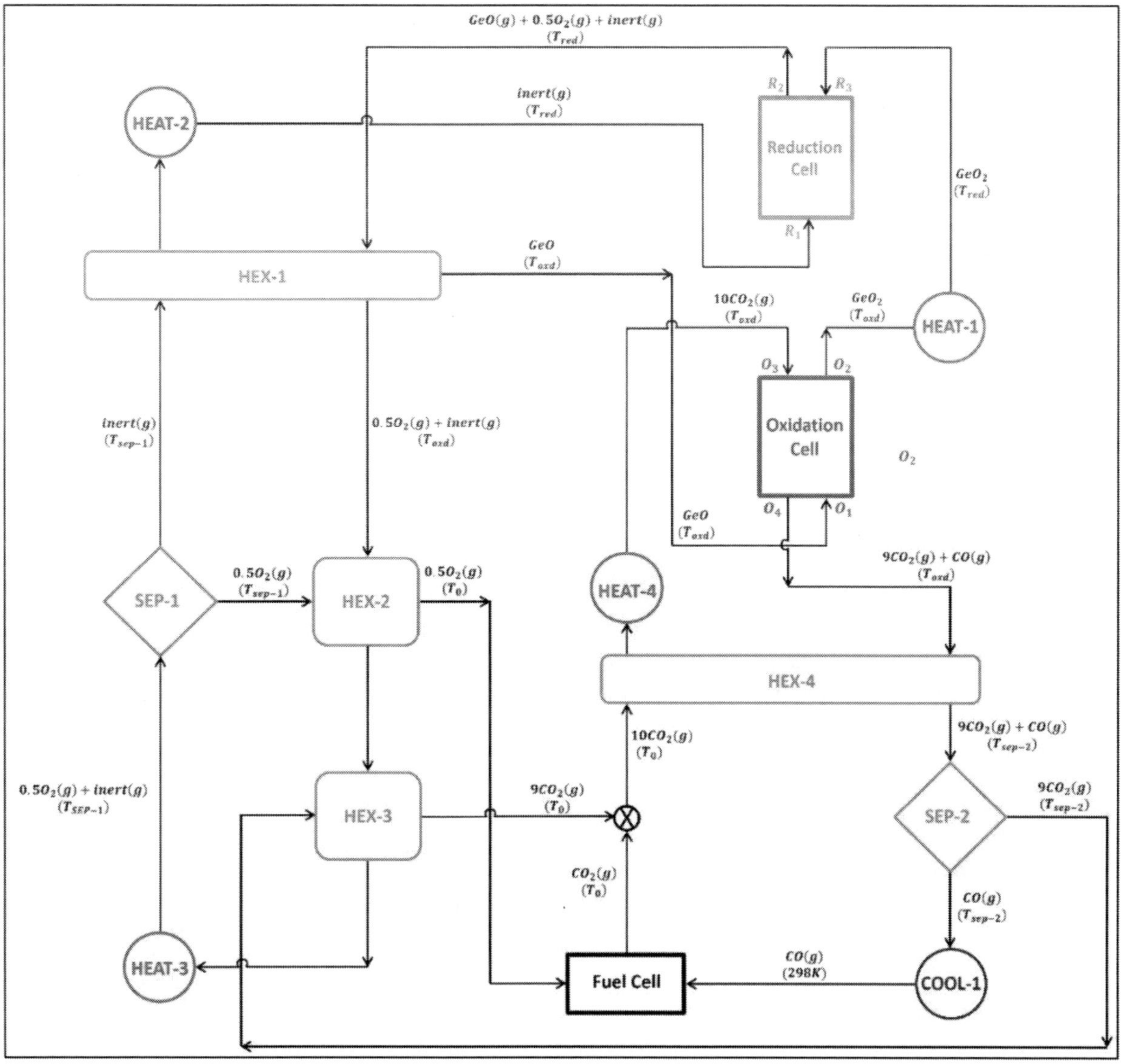

Fig. 1. A process model of the solar CO generation via GeO_2/GeO-based CO_2 Splitting cycle.

The thermal energy demand of the CO_2 heating process:

$$\dot{Q}_{CO_2-heat} = \varepsilon_{gg}\left[\dot{Q}_{(CO_2+\,CO)-cool}\right] + \dot{Q}_{heater-4} \tag{9}$$

Here,

$$\dot{Q}_{CO_2-heat} = \dot{n}_{CO_2}\Delta H|_{CO_2(g)@T_0\to CO_2(g)@T_{oxd}} \tag{10}$$

$$\dot{Q}_{(CO_2+\,CO)-cool} = \dot{n}_{CO_2}\Delta H|_{CO_2(g)@T_{oxd}\to CO_2(g)@T_{sep-2}} + \dot{n}_{CO}\Delta H|_{CO(g)@T_{oxd}\to CO(g)@T_{sep-2}} \tag{11}$$

The thermal energy required for the heating of inert/O_2 gas mixture:

$$\dot{Q}_{(inert+\,O_2)-heat} = \varepsilon_{gg}\left[\dot{Q}_{O_2-cool} + \dot{Q}_{CO_2-cool}\right] + \dot{Q}_{heater-3} \tag{12}$$

Here,

$$\dot{Q}_{(inert+\,O_2)-heat} = \dot{n}_{inert}\Delta H|_{inert(g)@T_{oxd}\rightarrow inert(g)@T_{sep-1}} + \dot{n}_{O_2}\Delta H|_{O_2(g)@T_{oxd}\rightarrow O_2(g)@T_{sep-1}} \quad (13)$$

$$\dot{Q}_{O_2-cool} = \dot{n}_{O_2}\Delta H|_{O_2(g)@T_{sep-1}\rightarrow O_2(g)@T_0} \quad (14)$$

$$\dot{Q}_{CO_2-cool} = \dot{n}_{CO_2}\Delta H|_{CO_2(g)@T_{sep-2}\rightarrow CO_2(g)@T_0} \quad (15)$$

The thermal energy required for the separation of inert sweep gas and O_2:

$$\dot{Q}_{sep-1} = \dot{n}_{GeO_2}\left[\frac{T_{sep-1}}{\eta_{sep-1}}(\Delta S_{mix,R_2} - \Delta S_{mix,R_1})\right] \quad (16)$$

$$\Delta S_{mix,R_1} = -R\{n_{inert}\ln(1 - y_{O_2,R_1}) + n_{O_2,R_1} ln y_{O_2,R_1}\} \quad (17)$$

$$\Delta S_{mix,R_2} = -R\{n_{inert}\ln(1 - y_{O_2,R_2}) + n_{O_2,R_2} ln y_{O_2,R_2}\} \quad (18)$$

The auxiliary heater-1 will provide the additional thermal energy needed (after passing through HEX-1).

$$\dot{Q}_{inert-heat} = \varepsilon_{gg}\left[\dot{Q}_{(inert+\,O_2+GeO)-cool}\right] + \dot{Q}_{heater-2} \quad (19)$$

Here,

$$\dot{Q}_{inert-heat} = \dot{n}_{inert}\Delta H|_{inert(g)@T_{sep-1}\rightarrow inert(g)@T_{red}} \quad (20)$$

$$\dot{Q}_{(inert+\,O_2+GeO)-cool} = \dot{n}_{inert}\Delta H|_{inert(g)@T_{red}\rightarrow inert(g)@T_{oxd}} + \dot{n}_{O_2}\Delta H|_{O_2(g)@T_{red}\rightarrow O_2(g)@T_{oxd}} + \dot{n}_{Zn}\Delta H|_{GeO@T_{red}\rightarrow GeOT_{oxd}} \quad (21)$$

$\eta_{solar-to-fuel}$ of the cycle:

$$\eta_{solar-to-fuel} = \frac{\dot{n}_{CO}\times HHV_{CO}}{\dot{Q}_{solar}} \quad (22)$$

The amount of solar energy required ($\dot{Q}_{solar}$):

$$\dot{Q}_{solar} = \frac{\dot{Q}_{TC}}{\eta_{absoption}} \quad (23)$$

Here,

$$\eta_{absorption} = 1 - \left(\frac{\sigma T_{red}^4}{IC}\right) \quad (24)$$

The summative thermal energy required for driving the cycle:

$$\dot{Q}_{TC} = \dot{Q}_{GeO_2-red} + \dot{Q}_{heater-1} + \dot{Q}_{heater-2} + \dot{Q}_{heater-3} + \dot{Q}_{heater-4} + \dot{Q}_{sep-1} + \dot{Q}_{sep-2} + \dot{Q}_{surf} \quad (25)$$

Here,

$$\dot{Q}_{surf} = 0.2 \times \dot{Q}_{GeO_2-red} \quad (26)$$

Thermal losses due to the re-radiation:

$$\dot{Q}_{re-rad} = \eta_{absorption} \times \dot{Q}_{solar} \quad (27)$$

4. Conclusions

A process model for the GeO_2/GeO driven solar CO production via thermochemical CO_2 splitting cycle is developed. This process model comprises various units required for the operation of the cycle, including the reaction cells, heat exchangers, heaters and coolers, gas separation units, and an ideal fuel cell. Model equations were derived using the developed process model and assumptions, which can be used to calculate the process parameters.

5. Acknowledgments

The authors would like to acknowledge the support provided by Qatar University International Research Collaboration Co-funds Grant No. IRCC-2021-002 and the Ruth S. Holmberg Grant for Faculty Excellence, University of Tennessee, Chattanooga, TN, USA.

References

[1] Bhosale RR, Kumar A, AlMomani F, Ghosh U, AlNouss A, Scheffe J, et al. CO 2 Capture Using Aqueous Potassium Carbonate Promoted by Ethylaminoethanol: A Kinetic Study. Ind Eng Chem Res 2016;55:5238–46. https://doi.org/10.1021/acs.iecr.5b04398.

[2] Vaidya PD, Kenig EY. CO2-Alkanolamine Reaction Kinetics: A Review of Recent Studies. Chem Eng Technol 2007;30:1467–74. https://doi.org/10.1002/ceat.200700268.

[3] Gulied M, Al Momani F, Khraisheh M, Bhosale R, AlNouss A. Influence of draw solution type and properties on the performance of forward osmosis process: Energy consumption and sustainable water reuse. Chemosphere 2019;233:234–44. https://doi.org/10.1016/j.chemosphere.2019.05.241.

[4] Zhang QH, Yang WN, Ngo HH, Guo WS, Jin PK, Dzakpasu M, et al. Current status of urban wastewater treatment plants in China. Environ Int 2016;92–93:11–22. https://doi.org/10.1016/j.envint.2016.03.024.

[5] Mohamad S, Fares A, Judd S, Bhosale R, Kumar A, Gosh U, et al. Advanced wastewater treatment using microalgae: effect of temperature on removal of nutrients and organic carbon. IOP Conf Ser Earth Environ Sci 2017;67:012032. https://doi.org/10.1088/1755-1315/67/1/012032.

[6] Ahmed SN, Haider W. Heterogeneous photocatalysis and its potential applications in water and wastewater treatment: a review. Nanotechnology 2018;29:342001. https://doi.org/10.1088/1361-6528/aac6ea.

[7] Bhosale RR, Mahajani V V. Kinetics of thermal degradation of renewably prepared amines useful for flue gas treatment. Journal of Renewable and Sustainable Energy 2013;5. https://doi.org/10.1063/1.4831960.

[8] Moreira D, Pires JCM. Atmospheric CO2 capture by algae: Negative carbon dioxide emission path. Bioresour Technol 2016;215:371–9. https://doi.org/10.1016/j.biortech.2016.03.060.

[9] Bhosale RR, Mahajani V V. Kinetics of Absorption of Carbon Dioxide in Aqueous Solution of Ethylaminoethanol Modified with N-methyl-2-pyrolidone. Sep Sci Technol 2013;48:2324–37. https://doi.org/10.1080/01496395.2013.805228.

[10] Stolaroff JK, Ye C, Nguyen DT, Oakdale J, Knipe JM, Baker SE. CO2 Absorption Kinetics of Micro-encapsulated Ionic Liquids. Energy Procedia 2017;114:860–5. https://doi.org/10.1016/j.egypro.2017.03.1228.

[11] Almomani F, Bhosale R, Kumar A, Khraisheh M. Potential use of solar photocatalytic oxidation in removing emerging pharmaceuticals from wastewater: A pilot plant study. Solar Energy 2018;172:128–40. https://doi.org/10.1016/j.solener.2018.07.041.

[12] Toth AJ. Modelling and Optimisation of Multi-Stage Flash Distillation and Reverse Osmosis for Desalination of Saline Process Wastewater Sources. Membranes (Basel) 2020;10:265. https://doi.org/10.3390/membranes10100265.

[13] Almomani F, Bhosale RR. Bio-sorption of toxic metals from industrial wastewater by algae strains Spirulina platensis and Chlorella vulgaris: Application of isotherm, kinetic models and process optimization. Science of The Total Environment 2021;755:142654. https://doi.org/10.1016/j.scitotenv.2020.142654.

[14] Crini G, Lichtfouse E. Advantages and disadvantages of techniques used for wastewater treatment. Environ Chem Lett 2019;17:145–55. https://doi.org/10.1007/s10311-018-0785-9.

[15] Almomani F, Bohsale RR. Optimizing nutrient removal of moving bed biofilm reactor process using response surface methodology. Bioresour Technol 2020;305:123059. https://doi.org/10.1016/j.biortech.2020.123059.

[16] Tungal R, Shende R V. Hydrothermal liquefaction of pinewood (Pinus ponderosa) for H2, biocrude and bio-oil generation. Appl Energy 2014;134:401–12. https://doi.org/10.1016/j.apenergy.2014.07.060.

[17] Pugazhendhi A, Nagappan S, Bhosale RR, Tsai P-C, Natarajan S, Devendran S, et al. Various potential techniques to reduce the water footprint of microalgal biomass production for biofuel—A review. Science of The Total Environment 2020;749:142218. https://doi.org/10.1016/j.scitotenv.2020.142218.

[18] Rodionova MV, Poudyal RS, Tiwari I, Voloshin RA, Zharmukhamedov SK, Nam HG, et al. Biofuel production: Challenges and opportunities. Int J Hydrogen Energy 2017;42:8450–61. https://doi.org/10.1016/j.ijhydene.2016.11.125.

[19] Shurair M, Almomani F, Bhosale R, Khraisheh M, Qiblawey H. Harvesting of intact microalgae in single and sequential conditioning steps by chemical and biological based – flocculants: Effect on harvesting efficiency, water recovery and algal cell morphology. Bioresour Technol 2019;281:250–9. https://doi.org/10.1016/j.biortech.2019.02.103.

[20] Voloshin RA, Rodionova M V., Zharmukhamedov SK, Nejat Veziroglu T, Allakhverdiev SI. Review: Biofuel production from plant and algal biomass. Int J Hydrogen Energy 2016;41:17257–73. https://doi.org/10.1016/j.ijhydene.2016.07.084.

[21] Rajesh Banu J, Yukesh Kannah R, Kavitha S, Ashikvivek A, Bhosale RR, Kumar G. Cost effective biomethanation via surfactant coupled ultrasonic liquefaction of mixed microalgal biomass harvested from open raceway pond. Bioresour Technol 2020;304:123021. https://doi.org/10.1016/j.biortech.2020.123021.

[22] Matin MA, Kumar A, Bhosale RR, Saleh Saad MAH, Almomani FA, Al-Marri MJ. PdZn nanoparticle electrocatalysts synthesized by solution combustion for

methanol oxidation reaction in an alkaline medium. RSC Adv 2017;7:42709–17. https://doi.org/10.1039/C7RA07013F.
[23] Almomani F, Shawaqfah M, Bhosale RR, Kumar A, Khraisheh MAM. Intermediate ozonation to enhance biogas production in batch and continuous systems using animal dung and agricultural waste. Int Biodeterior Biodegradation 2017;119:176–87. https://doi.org/10.1016/j.ibiod.2016.11.008.
[24] Abanades S, Abbaspour H, Ahmadi A, Das B, Ehyaei MA, Esmaeilion F, et al. A critical review of biogas production and usage with legislations framework across the globe. International Journal of Environmental Science and Technology 2022;19:3377–400. https://doi.org/10.1007/s13762-021-03301-6.
[25] Almomani F, Bhosale RR, Khraisheh MAM, Shawaqfah M. Enhancement of biogas production from agricultural wastes via pre-treatment with advanced oxidation processes. Fuel 2019;253:964–74. https://doi.org/10.1016/j.fuel.2019.05.057.
[26] Kabeyi MJB, Olanrewaju OA. Biogas Production and Applications in the Sustainable Energy Transition. Journal of Energy 2022;2022:1–43. https://doi.org/10.1155/2022/8750221.
[27] Rodriguez JA, Liu P, Stacchiola DJ, Senanayake SD, White MG, Chen JG. Hydrogenation of CO_2 to Methanol: Importance of Metal–Oxide and Metal–Carbide Interfaces in the Activation of CO_2. ACS Catal 2015;5:6696–706. https://doi.org/10.1021/acscatal.5b01755.
[28] Ashok A, Kumar A, Bhosale R, Saleh Saad MA, AlMomani F, Tarlochan F. Study of ethanol dehydrogenation reaction mechanism for hydrogen production on combustion synthesized cobalt catalyst. Int J Hydrogen Energy 2017;42:23464–73. https://doi.org/10.1016/j.ijhydene.2017.01.175.
[29] GO K, SON S, KIM S. Reaction kinetics of reduction and oxidation of metal oxides for hydrogen production. Int J Hydrogen Energy 2008;33:5986–95. https://doi.org/10.1016/j.ijhydene.2008.05.039.
[30] Kumar A, Ashok A, Bhosale RR, Saleh MAH, Almomani FA, Al-Marri M, et al. In situ DRIFTS Studies on Cu, Ni and CuNi catalysts for Ethanol Decomposition Reaction. Catal Letters 2016;146:778–87. https://doi.org/10.1007/s10562-016-1706-9.
[31] Li J, Kazakov A, Dryer FL. Experimental and Numerical Studies of Ethanol Decomposition Reactions. J Phys Chem A 2004;108:7671–80. https://doi.org/10.1021/jp0480302.
[32] Almomani F, Bhosale R, Khraisheh M, Kumar A, Tawalbeh M. Photocatalytic conversion of CO2 and H2O to useful fuels by nanostructured composite catalysis. Appl Surf Sci 2019;483:363–72. https://doi.org/10.1016/j.apsusc.2019.03.304.
[33] Dhabarde N, Selvaraj J, Yuda A, Kumar A, Subramanian VR. Review of photocatalytic and photo-electrocatalytic reduction of CO2 on carbon supported films. Int J Hydrogen Energy 2022;47:30908–36. https://doi.org/10.1016/j.ijhydene.2022.02.124.
[34] Ashok A, Kumar A, Bhosale RR, Almomani F, Saleh Saad MAH, Suslov S, et al. Influence of fuel ratio on the performance of combustion synthesized bifunctional

cobalt oxide catalysts for fuel cell application. Int J Hydrogen Energy 2019;44:436–45. https://doi.org/10.1016/j.ijhydene.2018.02.111.
[35] Hao Y, Jin J, Jin H. Thermodynamic analysis of isothermal CO2 splitting and CO2-H2O co-splitting for solar fuel production. Appl Therm Eng 2020;166:113600. https://doi.org/10.1016/j.applthermaleng.2019.04.010.
[36] Ashok A, Kumar A, Bhosale RR, Saleh MAH, Ghosh UK, Al-Marri M, et al. Cobalt oxide nanopowder synthesis using cellulose assisted combustion technique. Ceram Int 2016;42:12771–7. https://doi.org/10.1016/j.ceramint.2016.05.035.
[37] Bhosale RR, Kumar A, AlMomani F, Ghosh U, Dardor D, Bouabidi Z, et al. Solar co-production of samarium and syngas via methanothermal reduction of samarium sesquioxide. Energy Convers Manag 2016;112:413–22. https://doi.org/10.1016/j.enconman.2016.01.032.
[38] Steinfeld A. Solar hydrogen production via a two-step water-splitting thermochemical cycle based on Zn/ZnO redox reactions. Int J Hydrogen Energy 2002;27:611–9. https://doi.org/10.1016/S0360-3199(01)00177-X.
[39] Bhosale RR, Alxneit I, van den Broeke LLP, Kumar A, Jilani M, Gharbia SS, et al. Sol-Gel Synthesis of Nanocrystalline Ni-Ferrite and Co-Ferrite Redox Materials for Thermochemical Production of Solar Fuels. MRS Proceedings 2014;1675:203–8. https://doi.org/10.1557/opl.2014.866.
[40] Diver RB, Miller JE, Allendorf MD, Siegel NP, Hogan RE. Solar Thermochemical Water-Splitting Ferrite-Cycle Heat Engines. J Sol Energy Eng 2008;130. https://doi.org/10.1115/1.2969781.
[41] Bhosale RR, Kumar A, Ashok A, Sutar P, Takalkar G, Khraisheh M, et al. La-Based Perovskites as Oxygen-Exchange Redox Materials for Solar Syngas Production. MRS Adv 2017;2:3365–70. https://doi.org/10.1557/adv.2017.300.
[42] Bare ZJL, Morelock RJ, Musgrave CB. A Computational Framework to Accelerate the Discovery of Perovskites for Solar Thermochemical Hydrogen Production: Identification of Gd Perovskite Oxide Redox Mediators. Adv Funct Mater 2022;32:2200201. https://doi.org/10.1002/adfm.202200201.
[43] Takalkar G, Bhosale RR. Solar thermocatalytic conversion of CO2 using PrxSr(1−x)MnO3−δ perovskites. Fuel 2019;254:115624. https://doi.org/10.1016/j.fuel.2019.115624.
[44] Li S, Wheeler VM, Kumar A, Venkataraman MB, Muhich CL, Hao Y, et al. Thermodynamic Guiding Principles for Designing Nonstoichiometric Redox Materials for Solar Thermochemical Fuel Production: Ceria, Perovskites, and Beyond. Energy Technology 2022;10:2000925. https://doi.org/10.1002/ente.202000925.
[45] Bhosale RR. Thermodynamic analysis of Ni-ferrite based solar thermochemical H2O splitting cycle for H2 production. Int J Hydrogen Energy 2019;44:61–71. https://doi.org/10.1016/j.ijhydene.2018.03.145.
[46] Scheffe JR, Li J, Weimer AW. A spinel ferrite/hercynite water-splitting redox cycle. Int J Hydrogen Energy 2010;35:3333–40. https://doi.org/10.1016/j.ijhydene.2010.01.140.
[47] Takalkar G, Bhosale R, AlMomani F. Combustion synthesized A0.5Sr0.5MnO3-δ perovskites (where, A = La, Nd, Sm, Gd, Tb, Pr, Dy, and Y) as redox materials for

thermochemical splitting of CO2. Appl Surf Sci 2019;489:80–91. https://doi.org/10.1016/j.apsusc.2019.05.284.
[48] Jacot R, Moré R, Michalsky R, Steinfeld A, Patzke GR. Trends in the phase stability and thermochemical oxygen exchange of ceria doped with potentially tetravalent metals. J Mater Chem A 2017;5:19901–13. https://doi.org/10.1039/C7TA04063F.
[49] Luciani G, Landi G, Di Benedetto A. Syngas Production Through H2O/CO2 Thermochemical Splitting Over Doped Ceria-Zirconia Materials. Front Energy Res 2020;8. https://doi.org/10.3389/fenrg.2020.00204.
[50] Bhosale RR, Shende RV, Puszynski JA. Sol-Gel Synthesis of Ferrite Foam Materials for H2 Generation from Water-Splitting Reaction. Proceedings of NSTI–Nanotech Conference & Expo., 2010, p. 368–71.
[51] Tamaura Y, Hasegawa N, Kojima M, Ueda Y, Amano H, Tsuji M. Water splitting with the Mn(II)-ferrite–CaO–H2O system at 1273K. Energy 1998;23:879–86. https://doi.org/10.1016/S0360-5442(98)00030-9.
[52] Dardor D, Bhosale RR, Gharbia S, Kumar A, AlMomani FA. Solar carbon production via thermochemical ZnO/Zn carbon dioxide splitting cycle. Journal of Emerging Trends in Engineering and Applied Sciences 2016;7:129–35.
[53] Chambon M, Abanades S, Flamant G. Solar thermal reduction of ZnO and SnO2: Characterization of the recombination reaction with O2. Chem Eng Sci 2010;65:3671–80. https://doi.org/10.1016/j.ces.2010.03.005.
[54] Bhosale RR, Shende R V, Puszynski JA. H 2 Generation From Thermochemical Water-Splitting Using Sol-Gel Synthesized Zn/Sn/Mn-doped Ni-Ferrite. International Review of Chemical Engineering (IRECHE) 2010;2.
[55] Kaneko H, Gokon N, Hasegawa N, Tamaura Y. Solar thermochemical process for hydrogen production using ferrites. Energy 2005;30:2171–8. https://doi.org/10.1016/j.energy.2004.08.020.
[56] Shende RV, Puszynski JA, Opoku MK, Bhosale RR. Synthesis of novel ferrite foam material for water-splitting application. Proceedings of NSTI Nanotech conference & expo., 2009, p. 201–4.
[57] Scheffe JR, McDaniel AH, Allendorf MD, Weimer AW. Kinetics and mechanism of solar-thermochemical H2 production by oxidation of a cobalt ferrite–zirconia composite. Energy Environ Sci 2013;6:963. https://doi.org/10.1039/c3ee23568h.
[58] Bhosale RR, Shende R V, Puszynski JA. H 2 Generation from Thermochemical Water-Splitting Using Sol-Gel Derived Ni-Ferrite. Journal of Energy and Power Engineering 2010;4.
[59] Mostrou S, Büchel R, Pratsinis SE, van Bokhoven JA. Improving the ceria-mediated water and carbon dioxide splitting through the addition of chromium. Appl Catal A Gen 2017;537:40–9. https://doi.org/10.1016/j.apcata.2017.03.001.
[60] Bhosale R, Khadka R, Puszynski J, Shende R. H2 generation from two-step thermochemical water-splitting reaction using sol-gel derived SnxFeyOz. Journal of Renewable and Sustainable Energy 2011;3. https://doi.org/10.1063/1.3659684.
[61] Agrafiotis C, Zygogianni A, Pagkoura C, Kostoglou M, Konstandopoulos AG. Hydrogen production via solar-aided water splitting thermochemical cycles with nickel ferrite: Experiments and modeling. AIChE Journal 2013;59:1213–25. https://doi.org/10.1002/aic.13882.

[62] Bhosale RR, Kumar A, AlMomani F, Ghosh U, Sutar P, Takalkar G, et al. Effectiveness of Ni incorporation in iron oxide crystal structure towards thermochemical CO2 splitting reaction. Ceram Int 2017;43:5150–5. https://doi.org/10.1016/j.ceramint.2017.01.034.

[63] Haeussler A, Julbe A, Abanades S. Investigation of reactive perovskite materials for solar fuel production via two-step redox cycles: Thermochemical activity, thermodynamic properties and reduction kinetics. Mater Chem Phys 2022;276:125358. https://doi.org/10.1016/j.matchemphys.2021.125358.

[64] Bhosale RR, Sutar P, Kumar A, AlMomani F, Ali MH, Ghosh U, et al. Solar hydrogen production via erbium oxide based thermochemical water splitting cycle. Journal of Renewable and Sustainable Energy 2016;8:034702. https://doi.org/10.1063/1.4953166.

[65] Bhosale RR. Thermodynamic efficiency analysis of zinc oxide based solar driven thermochemical H2O splitting cycle: Effect of partial pressure of O2, thermal reduction and H2O splitting temperatures. Int J Hydrogen Energy 2018;43:14915–24. https://doi.org/10.1016/j.ijhydene.2018.06.074.

[66] Jacot R, Naik JM, Moré R, Michalsky R, Steinfeld A, Patzke GR. Reactive stability of promising scalable doped ceria materials for thermochemical two-step CO_2 dissociation. J Mater Chem A Mater 2018;6:5807–16. https://doi.org/10.1039/C7TA10966K.

[67] Bhosale RR, Takalkar GD. Nanostructured co-precipitated Ce0.9Ln0.1O2 (Ln = La, Pr, Sm, Nd, Gd, Tb, Dy, or Er) for thermochemical conversion of CO2. Ceram Int 2018;44:16688–97. https://doi.org/10.1016/j.ceramint.2018.06.096.

[68] Bhosale RR, Kumar A, AlMomani F, Alxneit I. Propylene oxide assisted sol–gel synthesis of zinc ferrite nanoparticles for solar fuel production. Ceram Int 2016;42:2431–8. https://doi.org/10.1016/j.ceramint.2015.10.043.

[69] Dey S, Rao CNR. Splitting of CO_2 by Manganite Perovskites to Generate CO by Solar Isothermal Redox Cycling. ACS Energy Lett 2016;1:237–43. https://doi.org/10.1021/acsenergylett.6b00122.

[70] Bhosale RR, Kumar A, AlMomani F, Ghosh U, Khraisheh M. A comparative thermodynamic analysis of samarium and erbium oxide based solar thermochemical water splitting cycles. Int J Hydrogen Energy 2017;42:23416–26. https://doi.org/10.1016/j.ijhydene.2017.03.172.

[71] Padella F, Alvani C, Barbera A La, Ennas G, Liberatore R, Varsano F. Mechanosynthesis and process characterization of nanostructured manganese ferrite. Mater Chem Phys 2005;90:172–7. https://doi.org/10.1016/j.matchemphys.2004.10.033.

[72] Bhosale R, Kumar A, AlMomani F. Solar Thermochemical Hydrogen Production via Terbium Oxide Based Redox Reactions. International Journal of Photoenergy 2016;2016:1–9. https://doi.org/10.1155/2016/9727895.

[73] Chueh WC, Falter C, Abbott M, Scipio D, Furler P, Haile SM, et al. High-Flux Solar-Driven Thermochemical Dissociation of CO_2 and H_2O Using Nonstoichiometric Ceria. Science (1979) 2010;330:1797–801. https://doi.org/10.1126/science.1197834.

[74] Takalkar GD, Bhosale RR, Kumar A, AlMomani F, Khraisheh M, Shakoor RA, et al. Transition metal doped ceria for solar thermochemical fuel production. Solar Energy 2018;172:204–11. https://doi.org/10.1016/j.solener.2018.03.022.

[75] Roeb M, Säck J-P, Rietbrock P, Prahl C, Schreiber H, Neises M, et al. Test operation of a 100kW pilot plant for solar hydrogen production from water on a solar tower. Solar Energy 2011;85:634–44. https://doi.org/10.1016/j.solener.2010.04.014.

[76] Bhosale RR, Shende R V., Puszynski JA. Sol-Gel Derived $NiFe_2O_4$ Modified with ZrO_2 for Hydrogen Generation from Solar Thermochemical Water-Splitting Reaction. MRS Proceedings 2012;1387:mrsf11-1387-e09-07. https://doi.org/10.1557/opl.2012.772.

[77] Fresno F, Fernández-Saavedra R, Belén Gómez-Mancebo M, Vidal A, Sánchez M, Isabel Rucandio M, et al. Solar hydrogen production by two-step thermochemical cycles: Evaluation of the activity of commercial ferrites. Int J Hydrogen Energy 2009;34:2918–24. https://doi.org/10.1016/j.ijhydene.2009.02.020.

548g Unifying CO_2 Capture and Utilization through Process Modeling: Sensitivity Analysis of Process Performance on Bifunctional Materials in a Single Bed Reactor

Abdullah Al Moinee, Ali Rownaghi, Fateme Rezaei

Fischer-Tropsch to olefin process (FTO) coupled with the reverse water gas shift reaction (RWGS) offers a compelling pathway to directly convert captured CO_2 into valuable light olefins while mitigating greenhouse gas emissions. In our study, we explored an integrated approach using a single plant with an integrated adsorber and reactor, compared to a separated system with distinct capture and utilization units. CO_2-containing flue gas from a 500 MW power plant serves as the carbon source for the process. The CO_2 undergoes adsorptive capture and hydrogenation reactions to produce light olefins via FTO. A comprehensive sensitivity analysis is conducted to evaluate the process performance indicators (PPIs) for both the unified adsorber-reactor system and the separated system. Various operational parameters, material properties, and downstream treatment units are considered in this analysis, including pressure, temperature, H_2/CO_2 molar ratio, catalyst and adsorbent activity, deactivation rate, and heat integration. The results reveal the superior performance of the integrated adsorber-reactor process in terms of PPIs, highlighting its enhanced cost-effectiveness and efficiency compared to the separated system, which relies on separate PSA units for capture and hydrogenation. By focusing on the sensitivity analysis, this study provides valuable insights into optimizing the CO_2 conversion process, emphasizing the significance of the integration in achieving improved overall performance.

548a Techno-Economic Analysis and Life Cycle Assessment of Hydroxylamine Eco-Manufacturing Via Wastewater Electrochemical Reduction

Manish Mosalpuri, Mark Mba Wright, Wenzhen Li

The manufacturing of nitrogen-based chemicals is dependent mainly on the Haber-Bosch process, which produces cheap ammonia (NH_3) from hydrogen (H_2) and atmospheric nitrogen (N_2) gas. However, this process has high energy requirements due to the high process temperature (500^0 C) and pressure (100-150 bar). It is also responsible for significant CO_2 emissions (400 metric tons annually) and is energy-intensive- it consumes around 2% of the world's primary energy.

Hydroxylamine (NH_2OH) is an important chemical intermediate currently produced from NH_3 formed by the Haber-Bosch process, followed by the Ostwald process (which generates nitric oxide, NO from NH_3) and NO reduction by H_2. We propose an alternative reaction mechanism to the conventional method described above, which involves the electroreduction of nitrate ions (NO_3^-) to NH_2OH, as depicted in figure 1. Wastewater from fertilizer run-offs, nuclear power plants, and slaughterhouses is a rich source of NO_3^-, which harms the ecosystem.

Many researchers are now focusing on manufacturing chemicals via electrochemical processes. Electrochemical processes are typically executed at ambient temperature and pressure, thus requiring mild operating conditions and low process energy.

We conducted a techno-economic analysis (TEA) and a life cycle assessment (LCA) of this electrochemical system converting NO_3^- to NH_2OH. We conducted TEA using a general model by Orella et al., 2020.[1] We performed a sensitivity analysis to calculate the maximum possible reduction in NH_2OH production cost and discover its influencing parameters. The most dominant parameter influencing the cost was the separation factor, followed by lifetime, product Faradaic Efficiency, electricity price, and catalyst loading. We obtained the optimum projected cost of NH_2OH as $1.65/kg, lower than its market price estimated at $1.72/kg. We obtained the base cost, or the current production cost, as $6.03/kg. Our analysis suggests an immense potential for cost reduction, which motivates further research.

We also conducted an LCA of the process system depicted in figure 1. The goal of this LCA study is to calculate the environmental impacts of the proposed eco-manufacturing pathway to produce NH_2OH in a wastewater treatment plant in Iowa. We determined if the proposed alternate method of producing NH_2OH is a sustainable option compared to the most prevalent method of producing almost every N-based chemical, the Haber-Bosch process.

This study's life cycle inventory (LCI) data was obtained from balanced chemical reactions, the preliminary TEA model, and literature, scaled to the functional unit: 1 kg of NH_2OH produced. We performed a 'Cradle to Gate' LCA study, focusing on the system from resource extraction to transport and not including the use and waste management phases of NH_2OH. We used openLCA 1.11.0 to perform LCA using Tool for Reduction and Assessment of Chemicals and Other Environmental Impacts (TRACI) 2.1 as the impact assessment method. It evaluates these impact categories- acidification, carcinogenics, ecotoxicity, eutrophication, fossil fuel depletion, global warming, non-carcinogens, ozone depletion, respiratory effects, and smog. We modeled the NO_3^- reduction unit process in openLCA based on Dominguez-Ramos et al. (2013) and Orella et al. (2020).[1,2]

We compared our LCA results for the proposed pathway with the conventional method of NH_2OH production. We found that the proposed method is environmentally friendly for most impact categories while using grid electricity to operate the electrical components. We also observed that the proposed pathway is better than the conventional method for all the impact categories considered if we use solar photovoltaic (PV) energy instead of grid electricity.

Figures 2 and 3 show the life cycle impacts comparison of producing 1 kg NH_2OH produced using the conventional and proposed route in Iowa using grid and solar PV electricity, respectively.

Our analysis shows that the proposed electrochemical pathway has less environmental impact than the conventional pathway. The economic feasibility is expected to improve based on future technological considerations. The proposed pathway has massive potential and could be used as an alternative to conventional processes, including the Haber-Bosch process, followed by the Ostwald process and hydrogenation to manufacture NH_2OH.

Acknowledgments: The authors gratefully acknowledge the funding for the project from the National Science Foundation (NSF) with the NSF Award number 2036944. We thank our collaborators- Dr. Shuang Gu, Dr. Wenzhen Li, and his students (Hengzhou and Yifu), for performing the experimental work for the project.

References

[1] Orella, M. J., Brown, S. M., Leonard, M. E., Román-Leshkov, Y., & Brushett, F. R. (2020). A General Technoeconomic Model for Evaluating Emerging Electrolytic Processes. Energy Technology, 8(11).

[2] Dominguez-Ramos, A., Singh, B., Zhang, X., Hertwich, E. G., & Irabien, A. (2015). Global warming footprint of the electrochemical reduction of carbon dioxide to formate. Journal of Cleaner Production, 104, 148–155.

548b Moving to a Zero Carbon Emissions World While Managing Key Transitions in the Food and Energy Supply Chains.

John Starr

The long-term movement to energy supplies with low greenhouse gas emissions will cause many transitions in our global economy impacting both our food and energy supply chains. Three potential transitions in food and energy supply chains need to be managed effectively by companies and governments over the next 30 years to mitigate risks in meeting basic needs of nutrition and transportation. These transitions include:

- Shifting ethanol supply chains from gasoline blending to a feedstock for sustainable aviation fuel
- Increased production of vegetable oil-based fuels while minimizing land use changes to expand soybean or palm production
- Expanded use of lignocellulosic materials as a primary feedstock for biofuels to address demand from aviation and maritime industries and to direct corn and soybean supply chains to meet nutritional needs of a growing population.

This presentation will look to the future on how investments in biofuel processing and infrastructure could develop based on using capital efficiently and minimizing economic disruptions through these transitions.

548c Process Simulation Coupled with Economic, Energy and Environmental Indicators As an Eco-Design Tool for Sustainable Products

Andrea Mio, Elena Barbera, Alessandro Massi Pavan, Alberto Bertucco, Maurizio Fermeglia

The sustainable design of novel processes and products requires the estimation of indicators to support decisions makers and investors. The sustainability concept accounts not only environmental issues, but also economic considerations and social aspects, such as the correct management of energy sources. Therefore, an informed selection of the most sustainable process or product to be developed should be based on well-established economic, energetic and environmental indicators. Additionally, they ought to be rather simple to calculate at an early design stage. These indicators frequently draw on data on production processes as well as chemo-physical characteristics that are closely tied to molecular structure. Unfortunately, the majority of the pertinent information for the computation of the indicators is not available experimentally, in the literature, or in the databases employed for the estimation of the indicators during the conceptual design phase of a new product or process. Furthermore, the accuracy of the data possibly present is missing or questionable: consequently, it is necessary to predict the properties of new products and perform material and energy balance for their production processes.

Process simulators have become effective instruments for resolving the material and energy balances for any manufacturing process as well as for energy integration and cost analysis. They are a crucial part of multiscale molecular modeling, along with molecular and mesoscale simulations. They can specifically cope with novel procedures including batch operations, difficult separations, and reactions in the manufacturing of new chemicals. Our purpose is offering a framework for coupling process modeling tools to the primary energy (EROEI), economic (Levelized Cost), and environmental indicators (LCA) to be used for process or product selection. Applications including carbon capture and storage (CCS) from a natural gas power plant, hydrogen production via different methods, LNG produced from various raw materials, and ammonia synthesis via alternative paths will support the adoption of this approach.

548d Towards the Development of Safe and Sustainable By Design Chemicals, Materials and Products

Dimosthenis Sarigiannis, Fotini Nikiforou, Achilleas Karakoltzidis, Antonis Gypakis, Spyros Karakitsios

The safe and sustainable by design (SSbD) concept is defined as a process to accelerate widespread market uptake of new and alternative chemical products and technologies that deliver greater consumer confidence in their safety, environmental and societal benefits and advance the transition towards a circular economy and climate-neutral society. Currently SSbD is at the core of institutional, process and technological systems innovation in the chemical industry in Europe. Given that the criteria for defining specifically the SSbD concept have been coined by the European Commission, the need to move towards the operationalization of SSbD is a key objective of the EU partnership on chemical risk assessment (PARC). Thus, an SSbD toolbox integrating tools for safety and sustainability assessment coming from different policy areas and strategies as well as new tools developed in PARC is being designed.

The overall PARC toolbox comprises interoperable models building in the EC SSbD framework, covering both the requirements for (a) newly developed chemicals and materials and (b) existing ones. The framework supports the redesign of their production processes by evaluating alternative processes to improve their safety and sustainability performance on the basis of SSbD criteria.

Towards this aim, a comprehensive framework that accounts for both the safety and sustainability criteria will be built, addressing the steps proposed by the EC SSbD framework, pertaining to:

- Step 1: Hazard assessment (intrinsic properties):
- Step 2: Human health and safety aspects of production and processes
- Step 3: Human health and environmental aspects in the final application phase
- Step 4: Environmental sustainability assessment
- Step 5: Social and economic assessment

Based on the above and aiming at the operationalisation of the EC SSbD framework, the PARC toolbox proposes a methodological framework that builds technically on innovative and beyond-the-state-of-the-art methods for hazard, exposure (both occupational and for the general population), human health risk assessment, and sustainability assessment. Starting from the user input where parameters related to chemical structure and a list of parameters, tools of different levels of complexity will be used to support the SSbD assessment.

Regarding the safe-by-design aspects, the source-to-impacts continuum is followed, aiming to provide the functional integration of environmental releases, environmental concentrations, multi-pathway and multi-route exposure, internal dose assessment, early biological effects, adverse outcome pathways, risk assessment and health impact assessment. With regard to impact sustainability, Life Cycle Assessment (LCA), protocols (ISO 14040, ISO 14044 and the International Reference Life Cycle Data System) is followed. Several methodological choices need to be made to assess the impacts associated to a certain product or process and to predict system performance upon upscaling, especially when prospective LCA is applied based on lab-scale data. The criteria to assess the quality of data and apply cut-off points, the methodology to be used to handle material recycling at the end of lifetime, and the way the environmental burdens and benefits are allocated among the different value chains, must be carefully selected, while different methodological options exist for the impact assessment, since these methodological choices can drastically influence the results of an LCA.

It has to be noted that:

- Tools of different level of complexity will be implemented for the assessment of chemicals, materials and processes across the different stages of innovation.
- An open architecture is followed in the computational toolbox allowing for the ready integration and use of new tools when appropriate; the overall methodology is built on the concept of concurrent engineering, able to integrate the methodological advances (e.g. omics, NAMS) that are of regulatory relevance.

A broad array of models is available within PARC, able to address both safety and sustainability issues. The aim of this work is the development of functional workflows (model pipelines) that involve the best suited tools for the specific user request. These pipelines are delivered automatically, either in the form of a new "pipeline tool", (e.g. linking portal) or a dashboard. The automated pipeline requires that the tools have interfaces for automated running of the models. The pipelines will be tested and refined with the case studies through a continuous iterative process following a concurrent engineering approach. This is critical in order to continuously adapt the workflows based on the emerging needs of the SSbD community, as well as for the integration of novel tools.

Equally important is to ensure link with databases that relate to various aspects of chemical safety and sustainability such as data on chemical structures, tonnage, phys-chem properties, toxicity, ancillary exposure data, carbon footprint and other sustainability indexes. Databases considered include the ZZS similarity tool, ECHA CLP C&L Inventory, PubChem, eChemportal, ECOTOX, RISCTOX, NORMAN, CompTox Chemicals Dashboard, Registry of Toxic Effects of Chemical Substances (RTECS), CAMEO Chemicals, Chemical Entities of Biological Interest (CheBI), ChemSpider, ChemView, OpenFoodTox, ChEMBL, OSHA Occupational Chemical Database, openLCA nexus and Ecoinvent.

Another important feature of the developed workflows is that they follow a tiered approach regarding the complexity of tools and methods, relying on the stage gate innovation model. In this sense, models of different level of complexity are employed based on the innovation stage and data availability. Various models will be interconnected through APIs. Models for safety aspects include INTEGRA, OECD SAAToolbox, ConsExpo, OECD QSAR Toolbox, QSARs lab database, VEGAHUB, GreenScreen for Safer Chemicals AMBIT tool (LRI) and ECETOC TRA. With regard to sustainability, models in the pipelines include ProScale, EF-LCIA Method, SubSelect, USEtox and INTEGRA-LCA.

The study presented herein concerns the development of an integrated solution with a user-friendly interface as well as user requirements for minimum data needs. A central interface is created aiming at the description of the pipeline of the SSbD toolbox. Initially, this provides links to the individual model user interface included in the toolbox for early and late stages of innovation. Later on, user interfaces will be created for model pipelines that will be tailored for specific SSbD assessment requests based on the specification of potential scenarios. The user interface will be web-based; in some cases (for specific tools) they may be a client-side application, executed from the web browser window that will guide the user through the pipeline of the SSbD toolbox.

Compounds of interest in the case study presented herein include bisphenol A, F, S and bisphenol alternatives. All human exposure routes and pathways to the toxic compounds were integrated at the level of systemic internal dose, that was linked with adverse health outcomes on the endocrine system with cascade impacts on human reproduction, metabolic syndrome and, even, neurotoxicity after chronic exposure to the persistent organic chemicals found in plastic products and waste to quantitatively assess the related health risk. On top of that, the various dimensions of sustainability have been investigated as well. The toolbox is also empowered with quantitative structure-activity relationship models for property prediction supported by a machine learning-based search engine that explores the relevant chemical space for alternative molecules that would eventually minimize the environmental and human health hazards associated with their use in consumer products. Thus, an integrated framework for selection of chemicals for product redesign has been coined; the computer-aided workflow we present supports the redesigning of these chemicals to achieve the sought of environmental and human health objectives using in silico generated structure suggestions.

548e Electrochemical Systems and Machine Learning for Cost-Effective Wastewater Treatment to Maximize Renewable Energy

Abdullah Maghfuri Sr., Manish Mosalpuri Sr., Wenzhen Li Sr., Mark Mba Wright

Abstract

Clean water is an essential and limited resource. Growing food demand and fertilizer use has increased the need for wastewater treatment systems. In Iowa, wastewater nitrogen levels range from 2.0 to 360 mg/L, while allowable nitrate levels in bottled water are 10 mg/L. Wastewater treatment facilities can lower nitrogen levels by more than 66%, but they require energy input and are expensive to operate. This study investigates utilizing renewable energy and electrochemical systems to lower wastewater treatment costs.

Renewable energy is abundant and inexpensive but intermittent. Because of this, we developed machine learning (ML) predictive models to operate the system during excess energy availability or high-water nutrient loading times. To further reduce costs, we investigate the sale of hydroxycinnamic acid (HCA) as a high-value by-product from the electrochemical treatment of wastewater streams.

For this project, a deep neural network (DNN) ML model was trained using US geological survey (USGS) data on nitrate levels for rivers across the US. The DNN predicts hourly nitrate levels within a 2% accuracy. An hourly timeframe allows the electrochemical system to switch between solar, battery, or grid electricity depending on availability and cost. At nitrate levels below 60 mg/L, the electrochemical system stores excess solar electricity using a battery unit. Above 60 mg/L, the system treats the wastewater stream to produce HCA. The HCA and sale of excess electricity allows the system to be profitable.

There is a required balance between the benefits of renewable energy options and the costs. This presentation will discuss the potential of using electrochemical systems and chemicals to reduce wastewater treatment costs from point sources by optimizing renewable energy use. We will also discuss how ML models can help address the critical technical parameters and challenges of electrochemical wastewater systems. This intersection of technology and chemistry could solve many of the problems posed by wastewater treatment, possibly even making it a profitable effort in order to drive interest in the process.

548f Grain Storage Goes Green: Innovations in Sustainable Processes and Products

Vinod Amar

Grain storage is a crucial aspect of food production and agriculture, and it is also an area where sustainable practices can have a significant impact. Green Process and Product Engineering offers innovative solutions to improve the sustainability of grain storage processes and products. This interdisciplinary field focuses on developing sustainable processes and products that minimize environmental impact. Energy-efficient storage facilities, green pest control methods, sustainable packaging, advanced monitoring and control systems, and the development of value-added products are some of the frontiers in Green Process and Product Engineering in grain storage. By implementing these practices, engineers and researchers can make a positive impact on the environment and society while supporting the critical role of grain storage in agriculture and food production. This abstract provides an overview of the potential of Green Process and Product Engineering in grain storage to improve sustainability and promote a more responsible approach to food production.

86a Selective and Efficient Valorization of Industrial Wastes Via Carbon Mineralization

Ning Zhang, Danny Huang, Aaron Moment, Ah-Hyung Park

Industrial waste treatment and utilization can efficiently address perspectives of sustainability and cost-effectiveness, in the context of green chemistry and engineering. Specifically, the integrated process of CO_2 mitigation technologies and valuable materials production is appealing to the industrial sectors, and is the subject of our presentation. In our studies, we used construction and demolition waste, steel slags and coal fly ash to develop multi-stage of processes to selectively extract out Ca first from waste materials by acid leaching and followed by indirect mineralization to produce high purity of calcium carbonates. We describe how vaterite, aragonite and calcite can be formed with controlled operating processes. Blaze900 probe from BlazeMetrics was integrated with the reactor to monitor the particle size distribution and the shape variations of particles/carbonates in real-time which provided optical information facilitating us to better understand the mechanisms of the leaching and crystallization processes. Multi-step of precipitation of impurities, prior to carbonation, was developed to improve the quality of carbonates by removing impurities from dissolution process (Fe, Al and Si). We applied elements of this process to a larger scale reactor (30L) after optimizing the lab-scale operating conditions and we aim to fix a large amount of CO_2 under mild conditions and produce high-value products. To achieve a more sustainable process, salt electrolysis is used to produce acid and base to offset chemical costs. Finally, we report using solid residues as supplementary cementitious materials in the construction materials.

86b Unique Synergies between Carbon Mineralization and Biomass Conversion: Alkaline Thermal Treatment for Combined Carbon Capture and Hydrogen Production

Jonah Williams, Ah-Hyung Park

Alkaline thermal treatment (ATT) is a promising and robust BECCS method which can run at mild temperatures and pressures (<500°C, 1 bar) and produces carbon-free high-purity H_2 (>90% v/v) in a single reactor. Hydroxide salts are utilized during the ATT reaction (e.g., NaOH, KOH, LiOH, etc.), facilitating the degradation and conversion of organic matter. Tandemly, carbon from the biomass is captured in the form of stable carbonate salts (e.g., Na_2CO_3). The type of hydroxide utilized will greatly alter the reactivity of the system, with Group I hydroxides operating differently than Group II hydroxides due to differences in melting point and basicity. Herein, we explore the conversion of different types of biomass wastes using ATT to make high purity hydrogen and sequester carbon in the process. Discussion about the ATT reactive mechanism is also provided in the context of conventional steam gasification processes. Unique synergies are also explored between alkaline thermal treatment and carbon mineralization processes to maximize both carbon conversion from biomass and also carbon storage; ideally, as the most stable carbonates of Mg or Ca.

86c Evaluation of Bioenergy Production from Advanced MFC-Fermentation Technology

Kundan Kumar, Ming-Hsun Cheng

Biobased energy and product production via biomanufacturing has gained a lot of attention, including advanced sources of energy, waste management, and decarbonization technologies. Integrating different types of organic waste, mostly including municipal solid waste (MSW) and agricultural residues (AR) for bioenergy conversions provides opportunities to secure sustainable waste management and strengthen the circular economy. Biofuel yield is achieved from feedstock deconstruction and upgradation of intermediates products to finished products. This study provides an overview of waste sources and composition and different bioconversion processes, including fermentation process with a special focus on microbial fuel cell (MFC) process and design for energy production from biomass (Dilip Kumar et al., 2022). The recent technology of microbial fuel cell involves various design considerations such as single chambered, double chambered, and integrated MFC coupled with various operational parameters, such as pH, temperature, electrode, substrate type etc. No microbial fuel cell does work on the same operating parameters and there is always compatibility problem in treating or using complex biomass. MFC integrated with fermentation technology to degrade the biomass parallel could be a promising option to facilitate the production of various types of bioenergy, such as liquid biofuel, hydrogen, and bioelectricity in a single pipeline. Along with the consolidated MFC-fermentation process, it is necessary to choose efficient microbe species and process design based on the feedstock properties. The consolidated MFC-fermentation process utilizing MSW and AR can be a good strategy for bioenergy conversions. This new process is expected to be a cost-efficient route improving the efficiency of bioenergy conversion, having lower CO_2 emission (40 - 60% faraday Efficiency), and facilitating waste and resources management (Santoro et al., 2017; Srivastava et al., 2022).

Reference

Dilip Kumar, S., Yasasve, M., Karthigadevi, G., Aashabharathi, M., Subbaiya, R., Karmegam, N., & Govarthanan, M. (2022). Efficiency of microbial fuel cells in the treatment and energy recovery from food wastes: Trends and applications - A review. *Chemosphere*, 287, 132439. https://doi.org/10.1016/j.chemosphere.2021.132439.

Santoro, C., Arbizzani, C., Erable, B., Leropoulos, L., 2017. Microbial fuel cells: From fundamentals to applications. A review. Journal of Power Sources, pp. 225-244, 356. http://dx.doi.org/10.1016/j.jpowsour.2017.03.109.

Srivastava, R. K., Boddula, R., Pothu, R., (2022). Microbial fuel cells: Technologically advanced devices and approach for sustainable/renewable energy development. *Energy Conversion and Management*, *13*, 1–15. https://doi.org/10.1016/j.ecmx.2021.100160.

86e *Chemical Sector Risk Management Agency Resources in Support of a Sustainable and Resilient Chemical Sector*

Janine Mason, Ashley Pennington, Cheryl Louck, Jacob Mehl

Risk management enhances sustainability. The US Department of Homeland Security's Cybersecurity and Infrastructure Security Agency (CISA) functions as the Sector Risk Management Agency (SRMA) for the chemical sector and the National Coordinator for critical infrastructure security and resilience, in this role, CISA leads the effort to understand, manage, and reduce risk to the physical and cyber infrastructure that Americans rely on every hour of every day. An infrastructure which, when it comes to the Chemical Sector, spans national and sector boundaries. Chemicals are used everywhere, and in an effort to better understand, manage, and mitigate risk, with CISA resources chemical sector partners can enhance site sustainability and resilience.

This poster will outline the Chemical SRMA and its roles and provide information and resources related to capacity building, training, publications, supply chain, public-private partnerships, voluntary programs, and no-cost resources available to the chemical sector and within the four industry segments (basic chemicals, specialty chemicals, agricultural chemicals, and consumer products). The Chemical SRMA focuses on coordination and collaboration across the sector and with cross-sector critical infrastructure partners, both at the headquarters level and at the individual site level through the CISA regional offices. Among the resources related to supply chain discussed will be an overview of the 2022 Effective All-Hazards Chemical Sector Response Playbook(2) and webinar series which provides a Standard Operating Procedure (SOP) to assist the Chemical Sector in preparing for, responding to, and recovering from an all-hazards emergency; as well as information pertinent to the supply chain of water treatment chemicals. The team will also discuss how to contact your regional office for interaction and resources, an overview of the Homeland Security Information Network – Critical Infrastructure (HSIN-CI) Chemical Portal information, attendance at classified and unclassified briefings, and how to receive pertinent Chemical SRMA updates will be covered.

- *(1) Critical Infrastructure Sectors | CISA; (2) Chemical Sector Playbook | CISA*

86g Landfill Gas Converted into Renewable Natural Gas: A Case Study in Texas

Helen Lou, Elizabeth Hastings, Jian Fang, Jessica Ledwith, Samantha Adelan

Landfill gas (LFG), a byproduct of decomposing organic waste in landfills, can be converted into renewable natural gas (RNG), a clean and sustainable source of energy. This conversion process aligns with the goal of the US Environmental Protection Agency's (EPA) Landfill Methane Outreach Program (LMOP) to reduce greenhouse gas emissions and promote renewable energy sources. The process of converting landfill gas to RNG goes through a series of steps, including primary treatment (collection), secondary treatment (purification), advanced treatment, and compressing injection into natural gas pipelines. In the case study, a municipal solid waste landfill was chosen based on its landfill gas flow rate, proximity to potential end users of the RNG, and relevant geography to the installation of a pipeline. A theoretical conversion process was then applied to the landfill, and the carbon footprint of the process was calculated based on data from similar projects obtained from state air permits and LMOP data to determine the benefit of implementing the process. Additionally, several chemical refineries in the vicinity of the landfill are interested in the RNG produced from the project, and according to LMOP data for Texas, most new landfill gas projects have been producing RNG rather than electricity or using the landfill gas directly since 2012. As more projects like this are implemented, the use of RNG can play an important role in reducing our reliance on fossil fuels and addressing climate change.

86h Synthesis and Tribological Properties of Bio-Based Lubricants from Vegetable Oils

Paulo Roberto C. F. Ribeiro Filho, Francisco Murilo Tavares de Luna, Celio Cavalcante Jr.

The demand for bio-based lubricants has grown considerably in recent years due to increasing environmental awareness among the public, governments, and industries. In this study, pequi, soybean and castor oils were used for obtaining biolubricants. From pequi oil, a sample of bio-based lubricant was obtained through hydrolysis (AGLP, Pequi Free Fatty Acids) and esterification (BLS, Synthetic Biolubricant) routes, using a long-chain alcohol (2- ethylhexanol). Hydrolysis (AGLS, Soybean Free Fatty Acids), esterification (BL1), epoxidation (BL2) and oxirane ring opening (BLOR) reactions using 2-ethylhexanol were used to obtain soybean oil-based biolubricants. Finally, samples based on castor oil fatty acids (AGLM, Castor Free Fatty Acids) were obtained through esterification (ERIC), epoxidation (EPOXIRIC) and oxirane ring opening routes using water (ORWATER) and 2-ethylhexanol (OR2EHRIC) as nucleophilic agents. The reaction products were characterized via Fourier-Transform Infrared Spectroscopy (FTIR) and Nuclear Magnetic Resonance (NMR). The main physicochemical properties were determined according to standard ASTM methods. Frictional characteristics of the samples were obtained through tribological tests performed on a four-ball tribometer. Commercial mineral lubricant samples were used as references to evaluate the tribological behavior of the bio-based lubricants. Friction coefficients and Wear Scar measurements were compared to the mineral oil results. The obtained tribological properties indicate that the bio-based samples obtained from pequi, soybean and castor oil may be used as lubricant base oil in commercial applications.

86i Materials for Direct Air Capture of CO_2 Via Particle Molecular Layer Deposition

Hailey Loehde-Woolard, Bergen Evans, Kent J. Warren, Alan Weimer

As the adverse impacts of anthropogenic climate change continue to become more prevalent and intense, adaptation and mitigation technologies are becoming vital. Among the mitigation technologies, direct air capture (DAC) of CO_2 addresses emissions from non-stationary sources and could aid in addressing the existing atmospheric concentration. Further implementation of DAC technology in conjunction with an array of other fields that use CO_2 such as solar thermal chemical processing or reverse water gas shift reactions could produce liquid fuels and other desirable products. However, DAC faces complications due to the high processing volume of air and the low concentration of CO_2. Improvements in the CO_2 sorbents are needed to reduce cost and politically motivate DAC implementation. Traditional CO_2 capture technology relied on liquid amines which require substantial energy, and therefore substantial cost, to regenerate. To lower sorbent regeneration cost, solid adsorbent materials have been a focus of research for the past few decades. In particular, supported amines are attractive for DAC as they can operate under ambient conditions, regenerate at relatively low temperatures, and are tolerant of humidity. The current techniques of functionalization through liquid methods face limitations; grafting tends to have lower adsorption capacities while impregnation faces poor regeneration stability (Unveren 2017). We aim to introduce a novel technique to generate supported amine adsorbents with high adsorption capacity and stable regeneration.

Here, we employ particle molecular layer deposition (MLD) as a vapor-phase functionalization method. MLD is closely related to atomic layer deposition (ALD), involving a series of stepwise reactions between gaseous reactants on a solid surface to create thin films (Weimer 2019). We use MLD to deposit a covalently bonded aminopropylsiloxane network on particle supports. Two MLD chemistries have been employed: (3-aminopropyl)triethoxysilane (mono-amine) and N1-(3-trimethoxysilylpropyl)diethylene triamine (tri-amine). Sorbent materials were generated on varying support materials with varying amine composition based on a design of experiments. For comparison, materials were also generated through traditional liquid phase grafting. Characterization of the sorbents included BET surface area analysis, LECO elemental analysis, CO_2 chemisorption, and thermogravimetric analysis (TGA). The adsorption capacity was shown to increase as the number of MLD cycles increased for both chemistries. To understand the efficacy of the MLD functionalization method, we took the adsorption capacity on the basis of surface area. For example, the 25 cycle tri-amine on fumed silica MLD sample achieved 0.005 mmol/m^2 which is competitive with other surface functionalization methods and well performing adsorbents in literature. Ambient adsorption and low regeneration temperatures, below 100°C, were demonstrated by the mono-amine and the tri-amine materials. These materials appeared robust over cycling conditions between 25°C and 80°C and appeared stable for adsorption/desorption cycling temperatures below the deposition temperature of 150°C. This work demonstrates a new functionalization method and capture material to the field of solid CO_2 capture research that could aid in widespread CO_2 DAC implementation.

Ünveren, E. E.; Monkul, B. Ö.; Sarıoğlan, Ş.; Karademir, N.; Alper, E. Solid amine sorbents for CO 2 capture by chemical adsorption: A review. Petroleum 2017, 3 (1), 37

Weimer, A.W. Particle atomic layer deposition. J. Nanopart. Res. 2019, 21:9

This page intentionally left blank.

86j Integration of Sustainable Extraction and Recovery of Energy-Relevant Metals Using Aldoxime-Based Ligands Systems from Unconventional Resources

Amanda Whai Shin Ooi, Hunter B. Vibbert, Ah-Hyung Park

The development of renewable energy infrastructure relies heavily on energy-relevant metals, but current industrial mining practices pose sustainability issues. To address this, new technologies are needed to explore alternative feedstocks and greener hydrometallurgical extraction processes. Although aldoxime-based ligands are currently used in industry to recover Cu in acidic conditions, this process has several environmental drawbacks and is not fully understood. Unconventional resources such as e-waste, in particular waste printed circuit boards and batteries, are particularly important as they contain many critical elements. In this study, we focus on the selective recovery of Ni using aldoxime-based ligands, even in the presence of other competing elements like Co, Mn, Zn, and Cu. We investigated the mechanisms of complexation for these metals and optimized operating conditions to extract them in a solvent extraction system. Finally, we explore the changes in electrochemical behavior using these ligand systems to selectively recover them as valuable products.

86k Guidelines for Designing Amine-Based CO_2 Capture Systems Onboard LNG Fuelled Ships

Anikesh Kumar, Preethi Sridhar, Farooq Shamsuzzaman, Iftekar Karimi

With the ever-growing concerns on the rise in Global Greenhouse (GHG) emissions, the International Maritime Organization has set timelines to reduce the GHG contribution of the maritime industry. Liquified Natural Gas (LNG) has emerged as a fuel of great interest, as it has 20-30% lower CO_2 emissions and negligible SOx emissions. Hence, a surge is expected in the manufacturing of LNG run ships (Al Baroudi et al., 2021). However, this alone is insufficient to meet the eventual target of 50 % reduction (with 2008 as benchmark). Therefore, Carbon Capture and Concentration (CCC) must still be studied and pursued as a potential remedy.

The existing literature has several studies on CO_2 capture onboard LNG fuelled ships. (Luo & Wang, 2017) and (Lee et al., 2021) studied amine-based aborption for engines of 15 MW and 18.2 MW. (Feenstra et al., 2019) studied both PZ and MEA as a solvent for a 3 MW ship. (Awoyomi et al., 2020) evaluated NH_3 based absorption as a technology for CCC for a 8.8 MW engine. (Oh et al., 2022) performed simulation studies on membrane-based capture for an 18.2 MW main engine. Our previous simulation studies on a 47,000 DWT MR Tanker suggested that amine-based absorption is the cheapest option for maritime capture.

All the above studies focused on specific ships, engines, or exhaust gas conditions only. To enhance the practical utility of their works, we develop a spectrum of designs for several combinations of flue gas conditions (flow rate, temperature, composition), maximum CO_2 storage time, and voyage length for LNG run ships. A rate-based model was simulated in Aspen Plus, and uses models/correlations, and numerical parameters were validated against the pilot plant data from (Notz et al., 2012) in our previous study. The design comprises the selection of key variables for separation, e.g., the absorber and regenerator column sizing, solvent flow rate, as well as selection of optimal CO_2 storage conditions. Additionally, the best configuration for cold energy integration to minimize the extra power demand for CO_2 compression and liquefaction is also assessed. The design is based on 90% recovery of CO_2 from the effective flue gas to be processed, including emissions stemming from extra fuel burnt for the fulfilment of the energy deficit for capture and the power demand for CO_2 compression. To this end, a novel non-iterative approach to calculate effective flue gas to be processed as a function of the flue gas conditions under optimized design conditions is also developed. Lastly, volume and weight losses incurred as a result of the installation of the capture unit are also presented.

References:

Al Baroudi, H., Awoyomi, A., Patchigolla, K., Jonnalagadda, K., & Anthony, E. J. (2021). A review of large-scale CO2 shipping and marine emissions management for carbon capture, utilisation and storage. Applied Energy, 287, 116510.

Awoyomi, A., Patchigolla, K., & Anthony, E. J. (2020). Process and Economic Evaluation of an Onboard Capture System for LNG-Fueled CO2 Carriers. Industrial & Engineering Chemistry Research, 59, 6951-6960.

Feenstra, M., Monteiro, J., van den Akker, J. T., Abu-Zahra, M. R. M., Gilling, E., & Goetheer, E. (2019). Ship-based carbon capture onboard of diesel or LNG-fuelled ships. International Journal of Greenhouse Gas Control, 85, 1-10.

Lee, S., Yoo, S., Park, H., Ahn, J., & Chang, D. (2021). Novel methodology for EEDI calculation considering onboard carbon capture and storage system. International Journal of Greenhouse Gas Control, 105, 103241.

Luo, X., & Wang, M. (2017). Study of solvent-based carbon capture for cargo ships through process modelling and simulation. Applied Energy, 195, 402-413.

Notz, R., Mangalapally, H. P., & Hasse, H. (2012). Post combustion CO2 capture by reactive absorption: Pilot plant description and results of systematic studies with MEA. International Journal of Greenhouse Gas Control, 6, 84-112.

Oh, J., Anantharaman, R., Zahid, U., Lee, P., & Lim, Y. (2022). Process design of onboard membrane carbon capture and liquefaction systems for LNG-fueled ships. Separation and Purification Technology, 282, 120052.

86m A Novel Integration of Solar Chimney Power Plant with Wind Farm for Green Hydrogen and Distilled Water Production

Fares Almomani, Jihad Eriekat, Hamza Alnawafah, Emad Abdelsalam

Jordan, like the rest of the globe, is working on securing its energy needs to satisfy the growing energy demands. Jordan is a hot and dry country with long, hot, dry summers and short cool winters. However, it is considered a sunny country with an average of 310 sunny days a year with high radiation capabilities between 4 to 8 kWh/, placing it among the top countries in the region to profit from this renewable source. According to NEPCO'S 2020 report, the total available capacity in Jordan is 5.424 GW. The available power is 4 GW for conventional energy and 1.424 GW for renewables. Aqaba has a high radiation capability. Also, its climate is considered a desert climate because there is barely any rainfall. As a result, the average temperature in Aqaba is 22.1°C. The necessity for using renewable energy has led to the development of numerous renewable energy technologies that can be applied to various purposes; one of these technologies is Solar Chimney Power Plant (SCPP). Solar Chimneys are passive solar heating and cooling systems used to control the building's temperature while also providing ventilation. Although SCPP is one of the most promising technologies, it has great potential for harnessing solar energy. Therefore, this technology is suitable in Aqaba due to its high solar radiation. *This work focuses on integrating the conventional solar chimney power plant (SCPP) with a wind farm. The integration aims to boost the production of green hydrogen and distilled water. The proposed integration was modeled after the wind farm located in the city of Al-Tafilah, Jordan. The SCPP was fitted with a seawater pool at the base. Furthermore, an electrolyzer was impeded in the seawater pool. The combined electricity production from the wind farm and the SCPP not only provides green power to the electrical grid but also provides power for the electrolyzer for generating green hydrogen. Additionally, the SCPP provides distilled water during the day due to condensation of the moist air at the chimney's inner walls. The results show that the integrated system can generate a combined 69311.86508 MWh of electrical energy and 68.3448 tons of distilled water annually. Furthermore, the results also showed that the system could produce 7593.87 tons of green hydrogen and 60750.9 tons of oxygen annually. Future work includes developing a control algorithm to optimize the efficiency of the integration.*

86n Electrochemical Performance of Hybrid Ferrite/(Mn, Ti)-Oxide Electrode for Asymmetric Supercapacitor

Khang Huynh, Bharathkiran Maddipudi, Rajesh Shende

Perovskite (ABX_3) materials have received significant attention for supercapacitors or asymmetric supercapacitors (ASCs) due to their high energy storage capacity. Ferrite materials have also been investigated as electrode materials for ASCs due to their high electrical conductivity, SSA, and excellent thermal stability. In this study, sol-gel derived ferrite materials such as $NiFe_2O_4$, $ZnFe_2O_4$, and $MnFe_2O_4$ were individually mixed at different weight ratio with the precursor salts of Mn and Ti, and sonicated in ethanol. To this dispersion, deionized (DI) water containing HCl was added to accomplish the gel formation. The gel was aged and calcined at 500-1200°C to accomplish the (Mn,Ti)-oxide phase formation. Hybrid ferrite/(Mn,Ti)-oxide electrode material was fully characterized by Brunauer-Emmett-Teller (BET) surface area analyzer, and powdered X-ray diffraction (XRD). The electrochemical performance of hybrid material was characterized using the 3-cell setup. ASCs were fabricated using the hybrid electrode materials with hydrothermal liquefaction (HTL)-derived porous carbon from corn stover. The performance of ASCs was measured to determine their specific capacitance and cyclic stability. Results obtained from the experimental findings will be presented.

243c System Level Analysis of Intermitent Photocatalytic Processes for the Production of Liquid Fuels

Luisdomingo Guzman, Juan Manuel Restrepo-Florez

Traditional chemical engineering process design has relied on the assumption that energy and resources are available in a continuous way. However, if new production paradigms based on renewable energy sources are to be deployed a radically new approach to the design of chemical processes needs to be developed. Specifically, to take advantage of renewable energy sources like wind or the sun it is fundamental to develop ways to deal with intermittency and variability at the process synthesis stage. Accounting for intermittency and variability requires incorporating these aspects explicitly in process models. In this work, we present a comprehensive discussion of current efforts, including our own, to develop alternative process synthesis frameworks for the design of photocatalytic systems accounting for intermittency and variability in sunlight. We discuss current approaches and challenges in the design of these processes. Special emphasis is placed on describing optimization-based frameworks leading to the formulation of mixed-integer non-linear programs (MINLP). We discuss how the use of stochastic programming enables us to account for variability in solar irradiance. From an application perspective, we focus on the photocatalytic synthesis of fuels based on CO_2 capture and reduction. A technology that may be of capital importance in mitigating the effects of global warming. Since there is a large set of technologies that can be used to go from CO_2 to liquid fuels (e.g., CO_2 reduction to methanol followed by multiple methanol chemistries available, CO_2 to syngas followed by Fischer Tropsch synthesis, and CO_2 methanation followed by methane chemistry), we develop a general superstructure-based optimization framework to explore this large design space and assess the energetic and economic feasibility of these processes. Additionally, historic data on solar irradiation at different regions in the US, available from public databases, is used to discuss the effect of geographic location on the process economics. Our work presents a roadmap for future research in the field of photocatalytic conversion systems for the production of CO_2-derived fuels, emphasizing the need for more integrated approaches that consider the variability of sunlight and the use of optimization tools to accelerate the implementation of this promising technology as a cost-effective solution.

399a Planning of Electricity Unit Commitment in Synergy with Nature's Ability to Mitigate Carbon Dioxide and Criteria Air Pollutants

Xinyu Zhen, Bhavik Bakshi

In 1970, the establishment of Clean Air Act (CAA) and Environmental Protection Agency (EPA) marked a breakthrough of the U.S. government to protect human health and environment by mandating the levels of air pollution.[1] The emissions of several precursor air pollutants have dropped over years of efforts, however, currently there are still 4.2 million people dying each year due to exposure to outdoor (or ambient) air pollution according to World Health Organization (WHO).[2] 2022 EPA Air Trends Report shows that most of the emissions are coming from stationary fuel combustion sources and industrial processes.[3] Additionally, CO_2 as a large contributor to greenhouse effect, its emission rate from fossil fuels has reached 37.12 billion tons in 2021, increased by 5.28% from year 2020.[4] CO_2 emissions from electricity generation and industry are still the largest portion over all sectors.[5] Unit Commitment (UC) is solved as an optimization problem to find the optimal hourly operations of multiple electricity generating plants. In recent years, researchers have started to incorporate environmental factors into this problem by imposing constraints on emissions and setting up separate goals for human health impacts.[6-8] However, their results are only based on additional objectives or constraints associated with emission control by using industrial pollutant removal technologies. Trees are natural air filters which can take up large amount of air pollutants[9] when properly planted, and they can be designed as unit operations to improve air quality and mitigate greenhouse gases. Some work[10–12] focused on Techno-Ecological Synergy (TES) which combines the operations of forests and modern emission removal technologies to treat air pollutants coming from manufacturing plants, however, they either did not consider the operations of the plant itself, or they only had one point source in the system boundary, and CO_2 emissions are not calculated and controlled in their study. In this work, we will introduce a new method which combines optimal electricity generation of multiple emission sources and the spatio-temporally varying TES framework to obtain the solution to air quality improvement, cost-saving operations, and social benefits. To the best of our knowledge, this is the first work to apply TES framework to electricity planning.

We design a framework called UC-TES (Unit Commitment with Techno-Ecological Synergy) to include ecosystem services into electricity unit commitment problem. Air pollution modeling was performed to obtain regional air quality maps and capacity of forests to take up air pollutants, and these results were then entered an optimization problem to explore the effect of vegetation on operations of electricity units and overall emissions and seek for a best solution to air pollution and monetary cost. A case study was conducted over the region of Louisville, KY. Nine power plants including coal, natural gas, and hydroelectric energy sources are optimized over a 24-hour time horizon together with decision variables for traditional air pollutants removal technologies and ecological tree planting design. CALPUFF version 7, an advanced non-steady-state meteorological and air quality modeling system, was used to generate concentration and dry deposition flux maps for three air pollutants (NO_2, SO_2, PM_{10}). Dry deposition flux is a measurement of the removal rate of forests in taking up air pollutants related to vegetation parameters.[13] To identify the effect of vegetation on air quality, we performed air pollution modeling for two types of land use: the original land use , and when land use changes to forests on some selected areas. Results show that concentration of three pollutants decreases if we change some land use to forests, and the dry deposition flux increases a lot with the presence of more trees. Then we formulate an optimization problem which integrates the unit commitment of nine power plants with decisions of traditional technologies and reforestation area and location. In this work, four technologies are considered: selective catalytic reducer, flue gas desulfurization, baghouse filter, and aqueous monoethanolamine (MEA) solution for removal of NO_2, SO_2, PM_{10}, CO_2, respectively. The results will decide a. whether to include a specific type of technology or not; b. if included, how much capacity will be needed; c. location and total area for tree planting; d. optimal power output from electricity units to meet projected demand. The optimization problem was formulated as a mixed-integer linear program (MILP) and was solved in Julia by Gurobi solver. We explored five scenarios with different combinations of features such as emission constraints, conventional technologies, ecological reforestation. This work incorporates spatial and temporal variations through maps from air pollution modeling and hourly operations of power plants.

Our results show that the TES method results in designs that are superior to designs without ecosystem services, because with the benefits of forests, the same emission goal can be met without setting up some traditional technologies. Moreover, TES method can achieve a cost-saving solution and provide additional social benefits to nearby populations, and overall carbon emissions will significantly decrease with presence of more trees. The best schedule of electricity generation was also obtained from the optimization problem with power output combination in energy sources of coal, natural gas, and hydroelectric power plants on an hourly basis. This spatial-temporal work demonstrates the benefits of vegetation, develops the methodological framework to include ecosystem into industrial designs as a solution to air pollution by multiple point sources, selects best location for tree planting that optimizes the benefits , and reduces the overall regional air pollution including CO_2 emissions. Comparing with traditional UC problem, UC-TES can achieve a flexible operation of electricity units and emissions will be naturally captured by trees in an economic and environmental-friendly way that benefits besides air quality improvement, such as a cooling effect[14], will last for a very long time. A sustainable way of pollutants removal and optimal power output will be explored and discussed which opens a new gate for system with both industrial system and ecological design.

Reference

1. Chay, K. & Greenstone, M. *Air Quality, Infant Mortality, and the Clean Air Act of 1970.* w10053 http://www.nber.org/papers/w10053.pdf (2003) doi:10.3386/w10053.
2. WHO fact sheets-Ambient(outdoor) air pollution. https://www.who.int/news-room/fact-sheets/detail/ambient-(outdoor)-air-quality-and-health.
3. EPA Air Trends through 2021. https://gispub.epa.gov/air/trendsreport/2022/#home.
4. Our World in Data CO2 emissions. https://ourworldindata.org/co2-emissions.
5. EPA Sources of Greenhouse Gas Emissions. https://www.epa.gov/ghgemissions/sources-greenhouse-gas-emissions.
6. Gjengedal, T. Emission constrained unit-commitment (ECUC). *IEEE Trans. On energy Conversion* **11**, 132–138 (1996).
7. Li, Y.-F., Pedroni, N. & Zio, E. A Memetic Evolutionary Multi-Objective Optimization Method for Environmental Power Unit Commitment. *IEEE Trans. Power Syst.* **28**, 2660–2669 (2013).
8. Kerl, P. Y. *et al.* New approach for optimal electricity planning and dispatching with hourly time-scale air quality and health considerations. *Proc. Natl. Acad. Sci. U.S.A.* **112**, 10884–10889 (2015).
9. Gopalakrishnan, V., Bakshi, B. R. & Ziv, G. Assessing the capacity of local ecosystems to meet industrial demand for ecosystem services. *AIChE J.* **62**, 3319–3333 (2016).
10. Shah, U. & Bakshi, B. R. Accounting for nature's intermittency and growth while mitigating NO $_2$ emissions by technoecological synergistic design—Application to a chloralkali process. *Jnl Adv Manuf & Process* **1**, (2019).

11. Shah, U. & Bakshi, B. R. Toward Nature-Positive Manufacturing by Adapting Industrial Processes to Pollution Uptake by Vegetation. *ACS Sustainable Chem. Eng.* **9**, 16709–16718 (2021).
12. Charles, M. & Bakshi, B. R. Designing industrial landscapes for mitigating air pollution with spatially-explicit techno-ecological synergy. *AIChE J* **67**, (2021).
13. García de Jalón, S. *et al.* Dry deposition of air pollutants on trees at regional scale: A case study in the Basque Country. *Agricultural and Forest Meteorology* **278**, 107648 (2019).
14. de la Paz, D. *et al.* Assessment of Air Quality and Meteorological Changes Induced by Future Vegetation in Madrid. *Forests* **13**, 690 (2022).

399b Economic and Environmental Analyses of Optimal Integrated Processes to Produce Chloromethanes Production from Wastes.

Diego Santamaría, Borja Hernández, Mariano Martin

Securing the drinking water supply is one of the most daunting challenges in the upcoming decades.[1] Population growth combined with the depletion of freshwater resources due to climate change is intensifying water scarcity. The percentage of the population without an affordable supply of water is estimated to rise from 41% nowadays to 52% in 2050.[2] To face this challenge, desalination technologies such as evaporation, pervaporation, electrolysis or reverse osmosis (RO) become necessary to ensure the supply and minimize the population affected. Among them, RO has been elected as the most popular one, around 85% of the plants in the world,[3] due to its better economic performance. However, RO plants generate environmental issues such as excessive energy use,[4] and the acidification and increase of salinity in the marine environment of the discharge.[5]

This work provides an economic and environmental (LCA) comparison of two processes (direct chlorination of methane and hydrochlorination of methanol that include two alternative configurations) to produce chloromethanes from waste brine and CO_2. The processes consist of the electrolysis of the brine to produce the chlorine and hydrogen. On the one hand, methane is produced from CO_2 hydrogenation and next, chlorine is used to obtain the chloromethanes. On the other hand methanol is produced also form Co2 hydrogenation. Next HCl is produced using a Deacon reactor or a burner. The HCl is used to synthesize the chloromethanes. Both processes are modeled using a mathematical optimization approach. The mathematical optimization framework is particularly interesting in the direct chlorination process where a chain reaction takes place and the residence time in the reactor defines the product distribution obtained for the chloromethanes. All the units are modelled using first principles, mass and energy balances, thermodynamic and phase equilibria and experimental data. In particular, the direct chlorination is modelled using an orthogonal collocation approach to capture the chain reaction that takes place since the residence time in the reactor defines the product distribution obtained for the chloromethanes.

Among the three alternatives, the hydrochlorination processes turned out to be more profitable and show lower environmental impacts than the direct chlorination process. The highest profitability is obtained with a hydrochlorination process based on a Deacon reactor for the production of HCl when Cl_2 is sold as by-product. However, if Cl_2 cannot be sold as a by-product, the use of a burner to produce the HCl is suggested as the technology of choice. The conversion in the reactor to chloromethanes in the direct chlorination process is low, being the process less profitable (a minimum selling price of $\$5/kg_{methyl\text{-}chloryde}$) than with hydrochlorination of methanol ($\sim\$1.8/kg_{methyl\text{-}chloryde}$). The hydrochlorination process presents the lowest environmental impacts, generating 3.6 kg_{CO2}/kg_{CH3Cl} produced versus 13.3 $kg_{CO2}/kg_{methyl\text{-}chloride}$ for direct chlorination of methane. The emissions of the hydrochlorination process are mainly due to the electricity consumption in the electrolyzer and can be reduced to avoidance of ~-1.5 kg_{CO2}/kg_{CH3Cl} if renewable energy is used as supply.

References

(1) United Nations. The Sustainable Development Goals Report 2022. Clean Water and Sanitation. . *Available in: https://www.un.org/sustainabledevelopment/water-and-sanitation/ Last access: 1st*

February 2023 **2022**.

(2) United Nations. The United Nations World Water Development Report 2018. *Available in: www.unwater.org/publications/world-water-development-report-2018/ Last access 1st February 2023* **2018**.

(3) Peter H. Gleick. *The World's Water. Volume 7*; 2011.

(4) Shahabi, M. P.; McHugh, A.; Anda, M.; Ho, G. Environmental Life Cycle Assessment of Seawater Reverse Osmosis Desalination Plant Powered by Renewable Energy. *Renew Energy* **2014**, *67*, 53–58. https://doi.org/10.1016/J.RENENE.2013.11.050.

(5) Zhou, J.; Chang, V. W. C.; Fane, A. G. Environmental Life Cycle Assessment of Reverse Osmosis Desalination: The Influence of Different Life Cycle Impact Assessment Methods on the Characterization Results. *Desalination* **2011**, *283*, 227–236. https://doi.org/10.1016/J.DESAL.2011.04.066.

399c Simulation of the Conversion of Japanese Waste Biomass to Light Olefins: A Comparative Revision of the Effect of Biomass Chemical Composition in the Yield of Olefins in PRO/II

Gabriel Talero, Yasuki Kansha

Japan has one of the best-stated and regulated waste management programs worldwide, befitting its geographical situation, shortage of disposal sites, sheltered resources, and environmental constraints. Only in 2020, a recycling ratio of 21% was endorsed with a disposal budget above USD 336 per waste ton. Still, Japan has a substantial capacity for Bioenergy and carbon capture, as its supply chain remains mostly non-renewable-based, ranking as the 4th largest consumer of petrochemicals and oil. Preliminary estimations with domestic wastes can forecast a theoretical substitution near 21% of the Japanese petrochemical olefins or BTX produced in 2020. A former revision on bio-waste availability suggests that Forest residue (leftover/thinning), rice waste (straw/husk), and cardboard waste are the most available lignified bio-waste in 2020. The last encourages a theoretical production of light olefins estimated at 2.16 Mt/y. While biomass for energy utilization may induce negative environmental effects, the production of biomass-based chemicals promotes an added-value approach, avoiding combustion and the production of biogenic CO2. The availability of Japanese waste hampers the conventional conversion via biochemical pathways like fermentation. For this reason, the thermochemical conversion of Biomass to Olefins (BTO) is recommended for lignified bio-waste, easily decomposed with pyrolysis or gasification. Still, the thermochemical conversion of Biomass to Olefins (BTO) remains on a conceptual level design. Industrially available technologies enclose the conversion via gasification, methanol synthesis, and catalytic methanol to olefins - MTO. Most revisions presented in the literature regarding BTO are oriented to the process design, by selecting and evaluating the platforms and chemistry concepts. Hence, the literature does not report an extensive revision of the role of feedstock in the process. Blocked studies compare the effect of different biomass only limited to the gasification or the methanol synthesis process. Still, no comprehensive revision is performed to date for the overall biomass to olefins bio-refinery.

The purpose of the current study is to quantify the effect of the biomass' ultimate composition on the yield of light olefins. A comparative revision of the olefins yield with a simplified global reaction model of the conversion of BTO, and a detailed process simulation model in PRO/II does this valorization. Biomass included in the study with a global reaction model is cardboard and paperboard, rice husk, rice straw, forest residues of Japanese cedar, Hinoki cypress, and red pine. On the other hand, the biomass included in the PRO/II simulation model is forest residues of the Japanese cedar, Hinoki cypress, and red pine. A significant difference between the global reaction model and the numerical simulation in PRO/II is observed during the comparative revision. The global reaction model predicts a maximum theoretical yield of olefins near 36.1% g/gBM, compared to a more realistic yield in the PRO/II model near 23.5% g/gBM. The model in PRO/II includes semi-empirical kinetic models that emulate more realistic yields of syngas, methanol synthesis, or MTO. Still, the global reaction model allows fast and simple comparison of the performance with different biomass, predicting a high yield of olefins for the use of Japanese cedar and pine, over a lower yield for rice husk or cardboard. The difference in terms of the biomass ultimate composition is linked to the higher content of oxygen in the biomass, compared to the availability of hydrogen. The last behavior is coherent with the BTO definition as an elimination process of oxygen throughout the conversion steps. This revision suggests for further studies the use of biomass in BTO conversion with a lower content of oxygen.

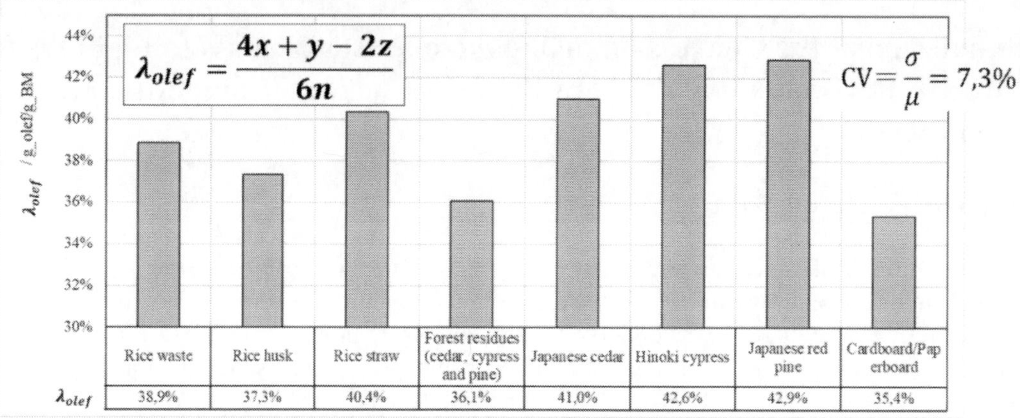

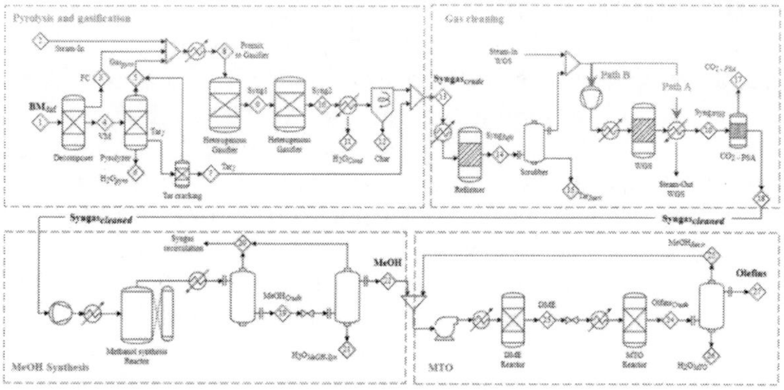

$$C_xH_yO_z + \boldsymbol{\lambda_{steam-in}}H_2O \rightarrow \boldsymbol{\lambda_{olef}}C_nH_{2n} + \boldsymbol{\lambda_{CO_2}}CO_2 + \boldsymbol{\lambda_{H_2O}}H_2O$$

$$C: \quad x = n\boldsymbol{\lambda_{olef}} + \boldsymbol{\lambda_{CO_2}}$$
$$H: \quad y + 2\boldsymbol{\lambda_{steam-in}} = 2n\boldsymbol{\lambda_{olef}} + 2\boldsymbol{\lambda_{H_2O}}$$
$$O: \quad z + 1\boldsymbol{\lambda_{steam-in}} = 2\boldsymbol{\lambda_{CO_2}} + 1\boldsymbol{\lambda_{H_2O}}$$

399d Process Operability Analysis of the Recovery of Rare Earth Elements from Coal Fly Ash

Beatriz Dantas, Deniz Talan, Fernando Lima

Rare Earth Elements (REEs) have been drawing the attention of government and industry sectors due to their applicability in diverse fields such as green energy, environmental protection technologies, military and defense industries. REEs are a group of seventeen metallic elements composed of scandium, yttrium, and lanthanides, characterized as critical materials due to uncertainties in their supply and prices. REE world resources are divided into primary (e.g., bastnaesite and monazite deposits) and secondary (e.g., coal and acid mine drainage) [1]. According to the United States Department of Energy [2], several investigations have identified coal containing a potentially significant amount of REEs. Additionally, REEs in coal ash and coal mine refuse/rejection, if released, can have significant adverse environmental impacts.

In this work, the modeling of a recovery process of REEs from coal fly ash (CFA) is developed in Aspen Plus®. This REE feedstock is considered to tackle a recurring issue of the utilization of post-combustion coal residues, acknowledging that this represents a promising alternative source for REE recovery [3]. Carbon mineralization is selected as the studied pathway due to the large availability of the mineral feedstock and its thermodynamically favored process. Moreover, this process represents a promising route for Carbon Capture and Utilization (CCU), converting captured CO_2 into carbonates. Among the several branches of carbon mineralization, this work considers the two-step carbonation process with a dissolution stage. In this process, the leaching section enables the release of metal ions by the acid, and the carbonation section occurs with the addition of CO_2 and the adjustment of pH with a base solution [4].

Systems analysis studies are performed to identify bounds for the desired region of process operation considering physical and chemical constraints. In particular, process operability analysis is carried out to assess process feasibility in early design stages [5]. The multimodel formulation adopted results in multiple linearized models that represent the process input-output mapping as a set of connected polytopes. This process mapping enables the subsequent evaluation of the inverse model to map the desired output region into feasible operational inputs. Furthermore, the Operability Index (OI) is evaluated to rank the designs given the operational inputs. The comparison for each scenario will be discussed considering the rescue of REEs, acid and base solutions usage, and the production of carbonates.

References

[1] Talan, D., & Huang, Q. (2022). A review of environmental aspect of rare earth element extraction processes and solution purification techniques. Minerals Engineering (Vol. 179). Elsevier Ltd. https://doi.org/10.1016/j.mineng.2022.107430

[2] United States Department of Energy. (2017). Report on Rare Earth Elements from Coal and Coal Byproducts.

[3] Huang, Z., Fan, M., & Tian, H. (2020). Rare earth elements of fly ash from Wyoming's Powder River Basin coal. Journal of Rare Earths, 38(2), 219–226. https://doi.org/10.1016/j.jre.2019.05.004

[4] Zhang, N., Chai, Y. E., Santos, R. M., & Šiller, L. (2020). Advances in process development of

aqueous CO2 mineralisation towards scalability. Journal of Environmental Chemical Engineering (Vol. 8, Issue 6). Elsevier Ltd. https://doi.org/10.1016/j.jece.2020.104453

[5] Gazzaneo, V., Carrasco, J. C., Vinson, D. R., & Lima, F. V. (2020). Process Operability Algorithms: Past, Present, and Future Developments. Industrial and Engineering Chemistry Research, 59(6), 2457–2470. https://doi.org/10.1021/acs.iecr.9b05181

EVALUATION OF POLYOL ESTER OILS AS ABSORBENTS IN A HYBRID ABSORPTION REFRIGERATION CYCLE

Aaron D. Liebold[*]
Florida Institute of Technology, Melbourne, FL

Paul Yelvington[†], *David Sykes, Ted Amundsen*[‡]
Mainstream Engineering Corp., Rockledge, FL

Introduction

Air conditioning and refrigeration are ubiquitous in many parts of the world. While this has provided comfort and improved health by preventing heat-related illness, it has also led to a large increase in energy consumption. The use of air conditioning is currently responsible for as much as 11% of all electrical energy consumption in the United States (EIA, 2023). This burden is felt even more in remote, off-grid applications where electrical power must be produced on-site by electric generators using internal combustion engines. Any efficiency improvements to air conditioning could provide significant progress toward energy conservation and sustainability.

The vapor-compression refrigeration cycle used in these applications is relatively efficient, having become a mature technology and having already realized many efficiency gains. However, this leaves little room for directly improving the efficiency of the cycle. Rather than trying to provide direct improvements to the vapor-compression cycle, this research sought to improve the overall efficiency of the cycle by utilizing waste heat in parallel with the primary electrical energy input. To accomplish this, the refrigeration cycle employed must be able to run on thermal energy rather than mechanical energy alone.

Refrigeration efficiency, described by a coefficient of performance (COP), is calculated using the energy flowing through the refrigeration cycle without regard to its source or type. The vapor-compression cycle is powered with mechanical energy, a high-value form of energy compared to heat. It is considered high-value energy because the conversion from heat to mechanical energy is only about 30% efficient, making mechanical energy about three times as valuable as heat energy. The remaining 70% of the heat energy used in electricity generation is lost as waste heat. Accounting for the efficiency at which the mechanical energy is produced is not necessary when comparing cycles that only run on mechanical energy. However, it is imperative to compare a cycle run on mechanical energy to a cycle run on heat (i.e., the vapor-absorption cycle). For this research, the common definition of COP will be denoted COP_{cycle}. A separate efficiency, referred to as COP_{heat}, will be defined as the heat accepted by the cold side of the

---

[*] Current address: Liebold Technologies, Melbourne, FL
[†] Current address: M2X Energy Inc., Rockledge, FL
[‡] Current address: Nestlé Nutrition (Gerber), Spring Lake, MI

cycle divided by the heat energy required to run the cycle, whether or not it is converted to mechanical energy. This new definition will assume that the thermal-to-electrical conversion is 30% efficient and that the reverse electrical-to-mechanical conversion (motor) efficiency is 90% efficient, resulting in an overall efficiency of 27% (0.27) from fuel energy to compressor shaft work. When the cycle efficiency is calculated with respect to the mechanical energy input, vapor-compression equipment commonly has a COP_{cycle} of about 2.5. However, if its efficiency is calculated using the original quantity of heat energy used to generate the mechanical energy, the same refrigeration cycle has a COP_{heat} equal to 2.5×0.27 = 0.68.

Hybrid Vapor-Compression/Vapor-Absorption Cycle

A thermodynamic cycle has been proposed where a vapor-absorption refrigeration cycle is placed in parallel with a vapor-compression refrigeration cycle, allowing the vapor-absorption cycle to utilize waste heat and supplement the cooling provided by the vapor-compression cycle. Figure 1 shows a basic implementation of this cycle.

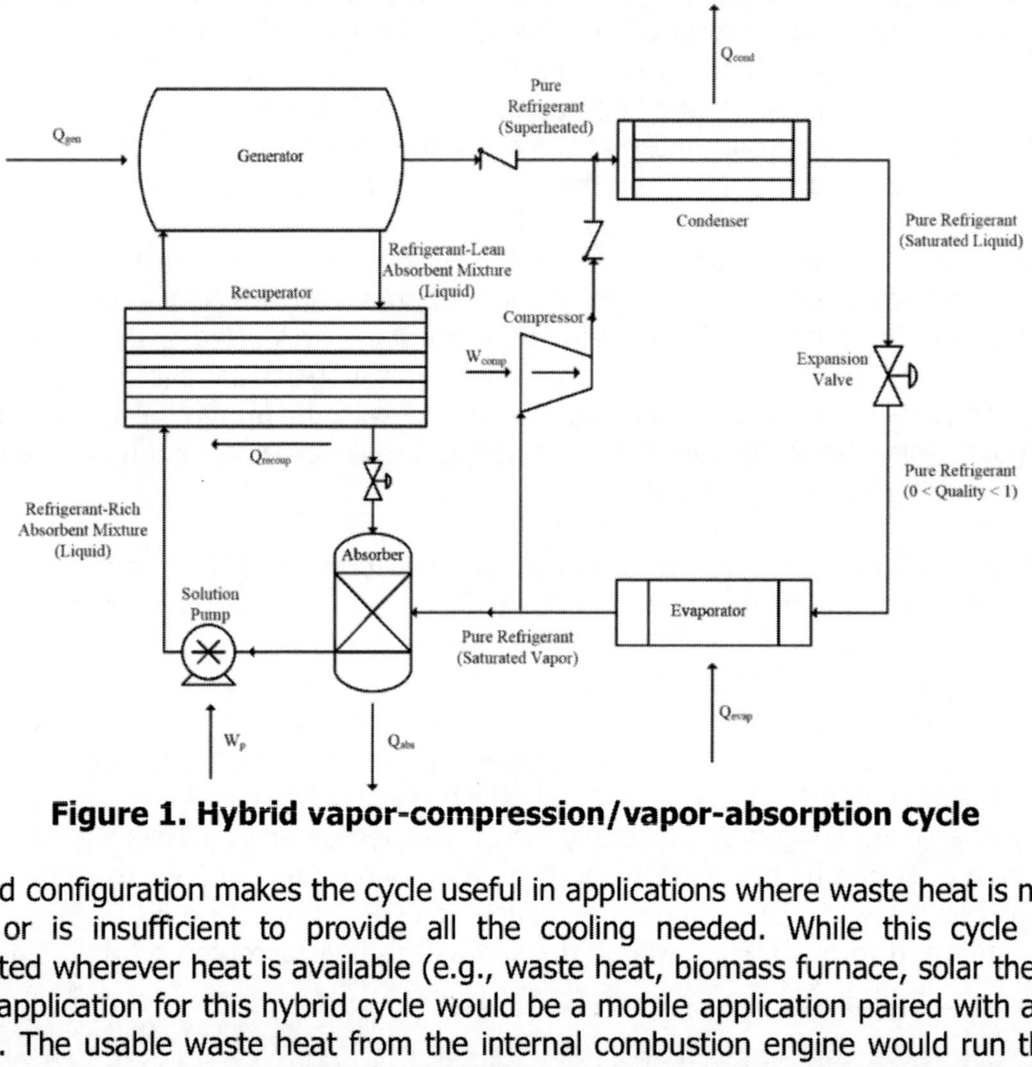

Figure 1. Hybrid vapor-compression/vapor-absorption cycle

This hybrid configuration makes the cycle useful in applications where waste heat is not always available or is insufficient to provide all the cooling needed. While this cycle could be implemented wherever heat is available (e.g., waste heat, biomass furnace, solar thermal), an excellent application for this hybrid cycle would be a mobile application paired with an electric generator. The usable waste heat from the internal combustion engine would run the vapor-absorption half of the cycle, while the vapor-compression half would run on the generated electricity. By using both the electrical and heat energy, the overall heat-to-cooling efficiency (COP_{heat}) would be increased. Because this new cycle integrates two established refrigeration cycles, the thermodynamic efficiency of each cycle remains unchanged. In this case, the COP of the hybrid cycle captures the ratio of refrigeration accomplished to energy "purchased." The components of the hybrid cycle that require an energy input include the compressor, liquid solution pump, and generator. However, the heat supplied to the refrigerant generator will come as "free" waste heat. Therefore, it is not included in the COP formula of the hybrid cycle.

$$\mathrm{COP}_{\mathrm{cycle}} = \frac{Q_{evap}}{W_p + W_{comp}} \tag{1}$$

To approximate the COP_{heat}, the mechanical energy requirements of the cycle, being the compressor and liquid solution pump, are divided by the assumed 27% efficiency of producing mechanical energy from heat.

$$\mathrm{COP}_{\mathrm{heat}} \cong \frac{0.27 Q_{evap}}{W_p + W_{comp}} \tag{2}$$

Figure 2 shows how the hybrid absorption cycle could be implemented with an electric generator in a mobile application (e.g., a disaster relief or military unit).

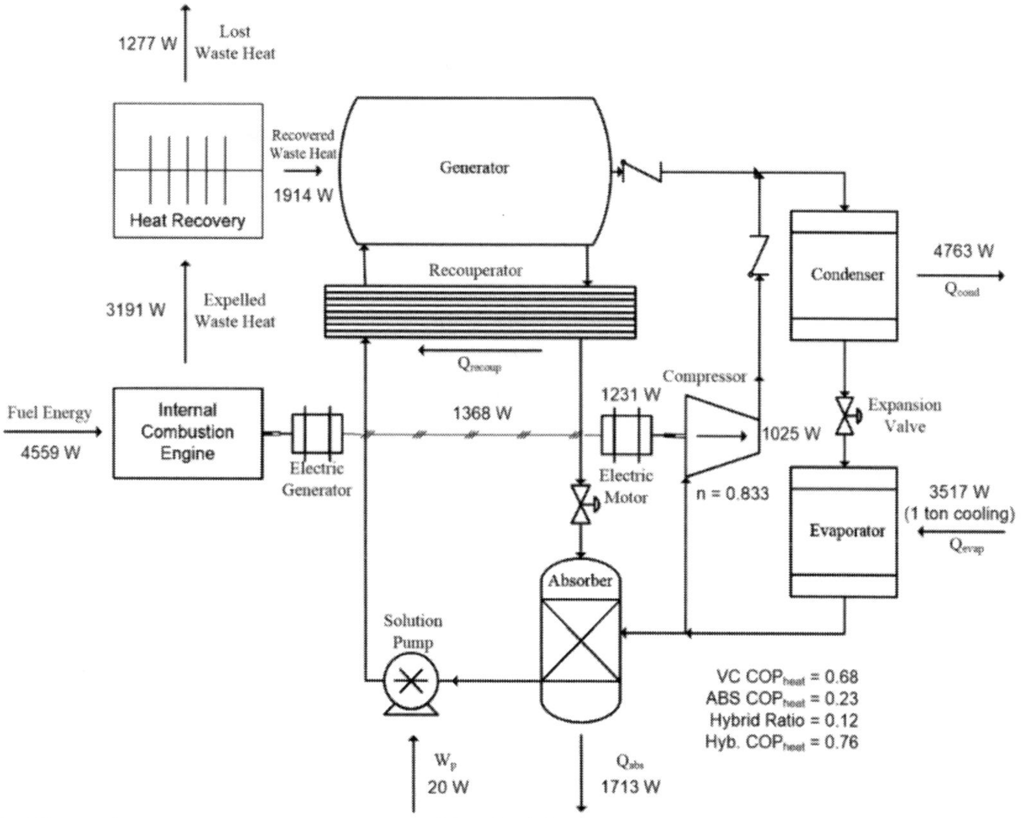

Figure 2. Electric generator-powered hybrid vapor-compression/vapor-absorption cycle

The values shown in Figure 2 are based on the measurement and simulation outlined herein. By harnessing the waste heat, the efficiency of the hybrid-absorption cycle is increased over the base vapor-compression cycle, achieving an overall fuel-to-cooling conversion ratio (COP_{heat}) of 0.76 when recovering only 60% of the waste heat. In the example in Figure 2, the heat and electrical energy produced by the electric generator are balanced to run the hybrid absorption cycle. In real-life applications, the electric generator often generates electrical energy for more than just the hybrid cycle. In these cases, even more waste heat is produced, increasing the amount of cooling achievable with the hybrid absorption cycle.

Working Fluids

In the hybrid cycle, the refrigerant is shared between the absorption and vapor-compression loops, imparting a unique set of requirements on the fluids. In addition to the refrigerant, this hybrid cycle will require other working fluids. The compressor in the vapor-compression cycle requires a lubricant, and the vapor-absorption cycle requires an absorbent. While these other fluids are separated from the refrigerant before the refrigerant outlets are joined, it is impossible to achieve perfect separation. In the hybrid cycle, cross-contamination is practically unavoidable. The simplest solution to this problem is to use the same chemical as the compressor lubricant and absorbent, as suggested by Scaringe et al. (2011).

Most modern vapor-compression refrigeration systems use the refrigerants R134a and R410a, which are compatible with polyol ester (POE) lubricants. Since these fluids are readily available and widely accepted, using one of these refrigerants in the hybrid cycle and the POE lubricant as the absorbent would be very convenient.

For a working fluid pair to be acceptable, both fluids must satisfy many criteria (ASHRAE, 2009):

- The absorbent must not become solid over the temperatures and pressures in the system to prevent blockages.
- The absorbent must be much less volatile than the refrigerant so the generator can easily achieve separation.
- The refrigerant and absorbent must demonstrate a high affinity for one another, meaning the refrigerant easily dissolves into the absorbent.
- The solubility of the refrigerant in the absorbent must show strong temperature dependence over the thermodynamic range of the cycle.

POE is already used as a compressor lubricant, suggesting that the freezing point is sufficiently low to avoid blockages. POE also has a negligible vapor pressure, indicating that it is much less volatile than the refrigerant, allowing easy separation (Nu-Calgon, 2012). The vapor-liquid equilibrium (VLE) interaction of R134a and R410 with the POE lubricant, demonstrating the affinity, was a primary focus of this research. The remaining qualities that the refrigerant and absorbent must exhibit are requirements for any refrigeration system, such as chemical stability, refrigerant saturation pressures, transport properties, and environmental soundness. Since both working fluids are already used within refrigeration cycles, it can be assumed that these properties are acceptable.

Scope of Research

In this research, the VLE of R134a with POE and R410a with POE were measured, the data was modeled with an equation of state, and finally, the hybrid refrigeration cycle was simulated. This study aimed to determine the feasibility of creating a hybrid vapor-compression/vapor-absorption refrigeration cycle and determine which working fluid pair would be more advantageous.

Previous Research

Vapor-Liquid Equilibrium Measurements

Several groups have previously studied the solubility behavior between refrigerants and polyol ester (POE) lubricants. These studies aimed not to determine their use as an absorption cycle working-fluid pair but to understand the interaction between the refrigerants and the lubricant within a vapor-compression cycle. However, the data is equally valid in both applications. The most prominent groups that have researched the interactions between refrigerants and POE lubricants include Burton et al. (1999), Huber et al. (2000, 2001), Martz et al. (1996a, 1996b), Takaishi et al. (1993), Tesser et al. (1999), as well as Wahlstrom and Vamling (1997, 1999). All these groups, except Wahlstrom and Vamling, studied the interaction of various refrigerants with commercially available POE lubricants. Wahlstrom and Vamling primarily studied the interactions of refrigerants with pure polyol esters rather than a blend to create a predictive model that could be used to model a POE oil and refrigerant mixture without gathering data for the unique pair.

The method of measuring VLE used by these groups was based upon having a known amount of refrigerant and oil in a vessel and indirectly determining the amount of dissolved refrigerant by measuring the amount of refrigerant that did not dissolve. Two methods were used among these groups to determine the amount of undissolved refrigerant in the vapor head space. The first method was to calculate the density of the vapor using an equation of state and find the mass of vapor using the headspace volume. This method was used by Takaishi et al. (1993), Martz et al. (1996b), Wahlstrom and Vamling (1997, 1999), Burton et al. (1999), and Tesser et al. (1999). The alternate method employed by Huber et al. (2000) was to use a vapor densimeter to measure the vapor density and calculate the vapor mass using the headspace volume, avoiding using an equation of state. These methods work on the assumption that a negligible amount of POE is present in the vapor phase, which is a valid assumption considering that POE has a negligible vapor pressure.

The current investigation extends previous measurements by including higher temperatures relevant to absorption-cycle generators. These previous studies are summarized in Table 1.

Table 1. Summary of the Fluids, as well as the Temperature and Pressure Ranges Studied by Previous Researchers

Research Group	Temperature Range, °C (°F)	Range of Refrigerant Mole Fraction	Refrigerants	Lubricants
Burton et al. (1999)	3-53 (37-127)	0.143-1.00	R32, R410a	POE
Huber et al. (2000)	7-82 (45-180)	0.97-0.98	R134a	POE
Martz et al. (1996b)	-20-70 (-4-158)	0.5-1.0	R134a, R22, R32, R125, R32/R125	POE
Takaishi et al. (1993)	-10-50 (14-122)	0.5-1.0	R134a	POE
Tesser et al. (1999)	-20-60 (-4-140)	0.5-1.0	R134a, R143, R236fa	Icematic SW32, POE
Wahlstrom and Vamling (1997, 1999)	30-70 (86-158)	0.0-0.7	R32, R125, R152a, R134a, R143a	PEB6[1], PEB8[2], PEC9[3]

Pure polyol esters: [1]Pentaerythritol tetra-2-ethylbutanoate, [2]Pentaerythritol tetra-2-ethylhexanoate, [3]Pentaerythritol tetranonanoate

Previous research lacked data sets spanning the entire composition range and at elevated temperatures, which motivated the present study.

Equation of State Modeling

When modeling the refrigerant/lubricant VLE data, researchers commonly employ the Flory-Huggins activity coefficient model or a fugacity coefficient approach employing a cubic equation of state (EOS). Wahlstom and Vamling (2000) and Tesser et al. (1999) investigated using the Flory-Huggins model, a semi-theoretical model that considers the large difference in size between polymer molecules and vapor molecules. Wahlstom and Vamling, and the other research groups mentioned previously, also modeled this data with equations of state such as the Redlich-Kwong (RK), Soave-Redlich-Kwong (SRK), Peng-Robbinson (PR), Van der Waals (VdW), and the Virial equations of state. The VdW and Virial EOS require fitting parameters for each fluid. At the same time, the RK, SRK, and PR equations can model pure fluids with the critical parameters alone, only requiring empirical fitting parameters for mixtures.

Methods and Equipment

Vapor-Liquid Equilibrium Measurement

The VLE measurement in this research was performed in a method similar to that used by Wahlstrom and Vamilng (1997). It was an isochoric technique in which refrigerant was dosed at a known temperature, pressure, and volume and then exposed to a known amount of POE. The refrigerant vapor in the headspace was then recalculated with the final temperature, pressure, and volume. The difference between the initial and final mass of refrigerant vapor was assumed to have been absorbed. In this experiment, it was assumed that the vapor was pure refrigerant. The concentration of the refrigerant dissolved in the POE was then calculated using the mass of POE and the mass of refrigerant that was absorbed.

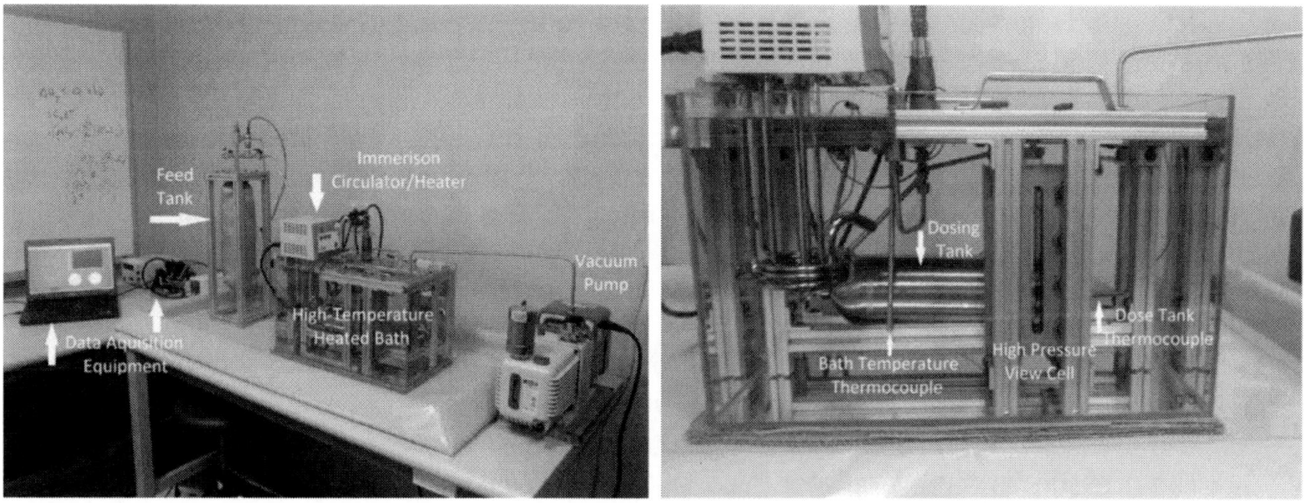

Figure 3 - Experimental VLE Measurement Apparatus

Equation of State Modeling

The resulting bubble-point data was then modeled with the Soave-Redlich-Kwong (SRK) equation of state (EOS), as demonstrated by Yokozeki et al. (2005). In the research performed by Yokozeki, the standard SRK EOS, along with modified mixing rules, was used to model refrigerant/absorbent fluid pairs. The temperature-dependent a term in the SRK equation is modeled empirically with four constants: β_i (i = 0,1,2,3). The modified mixing rules also use empirical mixing parameters, which are l_{12}, l_{21}, m_{12}, and C_{12}. The β_0 constant in the temperature-dependent a term for the absorbent is also treated as a fitting parameter. These five parameters are determined experimentally from the VLE data.

The SRK equation of state was then used to calculate the properties of the pure fluids and mixtures within the cycles, including the enthalpy and composition of each phase present. The enthalpy of the mixtures was calculated with the SRK equation of state using residual properties. The composition of the mixture in the different phases was calculated using fugacity. The phases are in equilibrium when the fugacity of one component is the same in all phases (i.e., the fugacity of the refrigerant in the liquid phase is equal to the fugacity of the refrigerant in the liquid phase). The fugacity of a component in the mixture is easily calculated with the equation of state as a fugacity coefficient, simply a ratio of the fugacity to the system pressure. Using the mixing rules provided by Yokozeki, the a_1' and b_1' coefficients in the fugacity coefficient calculation of Component 1 in a binary mixture were derived as Equations 3 and 4.

$$a_1' = \frac{\partial a_1}{\partial n_1} = 2\left(a_1 x_1 + x_1\sqrt{a_1 a_2}\left(1 - x_1 - fk_{12} + x_1\frac{fk_{12}^2}{l_{12}}\right)\right) - a_1 x_1^2 - a_2 x_2^2 \tag{3}$$

$$b_1' = \frac{\partial b_1}{\partial n_1} = 2b_1 x_1 + (b_1 + b_2)(1 - m_{12})x_2 - b \tag{4}$$

Hybrid Vapor-compression/Vapor-Absorption Cycle Simulation

The simulation of the hybrid cycle began with individual simulations of the vapor-compression and vapor-absorption cycles. The vapor-absorption cycle was simulated as presented by Yokozeki et al. (2005). The vapor-compression cycle was simulated similarly using the NIST (Boulder, CO) software, REFPROP, instead of the SRK equation of state (Smith, Van Ness, Abbott 2005). The vapor-compression and vapor-absorption cycle functions were then used together to create a function for the hybrid cycle. The assumptions used within these two models are as follows:

- The refrigerant exiting the condenser is a saturated liquid
- The refrigerant exiting the evaporator is a saturated vapor
- Flow through the expansion valve is isenthalpic
- The pressure in the condenser is equal to the pressure in the generator
- The pressure in the evaporator is equal to the pressure in the absorber
- The solutions leaving the absorber and generator are in chemical equilibrium
- The absorbent has a negligible vapor pressure
- The recuperative heat exchanger has an effectiveness of 1.
- There are no heat or pressure losses in the transfer lines.

For the hybrid simulation, the required inputs are the temperatures of the evaporator, condenser, generator, and absorber, the cooling capacity of the cycle, as well as the waste heat duty provided to run the generator. First, the vapor-absorption cycle simulation was solved to determine how much refrigeration can be accomplished with the waste heat generated from other electrical loads. The remaining refrigeration need was then simulated using the full hybrid cycle. Finally, the COP_{cycle} and COP_{heat} of the overall system were calculated using Equations 1 and 2.

Results

Vapor-Liquid Equilibrium

Data was collected for the R134a/POE and R410a/POE fluid pairs. The R134a/POE data is shown in Figure 4. The vertical uncertainty (pressure) is equal to the published errors of the pressure measurement equipment. The horizontal uncertainty (mole fraction) was determined for each point based on the uncertainty of the experimental measurements and the first derivative of each calculation required for the mole fraction calculation. The errors at x=0 (P=0) and x=1 (P=P_{sat}) of each curve were assumed to be zero because these points were taken from published fluid data.

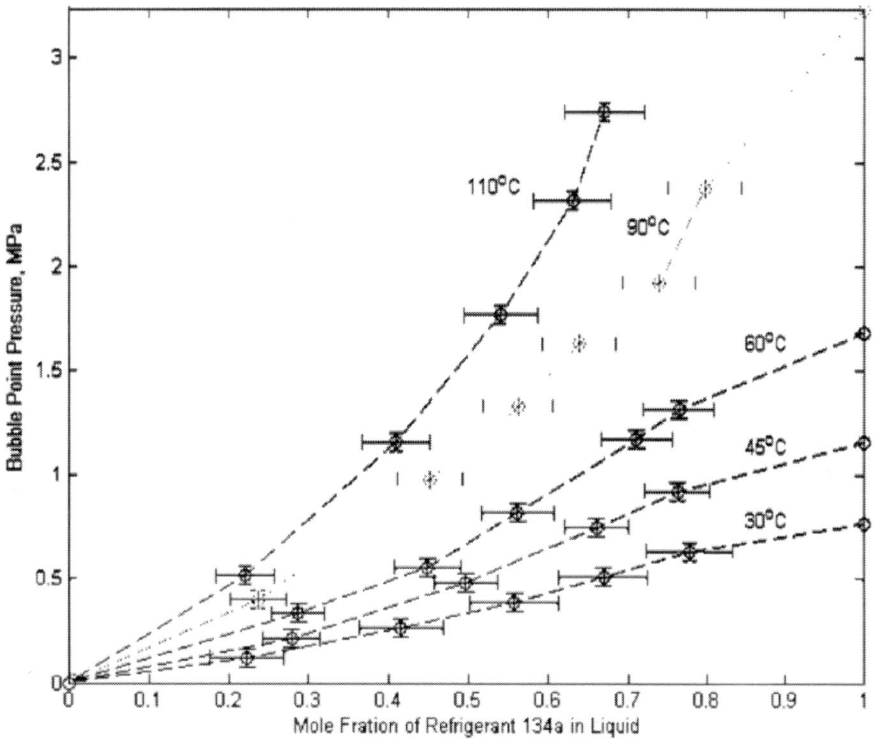

Figure 4. Experimental results of pressure versus mole fraction of R134a in POE

Upon an initial inspection, the refrigerant and POE mixtures demonstrate a slightly negative deviation from Raoult's law for ideal mixtures and a fairly strong temperature dependence. These are both favorable qualities in a vapor-absorption cycle (ASHRAE, 2009). All of the experimental data collected in the study are included in Table 6 and Table 7, located in Appendix A.

VLE Modeling

The experimental data was then modeled using the SRK equation of state. The pure fluid properties for R134a and PEB8 were taken from the work of Yokozeki et al. (2005). PEB8 is a pure polyol ester similar to the esters in the commercial POE blend used in the experiment. The PEB8 properties were chosen to model the POE blend based on similar molecular weight. Yokozeki found that the EOS calculations were not very sensitive to the critical parameters of the oil. In some research, Yokozeki took advantage of the calculation's insensitivity to the oil parameters and defined a set of oil properties that adequately modeled any oil given the average molecular weight of the oil. This model was dubbed UniOIL (Yokozeki, 2001). Considering the insensitivity of the oil parameters in the calculation, the accuracy of the critical parameters is not critical to the calculations. Because R410a is not a pure fluid, its properties were unavailable in Yokozek's work. Instead, these properties were regressed from and validated with data that REFPROP generated. R410a is a near azeotropic mixture of R32 and R125. Because there are two constituents, one can have a higher affinity for the POE than the other. This possibility was investigated with a simple experiment where R410a was allowed to come to equilibrium with POE, and the vapor in the mixture was sampled. This sample was then analyzed with a gas chromatograph and compared to an analysis of pure R410a. It was found that the proportion of R32 to R125 in the vapor had not changed significantly, indicating that the affinities of the individual refrigerants and POE were essentially the same, differing from previous research demonstrating that the R32 does have a slightly higher affinity for the oil than R125 (Yokozeki, 2001). However, because this behavior was not observed with the refrigerant/oil mixture used in this study and the VLE test stand in this study is not capable of sampling the vapor for compositional analysis, the refrigerant mixture R410a was modeled as a pseudo-pure fluid.

The constants used to calculate the state properties of the fluids in this study are shown in Table 2.

Table 2. Properties of Experimental Fluids

	MW	T_c (K)	P_c (Bar)	β_0	β_1	β_2	β_3	Source
R134a	101.0	374.21	40.59	1.0025	0.5053	-0.0498	0	(Yokozeki 2005)
R410a	72.6	345.95	48.8	1.0389	0.4291	0.0140	-0.0516	REFPROP Data Fit
PEB8	640	793	7.72	1.0	*	0	0	(Yokozeki 2005)

The constants for the ideal heat capacity of the fluids are shown in Table 3.

Table 3. Ideal Heat Capacity Coefficients of Experimental Fluids

	C_0	C_1	C_2	C_3	Source
R134a	12.89	0.3050	-2.342e-4	6.852e-8	(Yokozeki 2005)
R410a	25.38	8.741e-2	1.6080e-5	1.420e-7	REFPROP Data Fit
PEB8	2.594	2.7280	-1.42e-3	5.43e-7	(Yokozeki 2005)

These constants were used in Equation 5 to calculate the ideal enthalpy of a fluid or mixture, which was then combined with the residual enthalpy to obtain the actual enthalpy.

$$H_{ideal} = \int_{T_{ref}}^{T} C_0 + C_1T + C_2T^2 + C_3T^3 dt \tag{5}$$

Experimental Data

The binary interaction parameters for the SRK equation of state were solved using a least-squares regression analysis between a fugacity-based bubble point and the experimental bubble-point data. The regressed interaction parameters are shown in Table 4.

Table 4. SRK Binary Interaction Parameters for Yokozeki (2005) Mixing Rules Regressed from this Study's Experimental Data

	β_{1abs}	l_{12}	l_{21}	m_{12}	C
R134a/POE	0.3095	-0.0876	-0.2923	0.0876	-0.0050
R410a/POE	2.5745	-0.0036	-2.0e-6	0.0075	-0.5426

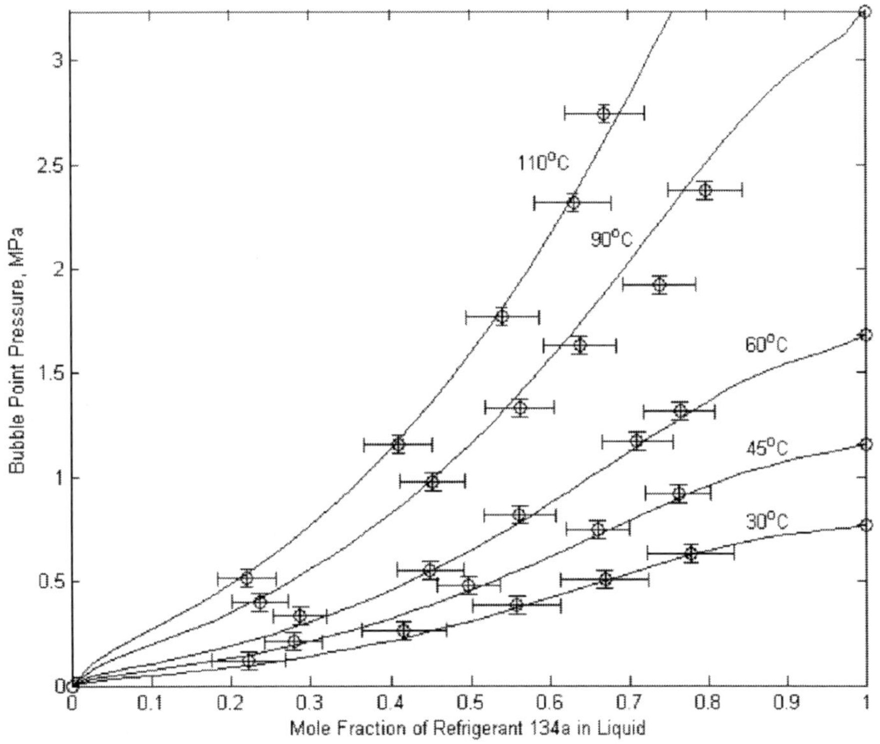

Figure 5. R134a/POE VLE Data with SRK model based on data from this study

From these plots, we can see that the compositional uncertainty dominates the uncertainty in the data, while the pressure uncertainty is relatively small. The error bars also show that the SRK model is consistent with the data within the experimental error. The coefficients for the EOS were regressed through a numerical nonlinear least squares method. The fit was similar for R410a/POE, with some deviation at high refrigerant concentrations ($0.9 < x < 1.0$).

Experimental and Literature Data

The interaction parameters were regressed again using the experimental data and the available literature data to improve the models, resulting in the coefficients shown in

Table 5.

Table 5. SRK Binary Interaction Parameters for Yokozeki (2005) Mixing Rules Regressed from All Available Data

	β_{1abs}	l_{12}	l_{21}	m_{12}	C
R134a/POE	0.9015	-0.0015	-0.0098	0.0549	-0.1080
R410a/POE	2.7999	-0.8685	-4.8e-06	0.0044	-0.4359

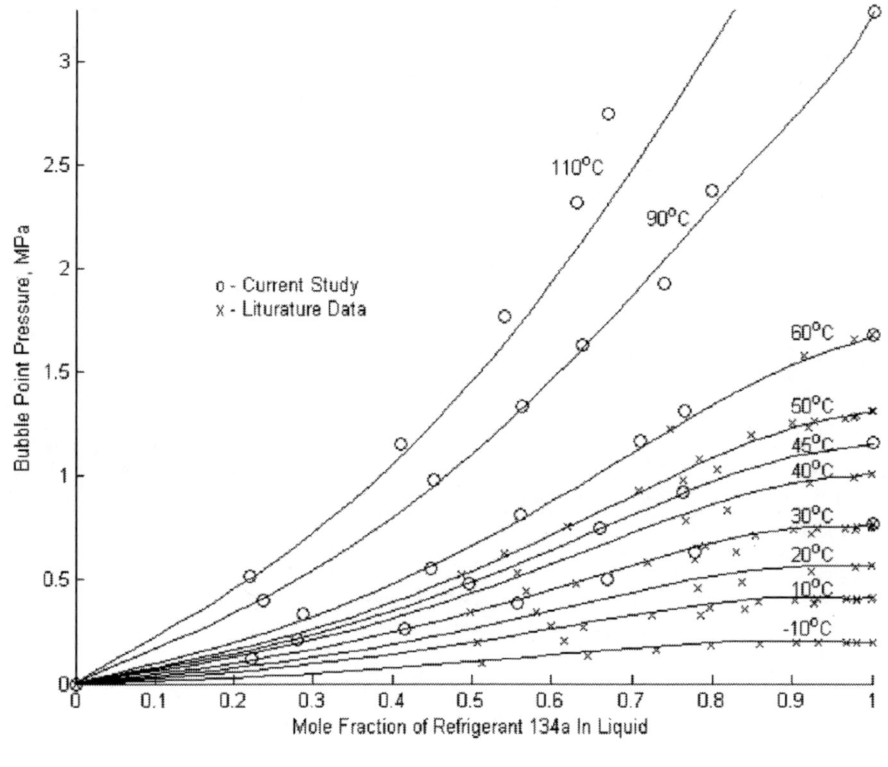

Figure 6. R134a/POE VLE Data with SRK model based on all available data [Huber et al. (2000), Takaishi et al. (1993), Tesser et al. (1999), Wahlstrom and Vamling (1999, 2000)]

The highest temperature at which R134a was studied, 110°C, is above the critical temperature and, therefore, does not have a saturation point (x = 1) to add to the data.

Hybrid Absorption Cycle Modeling

The refrigerants were simulated in three different cycles, and the resulting COPs were compared across ambient temperatures. The correlations between the ambient temperature and the system temperatures are shown below. All of the temperatures are calculated in degrees Celsius.

$$T_{cond}(°C) = T_{amb}(°C) + 15 \quad (6)$$

$$T_{indoor}(°C) = 0.233T_{amb}(°C) + 15.63 \quad (7)$$

$$T_{evap}(°C) = T_{indoor}(°C) - 15 \quad (8)$$

$$T_{abs}(°C) = T_{amb}(°C) + 5 \quad (9)$$

$$T_{gen}(°C) = bubblePointTemp(X_{abs}, P_{hi}) + 5 \quad (10)$$

Equations 6, 8, and 9 are based on a reasonable assumption for the temperature difference across an air-cooled heat exchanger. The generator temperature can be set by the amount of heat supplied. Figure 7 shows how the COP_{cycle} and recirculation ratio relates to the generator temperature.

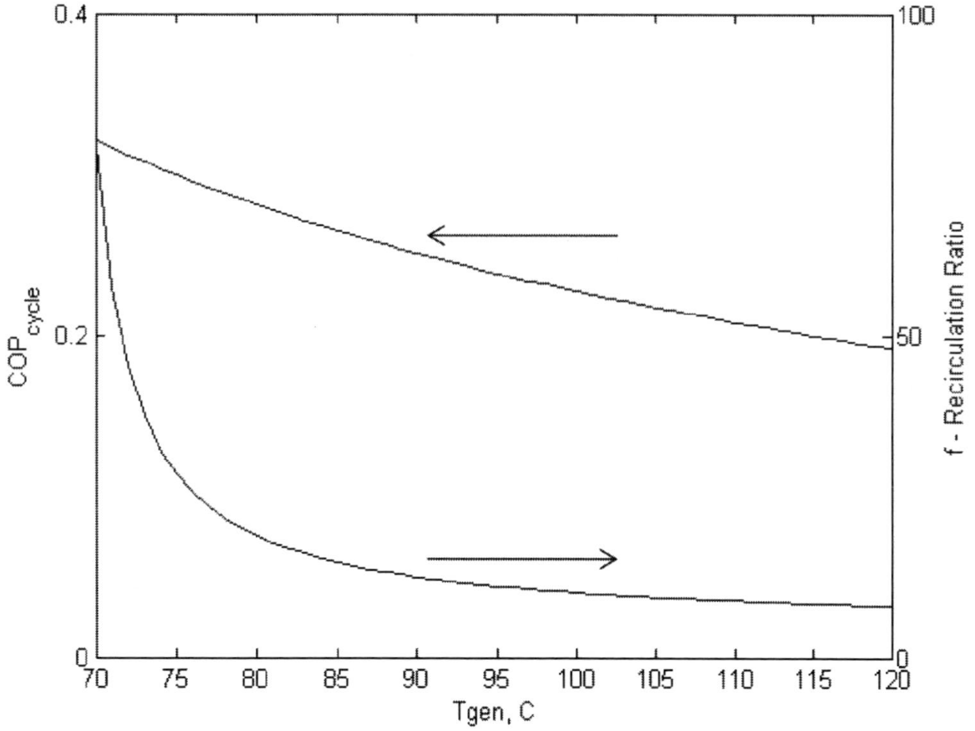

Figure 7. COP_{cycle} and recirculation ratio versus generator temperature for a simple vapor-absorption cycle

It can be seen that the recirculation ratio (i.e., the ratio between the mass of strong solution entering the generator to the mass of refrigerant generated) increases as the generator temperature decreases. Despite the increasing pumping power, the COP increases as the generator temperature decreases. The minimum generator temperature is the temperature at which the bubble-point pressure of the mixture is equal to the high side pressure. Therefore, the generator temperature (equation 10) was calculated as 10°C (18°F) greater than the minimum generator temperature to maintain the highest COP possible while avoiding an excessive recirculation ratio and the possibility of dropping below the minimum temperature. The indoor temperature (equation 7) was an assumed relationship between the desired and outdoor temperatures. The recommended indoor temperature was then reduced by 15°C (27°F) to represent the difference between indoor and evaporator temperatures. The compressor efficiency was assumed to be 0.28 in the vapor-compression and hybrid cycle. These temperature correlations were chosen to emulate a real-world scenario as best possible.

The two refrigerants were simulated in the vapor-compression and vapor-absorption cycle. In both of these applications, R134a performed more efficiently than R410a. The results of these simulations can be seen in Figure 8 and Figure 9. Being more efficient in each cycle would result in a higher efficiency in the combined hybrid cycle.

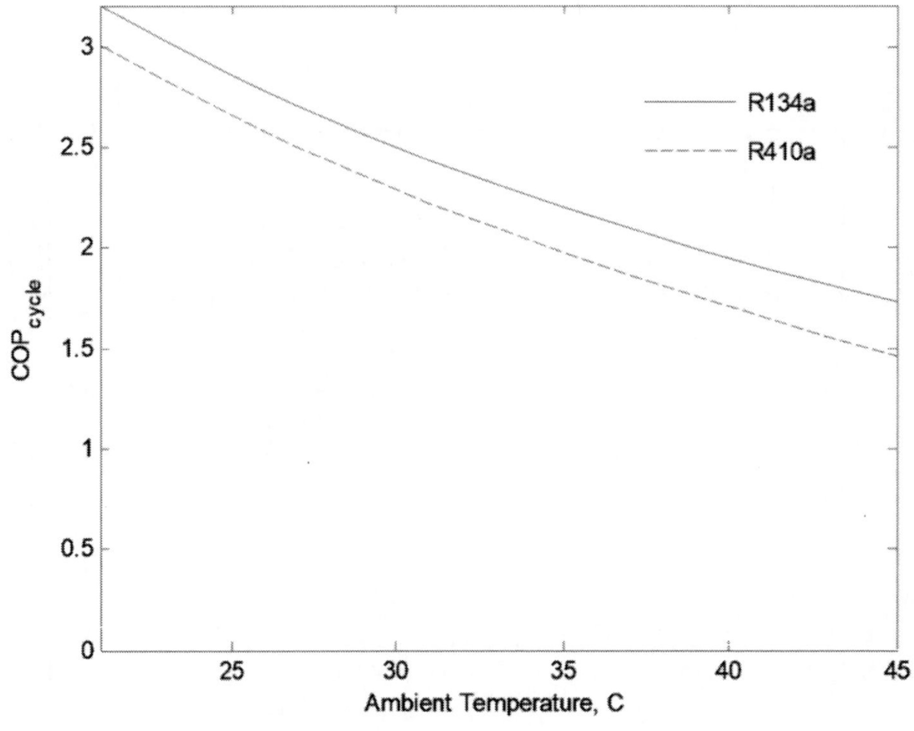

Figure 8. Comparison of the vapor-compression COP_{cycle} of R134a and R410a

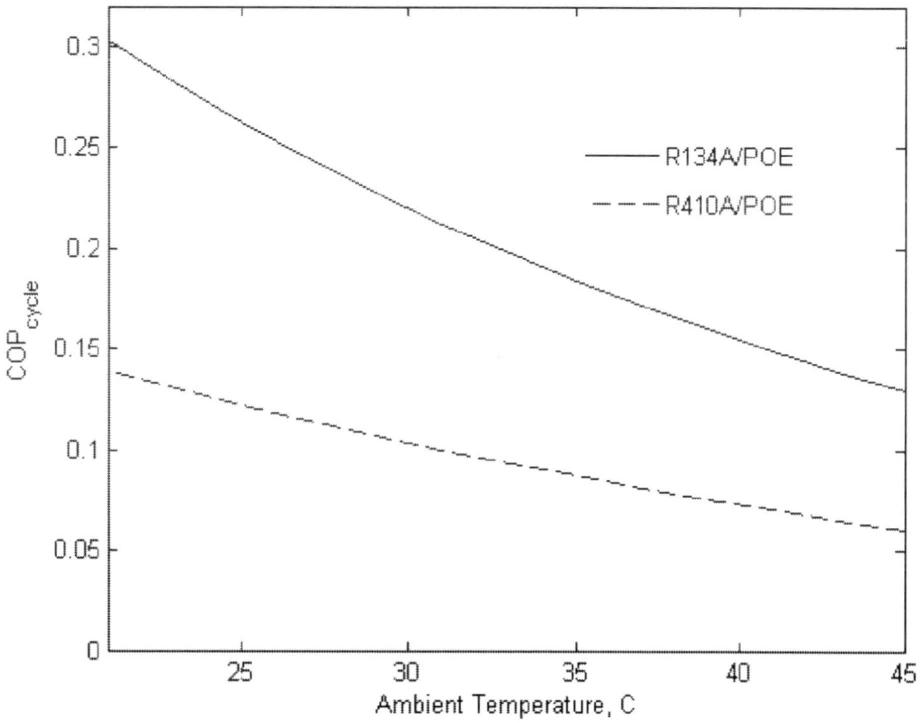

Figure 9. Comparison of the vapor-absorption COP_{cycle} of R134a and R410a

In the hybrid cycle, the efficiency increase is directly related to the amount of available waste heat. As more waste heat is provided, more refrigeration can be accomplished on the vapor-absorption side, reducing the compressor load. The hybrid-cycle simulation was performed with the same temperature dependencies used in the previous simulations. A hybrid cycle, with a total refrigeration capacity of 1 tonR (3.5 kW, 12,000 BTU/hr), was simulated as being paired with an electric generator that provides the required electrical energy and expelled waste heat. This simulation assumes that 30% of the fuel energy entering the electric generator is converted to mechanical energy, while 70% exits as waste heat. The model then assumes that 60% of the waste heat can be recovered from the engine exhaust and coolant. The results of this simulation are shown in Figure 10. The bottom curve shows the efficiency of the simple vapor-compression cycle. The second curve from the bottom ("Balanced Hyb") demonstrates the efficiency achieved by the hybrid cycle when the electrical and recovered heat energy (85% of waste heat) is supplied in a matched/balanced electric generator scenario. The other curves show the efficiencies resulting when there are other electrical loads on the electric generator in addition to the hybrid cycle. These extra electrical loads result in more waste heat availability, which aids the hybrid cycle, reducing the electrical energy requirement of the compressor.

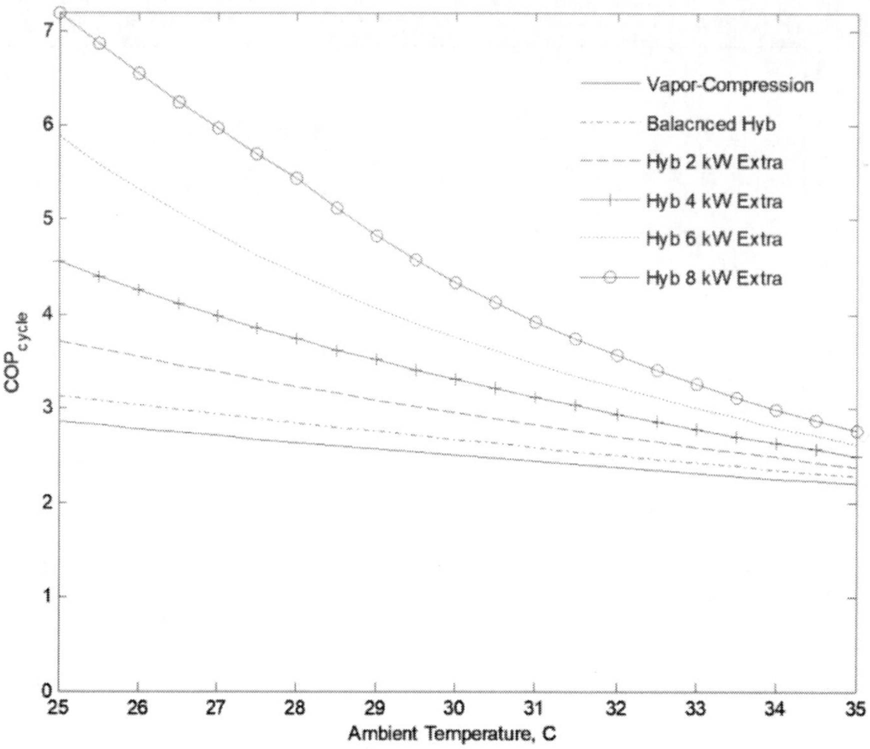

Figure 10. COP_{cycle} versus ambient temperature for a 1-tonR hybrid cycle paired with an electric generator with varying electrical loads

The simulation shows that the efficiency of the cycle is improved with the hybrid configuration. The improvement in COP is related to the amount of available waste heat. The efficiency of this cycle is highly dependent on the ambient temperature. The hybrid cycle becomes ineffective if the ambient temperature exceeds 36°C (96.8°F). As the ambient temperature increases, so does the pressure in the condenser and the temperature of the air-cooled absorber, causing the generator and absorber to operate less effectively.

Finally, the heat-to-cooling efficiency of the hybrid cycle, run from the electric generator, was compared to that of a typical vapor-compression cycle, shown in Figure 11.

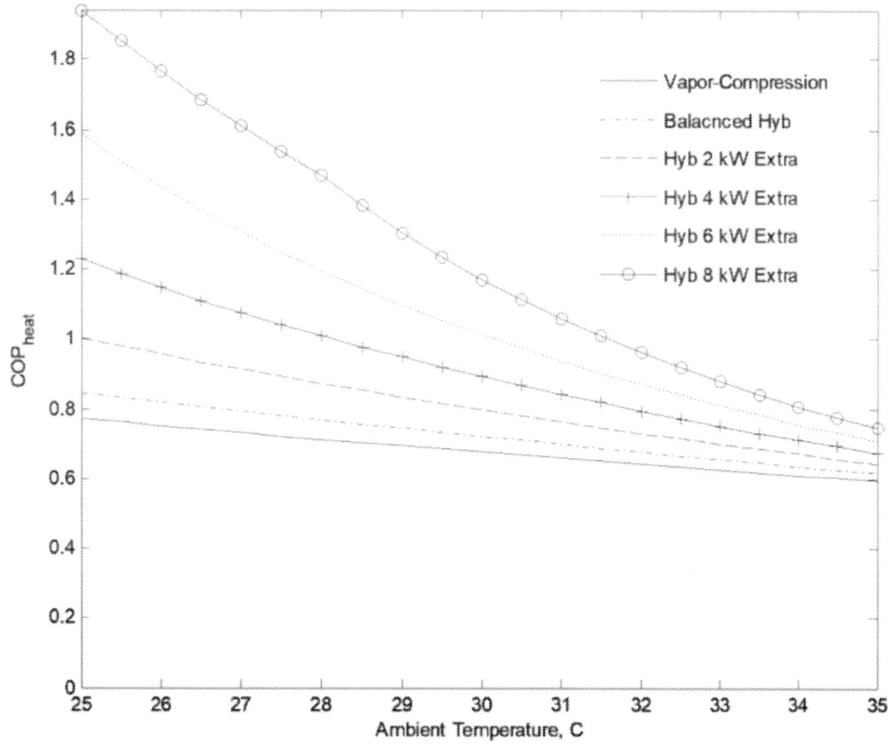

Figure 11. COP$_{heat}$ versus ambient temperature for a 1-tonR hybrid cycle paired with an electric generator with varying electrical loads

The hybrid absorption cycle achieves a greater COP$_{heat}$ than a simple vapor-compression cycle. For the balanced hybrid case, the improvement is 8.7%. At an ambient temperature of 30°C (86°F), the COPheat could be improved by up to 74% compared to the vapor-compression cycle when the electric generator produces 8 kW of electricity in addition to the electricity required by the hybrid absorption cycle.

Conclusion

Simulating the hybrid absorption cycle using the newly parameterized SRK model demonstrated that the R134a/POE working-fluid pair performs with 20% greater efficiency than the R410a/POE working-fluid pair in a matched case. Using refrigerant R134a and the POE oil as an absorbent at an ambient temperature of 30°C (86°F), the hybrid absorption cycle was capable of improving the COP$_{heat}$ by nearly 74% compared to a conventional vapor-compression cycle. Based on this study, it would seem thermodynamically feasible to create a hybrid cycle capable of harnessing waste heat to aid the refrigeration cycle and significantly increase the overall cycle efficiency.

Acknowledgments

We would like to thank the U.S. Office of Naval Research and Dr. Mark Spector for making this research possible with funding through SBIR contract N00014-11-C-0205. We would also like to thank Dr. Paul Jennings (Florida Institute of Technology), Dr. Manolis Tomadakis (Florida Institute of Technology), and Dr. Daniel Kirk (Florida Institute of Technology) for their contributions to this work.

Nomenclature

C_i – Ideal gas heat capacity coefficient
l_{12}, l_{21}, m, C – Binary interaction parameters
MW – Molecular Weight
P – Pressure
P_c – Critical pressure
Q – Heat flowrate
T – Temperature
T_c – Critical temperature
W – Work
x – Mole fraction in liquid phase
β_i – Coefficient for the temperature-dependent α term in the SRK equation

Appendix A – Experimental Data

Table 6. Experimental R134a/POE VLE data

Temperature, °C (°F)	Pressure, MPa (psia)	Error in Pressure, +/- MPa (psia)	Mole Fraction of R134a in Liquid	Error in Liquid Mole Fraction R134a, +/-
30.0 (86.0)	0.000 (0.0)	0.000 (0.0)	0.00	0.00
30.0 (86.0)	0.122 (17.7)	0.043 (6.2)	0.22	0.05
30.2 (86.4)	0.264 (38.3)	0.043 (6.2)	0.41	0.05
30.2 (86.4)	0.384 (55.7)	0.043 (6.2)	0.56	0.06
30.1 (86.2)	0.505 (73.3)	0.043 (6.2)	0.67	0.06
29.9 (85.8)	0.633 (91.8)	0.043 (6.2)	0.78	0.06
30.0 (86.0)	0.770 (111.7)	0.000 (0.0)	1.00	0.00
45.0 (113.0)	0.000 (0.0)	0.000 (0.0)	0.00	0.00
45.0 (113.0)	0.211 (30.6)	0.043 (6.2)	0.28	0.04
45.1 (113.2)	0.481 (69.8)	0.043 (6.2)	0.50	0.04
45.1 (113.2)	0.748 (108.5)	0.043 (6.2)	0.66	0.04
45.1 (113.2)	0.921 (133.6)	0.043 (6.2)	0.76	0.04
45.0 (113.0)	1.160 (168.3)	0.000 (0.0)	1.00	0.00
60.0 (140.0)	0.000 (0.0)	0.000 (0.0)	0.00	0.00
60.0 (140.0)	0.333 (48.3)	0.043 (6.2)	0.29	0.03
60.1 (140.2)	0.553 (80.2)	0.043 (6.2)	0.45	0.04
60.0 (140.0)	0.815 (118.2)	0.043 (6.2)	0.56	0.04
60.0 (140.0)	1.169 (169.6)	0.043 (6.2)	0.71	0.04
60.1 (140.2)	1.312 (190.3)	0.043 (6.2)	0.76	0.05
60.0 (140.0)	1.681 (243.9)	0.000 (0.0)	1.00	0.00
89.9 (193.8)	0.000 (0.0)	0.000 (0.0)	0.00	0.00
89.9 (193.8)	0.400 (58.0)	0.043 (6.2)	0.24	0.04
90.0 (194.0)	0.979 (142.0)	0.043 (6.2)	0.45	0.04
90.0 (194.0)	1.332 (193.2)	0.043 (6.2)	0.56	0.04
89.9 (193.8)	1.631 (236.6)	0.043 (6.2)	0.64	0.05
90.0 (194.0)	1.924 (279.1)	0.043 (6.2)	0.74	0.05
89.9 (193.8)	2.377 (344.8)	0.043 (6.2)	0.80	0.05
90.0 (194.0)	3.237 (469.6)	0.000 (0.0)	1.00	0.00
110.0 (230.0)	0.000 (0.0)	0.000 (0.0)	0.00	0.00
109.9 (229.8)	0.514 (74.6)	0.043 (6.2)	0.22	0.04
110.1 (230.2)	1.154 (167.4)	0.043 (6.2)	0.41	0.04
110.1 (230.2)	1.769 (256.6)	0.043 (6.2)	0.54	0.05

Table 7. Experimental R410a/POE VLE data

Temperature, °C (°F)	Pressure, MPa (psia)	Error in Pressure, +/- MPa (psia)	Mole Fraction of R410a in Liquid	Error in Liquid Mole Fraction R410a, +/-
30.0 (86.0)	0.000 (0.0)	0.000 (0.0)	0.00	0.00
30.2 (86.4)	0.348 (50.5)	0.047 (6.8)	0.22	0.03
30.1 (86.2)	0.759 (110.1)	0.047 (6.8)	0.41	0.04
30.0 (86.0)	0.976 (141.6)	0.047 (6.8)	0.56	0.04
30.0 (86.0)	1.254 (181.9)	0.047 (6.8)	0.67	0.04
29.9 (85.8)	1.650 (239.4)	0.047 (6.8)	0.78	0.04
30.0 (86.0)	1.883 (273.2)	0.000 (0.0)	1.00	0.00
45.0 (113.0)	0.000 (0.0)	0.000 (0.0)	0.00	0.00
45.0 (113.0)	0.416 (60.4)	0.047 (6.8)	0.20	0.04
45.0 (113.0)	0.865 (125.5)	0.047 (6.8)	0.38	0.04
45.0 (113.0)	1.421 (206.2)	0.047 (6.8)	0.53	0.05
44.9 (112.8)	1.904 (276.2)	0.047 (6.8)	0.65	0.05
44.9 (112.8)	2.454 (356.0)	0.047 (6.8)	0.78	0.05
45.0 (113.0)	2.725 (395.3)	0.000 (0.0)	1.00	0.00
60.0 (140.0)	0.000 (0.0)	0.000 (0.0)	0.00	0.00
60.0 (140.0)	0.466 (67.6)	0.047 (6.8)	0.16	0.04
60.1 (140.2)	0.913 (132.5)	0.047 (6.8)	0.30	0.05
60.0 (140.0)	1.417 (205.6)	0.047 (6.8)	0.44	0.06
60.1 (140.2)	1.906 (276.5)	0.047 (6.8)	0.53	0.06
60.1 (140.2)	2.262 (328.2)	0.047 (6.8)	0.58	0.06
60.1 (140.2)	2.725 (395.3)	0.047 (6.8)	0.66	0.06
60.0 (140.0)	3.833 (556.1)	0.000 (0.0)	1.00	0.00
75.0 (167.0)	0.000 (0.0)	0.000 (0.0)	0.00	0.00
75.0 (167.0)	0.456 (66.2)	0.047 (6.8)	0.16	0.04
75.1 (167.2)	0.885 (128.4)	0.047 (6.8)	0.26	0.06
75.1 (167.2)	1.366 (198.2)	0.047 (6.8)	0.34	0.06
75.1 (167.2)	1.785 (259.0)	0.047 (6.8)	0.40	0.07
75.0 (167.0)	2.277 (330.3)	0.047 (6.8)	0.49	0.07

References

1. ASHRAE. 2009. 2009. *ASHRAE handbook – fundamentals*. American Society of Heating, Refrigerating and Air-Conditioning Engineers.
2. Burton, C., A. M. Jacobi, and S. S. Mehendale. 1999. Vapor-liquid equilibrium for R-32 and R-410A mixed with a polyol ester: Non-ideality and local composition modeling. *International Journal of Refrigeration* 22 : 458,459-471.
3. EIA. (2023, April 20). Electricity explained - Use of electricity. https://www.eia.gov/energyexplained/electricity/use-of-electricity.php
4. Engineering Toolbox. Recommended indoor temperature versus outdoor temperature. [cited 12/28 2012]. Available from http://www.engineeringtoolbox.com/indoor-outdoor-temperature-d_1018.html.
5. Huber, Marcia L., Cynthia D. Holcomb, Stephanie L. Outcalt, and J. Richard Elliott. 2000. Vapor-liquid equilibria for an R134a/Lubricant mixture: Measurements and equation-of-state modeling. *ASHRAE Transactions* 106 : 768,789-773.
6. Huber, ML, EW Lemmon, and DG Friend. 2001. Modeling bubble points of mixtures of hydrofluorocarbon refrigerants and polyol ester lubricants. *Fluid Phase Equilibria* 194 (MAR 30): 511-9.
7. Martz, W. L., C. M. Burton, and A. M. Jacobi. 1996a. Local composition modeling of the thermodynamic properties of refrigerant and oil mixtures. *International Journal of Refrigeration* 19 (1): 25,26-33.
8. Martz, William L., Christopher Burton M., and Anthony M. Jacobi. 1996b. Vapor-liquid equilibria for R-22, R-134a, R-125, and R-32/125 with a polyol ester lubricant: Measurements and departure from ideality. *ASHRAE Transactions* 102 (1): 367,368-374.
9. Nu-Calgon. Emkarate RL 100S MSDS. https://www.nucalgon.com/sds/. Accessed 2012.
10. Scaringe, R. P., D. Sykes, T. Amundsen, and P. E. Yelvington. Combination electric-powered and waste-heat powered vapor-compression cooling system. Patent 13/153,644, issued 2011.
11. Takaishi, Yoshinori, Hajime Nakagawa, and Kosei Oguchi. 1993. Measurements of the vapor pressure for the solutions of HFC-134a and ester based oil. *Transactions of the Japanese Association of Refrigeration* 10 (1): 67,68-73.
12. Tesser, R., E. Musso, M. Di Serio, G. Basile, and E. Santacesaria. 1999. Description of the vapor-liquid equilibrium in binary refrigerant/lubrication oil system. *Journal of Fluorine Chemistry* 99 (1): 29,30-36.
13. U.S. Energy Information Administration. 2012. *Annual energy outlook 2012*. Washington, DC: U.S. Energy Information Administration.

14. Wahlström, Åsa, and Lennart Vamling. 1999. Solubility of HFC32, HFC125, HFC134a, HFC143a, and HFC152a in a pentaerythritol tetrapentanoate ester. *Journal of Chemical & Engineering Data* 44 (4) (07/01; 2012/02): 823-8.

15. Wahlström, Åsa, and Lennart Vamling. 1997. The solubility of HFC125, HFC134a, HFC143a and HFC152a in n-eicosane, n-hexadecane, n-tridecane and 2, 6, 10, 14-tetramethylpentadecane. *The Canadian Journal of Chemical Engineering* 75 (3): 544-50.

16. Yokozeki, A. 2005. Theoretical performances of various refrigerant–absorbent pairs in a vapor-absorption refrigeration cycle by the use of equations of state. *Applied Energy* 80: 384-399.

399f Spectroscopic Monitoring of Dissolved Carbon Dioxide for Ccus

Richa Sharma, T. S. Ramakrishnan

Decarbonization of the energy industry is an environment driven requirement. A sustained expansion and deployment of large-scale carbon capture, utilization and storage (CCUS), is needed to address this challenge, at least partly. Still, several technical barriers remain to be overcome for CCUS technologies to have commercial gains in field. CO_2 Geological storage (CGS) refers to long-term storage in suitable geological sites. Since CGS effectiveness is contingent upon developing miscibility through partitioning into water, this work is motivated by the need to reliably monitor dissolved CO2 under reservoir conditions of temperature and pressure. Although methods such as attenuated total reflection (ATR) have been proposed, a universal and reliable sensor for field application is not available. We approach this question by focusing on the use of spectroscopic techniques, particularly Raman Spectroscopy, in order to identify multiple gas components in aqueous. The proposed system complements the prevalent transmission infrared (IR) measurements and addresses some of their limitations. It can support continuous monitoring and quantitative identification of CO2 over a broad range of ambient conditions.We propose Raman spectroscopy-based sensor system for translating molecular interactions into molar concentrations. We propose the utility and feasibility of field deployment of our sensing system in aquifer reservoir CO2 injection projects. We also review a possible use-cases of monitoring system.

466a Effective Combination of Reactors for Biomass Feedstock for High-Efficiency Power Generation from Manure

Yukihiko Matsumura, Mengli Zhang, Takayuki Ichikawa, Wookyung Kim, Yutaka Nakashimada, Keiya Nishida

Utilization of manure for power generation is important for renewable energy supply. However, conventional methane fermentation has limitation due to ammonia inhibition. In this study, combination of reactors for biomass feedstock achieved improved power generation. Hydrothermal pretreatment allowed ammonia generation as well as improved fermentability. The treated feedstock is then fed to high-temperature biomethanation where ammonia is removed into gas phase and adsorbed onto ammonia adsorber. The ammonia is then electrolyzed with a part of produced electricity into nitrogen and hydrogen while biogas is used as engine fuel for power generation. Hydrogen is fed to the engine to enhance the efficiency. Four elementary topics for this system were studied separately, and process evaluation was made based on the finding. The mass and energy flow is shown together with the experimental results of each study. Combination of reactors biomass feedstock can achieve effective process that cannot be achieved by single reactors. This research was performed by the Environment Research and Technology Development Fund (JPMEERF20203001) of the Environmental Restoration and Conservation Agency of Japan, and three year project finished last March. This is the presentation of the total achievement through the three year project.

466b Ash Adhesion on a Metal Mesh during Combustion of Bamboo Powder with Inorganic Additives

Ken-ichiro Tanoue, Nao Umehara, Masayuki Fujita Masayuki Fujita, Yoshihiko Ninomiya Yoshihiko Ninomiya

If the high melting point oxides added to the bamboo powder, the ash adhesion on the metal mesh during combustion was counterproductive in some cases, but effective in others. For SiO 2 addition, although the shrinkage temperature (SHT) and part fusion temperature (PFT) were higher than that for only bamboo combustion due to the formed KAlSiO 4 whose melting temperature was 1750 °C, the amount of ash on the metal mesh decreased in some cases or increased in others with the additive mass percentage of SiO 2 . Because there was the ratio distribution of K 2 O to SiO 2 in the bamboo ash which decided the minimum melting point, the
amount of adhered ash had a large variation. For CaO addition, although the SHT and PFT were higher than that for only bamboo combustion, whether ash deposition was enhanced or reduced could be determined by the distribution of char present in the combusted ash, resulting in the adhesion on the Inconel mesh. Because the ash for only bamboo combustion had 13.6% char, not only CaCO 3 but also Ca(OH) 2 with a lower melting point than that of CaO were formed during combustion. On the other hand, for MgO powder addition, as the ash formed that could be adhered by MgO particles was fluffy, the adhesion mass decreased with the additive mass percentage. Then all fusion temperatures of SHT, PFT, and the full fusion temperature (FFT) for MgO addition were higher than that for only bamboo

466c Exhaust Gas Characteristics and NOx Formation Mechanism on Spark Ignition Engine Using By Bio-Syngas

Hossain Zakir, Naoya Kishi, Yukihiko Okumura

In this study, three types of syngas fuel (CO: 25 vol% + H_2: 75 vol%, CO: 50 vol% + H_2: 50 vol%, CO: 75 vol% + H_2: 25 vol%) were supplied to a small gas engine, and the optimum fuel composition, maximum mixing ratio of syngas and exhaust gas characteristics (i.e, NOx, CO emission etc.) were investigated. As a result, (1) the maximum mixing ratio of syngas increased as the H_2 ratio in syngas composition increased, (2) the NOx emission increased and the CO emission decreased as the syngas supplied.

466d Bio-Derived Gamma-Butyrolactone Production from Succinic Acid with Bio-Derived Solvents

Nanami Hirano, Ryu Ukawa-Sato, Ryuta X. Suzuki, Joerg Freitag, Hiroshi Hori, Masafumi Kono, Chihiro Fushimi

1. Introduction

Petrochemical resources are widely used in many industries in modern society. However, their use has caused environmental problems, such as global warming and pollution. As a result, many renewable resources are being studied as alternatives. The use of petrochemical resources can be divided into two categories: energy use and conversion to chemicals. Alternative resources for energy use include hydropower, wind power, solar power, biomass, and many others. In contrast, biomass is the only primary resource that can be used for chemical conversion. Therefore, bio-based chemicals are attracting attention. Among them, gamma-butyrolactone (GBL) is an important chemical compound with a wide range of applications in various fields, such as medicine, agriculture, and industry[1)]. Hydrogenation using bio-succinic acid (SA) as a raw material is attracting attention as an alternative to maleic acid derived from fossil resources, which is currently used as a raw material for GBL[2)]. Since bio-succinic acid TTis an important platform chemical and is already produced on a commercial scale, this study aims to produce GBL from SA. Most of the previous studies on the SA–GBL reaction used 1,4–dioxane as a solvent[3, 4)]. However, since 1,4–dioxane is a hazardous chemical with carcinogenic, mutagenic, and reproductive toxicity effects, its use in the production of GBL from SA is not sustainable. To achieve sustainable production, research using bio-based solvents for the generation of bio-gamma-butyrolactone from succinic acid is necessary. However, there is currently no literature available on this topic. Based on the above, in this study, we aimed to produce GBL from SA using not only raw materials, but also solvents derived from biomass. This research can contribute not only to the sustainable production of GBL but also to the production of biochemicals commonly produced by hydrogenation in industry.

2. Experimental

In this study, two types of experiments were conducted to produce GBL, Catalytic Transfer Hydrogenation (CTH) and direct hydrogenation.

The Materials were 5wt% Ru/C catalyst[*1], 5wt% Pd/Al_2O_3 catalyst[*2], SA (>99.5wt%)[*1], Solvents[*1,2], isopropanol (IPA) and tetrahydrofuran (THF), hydrogen gas (>99.999%)[*3], argon gas (>99.999%)[*3].

(*1 Fujifilm Wako Pure Chemicals Corporation, *2 Tokyo Kasei Kogyo, *3 Taiyo Nippon Sanso Gas & Welding Co.)

The reactor utilized in the experiments was a 100 mL stainless steel batch autoclave (TVS-N2 type, 100 mL, Taiatsu Techno. Corporation, Osaka, Japan).

2.1 Catalyst reduction

Ru/C catalyst or Pd/Al_2O_3 catalyst was reduced for 2.5 h at 240 °C under a hydrogen gas atmosphere.

2.2 Catalytic hydrogenation

2.2.1 Catalytic Transfer Hydrogenation (CTH)

Isopropanol (IPA) was used as the solvent and hydrogen donor in the CTH experiments. After the catalyst reduction (c.f., 2.1), the temperature was gradually lowered to room temperature, and the hydrogen gas was purged using argon gas. The hydrogenation experiments were then performed started by adding SA and IPA to the reactor followed by increasing the temperature to reaction temperature.

2.2.2 Direct hydrogenation

In the direct hydrogenation experiments, hydrogen gas was supplied externally. After the catalyst reduction (c.f., 2.1), the temperature was gradually lowered to room temperature, and the hydrogen gas was purged using argon gas. The direct hydrogenation experiments were then performed started by adding SA and the solvent to the reactor. In the direct hydrogenation process, the reactor was pressurized with hydrogen gas before the start of the temperature increase.

2.3 Analysis

After the catalytic hydrogenation reaction, the temperature was gradually lowered to room temperature, and the catalyst was filtered. Finally, the product solution was analyzed by GC-FID (Shimadzu, GC-2014).

3. Results & Discussion

In the CTH experiments conducted at 150, 180, and 200 °C with IPA, the maximum SA conversion of 98.1 mol% was achieved at 200 °C and the maximum GBL yield of 6.1mol% was achieved at 180 °C (at 79.7% SA conversion). In order to reduce the amount of SA ester produced, which was responsible for the low GBL yield, water was added. This resulted in a decrease of SA ester yield to 31.9mol%. Subsequently, the amount of water was increased to 5 and 10wt%, resulting in a decrease of SA ester yield to 24.7 and 23.0mol%, respectively. However, increasing the amount of water to 10wt% also resulted in a decrease of both SA conversion (67.4mol%) and GBL yield (5.1mol%). This indicates that while water addition greatly suppressed dehydration, it also made it difficult to suppress esterification alone.

On the other hand, in the direct hydrogenation experiment, the SA conversion and GBL yield were 57.7mol% and 9.5mol%, respectively, at a hydrogen pressure of 2.0 MPa before temperature increase and a reaction temperature of 200 °C for 4 h, using the Ru/C catalyst. However, at a reaction time of 2 h, the SA conversion and GBL yield dropped to 20.0 and 8.0mol%, respectively. At a hydrogen pressure of 1.5 MPa before temperature increase, the SA conversion and GBL yield dropped further to 17.9 and 4.7mol%, respectively. This indicates that the reaction time and hydrogen pressure have a significant effect on the conversion ratio. When the Pd/Al_2O_3 catalyst was used, the SA conversion decreased to 43.4mol%, while the GBL selectivity increased to 25.6mol%. This suggests that the Pd/Al_2O_3 catalyst forms fewer byproducts and that further improvement in GBL yield can be expected by positively influencing the SA conversion by increasing the reaction temperature and reaction time.

4. Conclusion

In this particular study, a series of experiments were carried out to produce GBL from bio-SA using a biomass-derived solvent and hydrogen. Firstly, CTH experiments were conducted using IPA at different reaction temperatures, and the maximum yield of GBL achieved was 6.1mol% at a reaction temperature of 180 °C and 79.7% SA conversion. This result indicates that SA underwent fast esterification during the CTH process.

Additionally, direct hydrogenation experiments were performed using THF as a solvent. In an experiment with THF utilizing Ru/C as catalyst, ~~a~~ with hydrogen pressure of 2.0 MPa before temperature increase, reaction temperature of 200 °C, and reaction time 4 h, the SA conversion reached 57.7mol%, and the corresponding GBL yield was 9.5mol%. Moreover, when the Pd/Al_2O_3 catalyst with THF as solvent was utilized under the same conditions, a high GBL selectivity of 25.6mol% was obtained. Those results indicate that further GBL yield improvement can be achieved by optimizing the reaction time and temperature.

This study about two catalytic hydrogenation approaches (CTH and direct hydrogenation) for GBL formation from bio-SA using bio-based solvents show the multiple tasks when it comes to sustainable biomass feedstock engineering, but it gives hope too, that finally we can do the transformation to a sustainable future.

References

1) Hong, U. G. *et al.*, *J. Ind. Eng. Chem.*, **17**, 316–320 (2011).

2) C. Zhang *et al.*, *Catal. Today*, **276**, 55–61 (2016).

3) Hong, U.G. *et al.*, *Catal. Letters*, **141(2)**, 332–338 (2011).

4) Shao, Z. *et al.*, *Ind. Eng. Chem. Res.*, **53**, 9638–9645 (2014).

428a Indicator-Based Supply Stability Analysis of Pharmaceutical Excipients: Method and Application

Kosuke Nemoto, Yusuke Hayashi, Hirokazu Sugiyama

Pharmaceuticals contain multiple excipients to enhance their functionality and manufacturability, e.g., corn starch as diluent, titanium oxide as colorant, magnesium stearate as lubricant. In R&D of pharmaceuticals, experts select excipients on an empirical basis[1] considering product efficacy as the major criteria. In order to stabilize the global supply chain issues in the pharmaceutical industry[2], systematic consideration of supply aspects is needed. Several studies have assessed the supply of drug materials, e.g., model-based minimization of cost/environmental impacts of pharmaceutical materials such as general pharmaceuticals[3] and vaccines[4]. However, a common approach is yet to be established for the evaluation of supply stability for pharmaceuticals.

This work presents an indicator-based method for supply stability analysis of pharmaceutical excipients and its application to commercial pharmaceutical products in Japan. The method quantifies the potential supply risk of the excipient compounds in a product mainly from the resource distribution perspective. The method defines the five steps: (i) list the excipient compounds of the drug products in scope, (ii) identify raw materials of each compound and collect data, (iii) calculate supply stability indicator at the element level, (iv) extend the indicator calculation to the compound level, and (v) evaluate the supply stability of the product. Herfindahl-Hirschman Index (HHI[5,6]) provides the basis of the supply stability indicator, which quantifies how evenly the materials are distributed in the world. We extended HHI to consider governance of the countries where the element is originated, and also newly defined the compound-level indicator. As a case study, the method was applied to all products (n > 14,000) in the Japanese pharmaceutical market. The analysis revealed potential risks of several specific excipients. It also elucidated that the supply stability of pharmaceuticals varied depending on the dosage forms or target diseases. These finding would be useful especially in the early phase of drug development (e.g., excipient selection or even dosage form determination), which can contribute to the sustainable supply of pharmaceuticals.

References

1. Bejarano, A., Hewa Nadungodage, C., Wang, F., Catlin, A. C. & Hoag, S. W. Decision Support for Excipient Risk Assessment in Pharmaceutical Manufacturing. AAPS PharmSciTech. 2019;20:1-16.

2. Sayarshad, H. R. Personal protective equipment market coordination using subsidy. Sustain Cities Soc. 2022;85:104044.

3. Goodarzian, F., Hosseini-Nasab, H. & Fakhrzad, M. B. A multi-objective sustainable medicine supply chain network design using a novel hybrid multi-objective metaheuristic algorithm. International Journal of Engineering, Transactions A: Basics. 2020;33:1986-1995.

4. Chowdhury, N. R., Ahmed, M., Mahmud, P., Paul, S. K. & Liza, S. A. Modeling a sustainable vaccine supply chain for a healthcare system. J Clean Prod. 2022;370:133423.

5. Herfindahl, O. C. CONCENTRATION IN THE STEEL INDUSTRY. 1950;p.9.

6. Hirschman, A. O. The Paternity of an Index. 1964;54:761.

428b Ecological Network Analysis and Information Theory to Evaluate the Performance of Circular Economy Designs

Farshid Nazemi, Brian D. Fath, Bhavik Bakshi

Circular Economy (CE) is an alternative business model to the conventional take-make-waste Linear Economy (LE) model, where products are reused, redistributed, refurbished, remanufactured, and recycled at their End of Life (EoL). Despite the huge efforts across the globe for moving from LE to CE designs, there is a lack of proper metrics that capture and evaluate the circularity, resilience, and robustness of different CE designs. In the present work, we develop a set of metrics based on Ecological Network Analysis (ENA) and Information Theory (IT) and apply these metrics on CE designs. We use eight metrics based on ENA and IT for trophic chains which have high potential to be used for a meaningful evaluation of the performance of CE systems, i.e., (1) Total System Throughput (TST), (2) Mean Trophic Level (MTL), (3) Detritivory : Herbivory ratio (D : H) (4) Ascendency (A), (5) Reserve (φ), (6) Development Capacity (DC), (7) Relative Ascendency (α), and (8) Robustness (R). In addition, we propose guidelines and/or modifications to develop metrics with meaningful insights for evaluating CE performance. We demonstrate the usefulness of these metrics in two "toy problem" examples. In the first toy problem, we develop one LE and four CE scenarios indicating different recovery strategies in CE systems, i.e., Reuse, Redistribute, Refurbish (or Remanufacture) and Recycle. Qualitatively speaking, in the case of equal recovery efficiency, CE strategies with inner (tighter) loops (e.g., Reuse) are more favorable than outer loops (e.g., recycle). Thus, in this toy problem, we study whether ENA-based metrics can be used to quantitatively capture the differences between circular strategies through implementing loops with different tightness in the system. We assume consumer's demand in all scenarios are the same, and we set a fixed 50% recovery efficiency for all CE scenarios (50% of the waste is recovered after use phase and is circulated). Moreover, we study whether metrics based on IT can demonstrate the difference between CE systems over LE systems with respect to the resilience and robustness of the system. Results indicate that all CE designs have higher resilience and robustness compared to the LE scenario, which indicates that by applying circular strategies, we can develop a more robust economy that is resilient to unprecedented perturbations. In addition, in CE scenarios, MTL and D:H ratio is the highest for the Reuse scenario, and it decreases as we implement outer cycles for recovery. This indicates that the Reuse scenario is less dependent on the primary resources and lower trophic levels compared to other CE scenarios, which shows advantage of the Reuse scenario over other CE scenarios in terms of resource consumption. Conversely, TST decreases as outer cycles are implemented in the CE design with the Reuse scenario having the lowest TST. This indicates that less system activity is needed in the reuse scenario in order to meet consumer's demand. This could indirectly indicate less resource consumption and environmental impacts since the scale of the industrial activity is lower. In the second toy problem, we study the importance of diversity in available upstream and EoL options for feedstock provision and waste handling and recovery of a product. More specifically, we study the possible tradeoffs between circularity efficiency and resilience in CE systems. In one scenario, we assume that there is only one upstream process and one EoL option that are highly efficient (we define efficiency as the amount of resource needed to produce one unit of flow of final product). In the second scenario, we introduce an alternative upstream and EoL options in addition to those that are available in the first scenario, which are less efficient compared to the already available options for upstream and EoL processes. In other words, in this scenario, there are two upstream processes for feedstock provision and two EoL processes for waste handling and treatment, and the total efficiency of this system is lower compared to the first scenario. However, despite the lower efficiency of the second scenario, this scenario demonstrates a higher resilience (φ) and robustness (R) due to the implementation of a diverse set of EoL and upstream processes. This indicates that efforts toward designing CE systems should not be only focused on implementing and developing emerging technologies with highest efficiency. Instead, implementation of a diverse set of emerging technologies at different life cycle stages of a CE can help build robust and resilient supply chain solutions. Finally, we apply the metrics on polyethylene terephthalate (PET) supply chains in order to first, demonstrate the applicability of the metrics for a real-world system, and secondly, to evaluate the circularity and resilience of PET supply chains and guide its development toward a sustainable and resilient circular economy.

428c Design of Resilient Supply Chain Networks for Pharmaceutical Products

Arun Sankar Eenhakkattu Mana, Smrithi Prakash, Venkat Venkatasubramanian

The disruption of global supply chains during the pandemic, and its continuing ill effects, has raised important questions about resilience in their design. For over four decades, the design and architecture of supply chains were hyper-focused on their efficient performance and cost minimization more than their robustness and resilience to attacks and failures. However, as the pandemic demonstrated, paying attention to robustness and resilience turns out to be important in the long run. A survey by Ernst & Young indicates that only 2% of the companies were prepared to respond to the covid-19 pandemic [1]. Another survey by Infosys Consulting reports that only 13% companies survived the disruptions to their SCs [2]. Towards that goal, we have developed a conceptual framework that helps us model and analyze efficiency-robustness trade-offs in network design [3,4]. Typical factors considered in the design are efficiency, robustness, uncertainty in supply and demand, unsatisfied demand [5] etc. Additional constrains on shelf life arise for perishable materials such as pharmaceutical products [5]. The elements contributing to resiliency are preparedness, response, and recovery [6]. In this work, we will present an overview of this conceptual framework and discuss the trade-offs we need to make among efficiency, robustness, and cost for the long-term sustainability of such systems for model network topologies such as star, circle, hub, small-world, scale-free, and von Neumann networks.

References

[1] 'https://www.ey.com/en_us/supply-chain/how-covid-19-impacted-supply-chains-and-what-comes-next,' [Online]. Available: https://www.ey.com/en_us/supply-chain/how-covid-19-impacted-supply-chains-and-what-comes-next.

[2] 'https://www.infosysconsultinginsights.com/wp-content/uploads/2020/08/covid-19-supply-chain-impact-survey_infosys-consulting.pdf,' [Online]. Available: https://www.infosysconsultinginsights.com/wp-content/uploads/2020/08/covid-19-supply-chain-impact-survey_infosys-consulting.pdf.

[3] Shukla, A., Agarwal. V.L. and Venkatasubramanian, V., "Optimizing Efficiency-Robustness Trade-offs in Supply Chain Design Under Uncertainty Due to Disruptions", International Journal of Physical Distribution & Logistics Management, 41(6), pp. 623-646, 2011.

[4] V. Venkatasubramanian, S. Katare, P. R. Patkar and F.-p. Mu, 'Spontaneous emergence of complex optimal networks through evolutionary adaptation,' *Computers & Chemical Engineering,* vol. 28, no. 9, pp. 1789-1798, 2004

[5] B. Zahiri, P. Jula and R. Tavakkoli-Moghaddam, 'Design of a pharmaceutical supply chain network under uncertainty considering perishability and substitutability of products,' *Information Sciences,* vol. 423, pp. 257-283, 2018.

[6] P. Chowdhury, P. Chowdhury and P. M. A. Chowdhury, 'COVID-19 pandemic related supply chain studies: A systematic review,' *Transportation Research Part E: Logistics and Transportation Review,* vol. 148, no. 102271, 2021.

428d An Agent-Based Simulation Model Towards Mitigating Domino Effects of Maritime Disruptions in LNG Supply Chains

Venkataramanan Swaminathan, Rajagopalan Srinivasan

Logistics is central in ensuring the uninterrupted functioning of supply chains (SC), enabling the transport of commodities from the suppliers to the various end customers. Maritime trade has evolved to serve as the backbone of logistics allowing for a globalized market through augmented SC linkages and timely delivery of products over long distances. Marine trade handles approximately 80% of global merchandise by volume (United Nations Conference on Trade and Development, 2021) owing to its sheer cargo-handling capacity and fleet size, which is the highest compared to any mode of transport utilized for the shipment of goods. Though it offers better interconnectivity among different SC entities and promotes economic interdependency, vulnerabilities do exist in maritime trade that can eventually manifest to disrupt normal SC operations. While perturbations can occur at any segment of the SC (for example, supply uncertainty due to political crisis at the export facility and operational and maintenance issues at the import facility), the transportation segment in itself has faced critical disruptions sending large supply shocks, the consequence (or impact) of which propagates as domino effects downstream across the SC causing significant economic downfalls along with damaging organizational reputations. **Hence, it becomes essential to study the effect of supply shocks induced in SCs due to maritime logistic disruptions.**

Maritime trade is susceptible to accidents or delays induced by turbulent weather conditions, piracy, terror attacks, and other human errors. It is understood that while weather-induced delays temporarily hamper normal operations, disruptions due to accidents or piracy can either result in delay or complete loss of shipment along with associated safety concerns. There have been several instances of marine disruptions, with the most recent being induced due to the CoVID-19 pandemic. In October 2021, container ships were seen maneuvering a few miles offshore in holding mode, unable to offload shipment at the stipulated time in Port of Los Angeles, California. This SC gridlock was attributed to a sudden surge in demand; however, an inadequate workforce at the port resulted in almost 200,000 shipping containers stationed off-coast over a span of a week (California Gridlock, October 2021). Additionally, global marine trade has a number of critical chokepoints, such as the Strait of Panama, Suez, and Malacca (US EIA Maritime Chokepoints, 2017) that can potentially disrupt supply chains. Recent examples of such disruptions include the Suez Canal Crisis in early 2021 due to Ever-Given blocking the narrow strait for about a week, making about 400 other shipments (oil tankers, gas carriers, container ships) to either wait in queue for clearance, or reroute around Cape of Good Hope taking up a more circuitous route. This compounded global SC woes already constrained due to the pandemic and resulted in a loss of around $95 million in revenue to the canal authorities, apart from inflicting about $54 billion in losses to global trade (Suez Canal nightmare, 2021). Further, cyclone Shaheen impacted the export of oil and gas from Oman and Qatar during its passage along the strait of Hormuz (Saadi and Anupam, 2021). The aforementioned real-world examples motivate us to develop mitigative strategies that can enhance SC resilience. Specifically, we consider LNG SC, given its potential to serve in global energy transitions being a low-carbon fossil. Further, with almost 73% of global LNG demand arising from the Asia-Pacific region (GIIGNL Annual Report, 2022), it is of greater importance to enhance resilience and improve energy security for Asian regions having higher import dependencies. Hence, **the objective of this work is to study and quantify the impact of unplanned events (disruptions) in maritime transport and the associated shock propagation across an LNG receiving terminal. This further encourages to design appropriate countermeasures that help alleviate its adverse consequences.**

From the perspective of an LNG importer, the LNG supply chain (LNG SC) comprises NG liquefaction, LNG shipment, storage & regasification, and distribution. Here, we develop an agent-based dynamic simulation model of the LNG supply chain that emulates the different participating entities like LNG suppliers (exporter), LNG carriers, receiving terminals (importer), end-customers (for example, regasified LNG, retail LNG, and reloading LNG). All entities and their internal departments are modeled as autonomous actors that are capable of making their own decisions. Each actor coordinates with one another and seeks to maximize their individualistic objectives. Several hazards that can potentially hamper normal operations were identified using the HAZOP technique and appropriate recovery options were proposed to be utilized during disruptive events. A number of countermeasures involving strategic safeguards, tactical planning, and operational policy-based decisions were considered depending on the severity and duration of the disruptive event. The dynamics of disruptions were initiated as stochastic factors affecting the normal SC dynamics through the activities of the respective agents, and the associated cascading impacts and emergent phenomena were analyzed.

In order to assess the impacts of disruptions, a case study involving maritime transportation delays of LNG carriers in transit was conducted for different delay magnitudes and durations. Due to the diverse operational decisions and management, an annual simulation horizon was considered with hourly operational time steps. In this case, the delay was introduced around the mean carrier transit duration, i.e., with a positive standard deviation of +3.5 days (high severity) and a duration lasting maximum for a month. It was observed to have significant impacts on the operations of the receiving terminal, specifically facing demurrage penalties due to increased waiting time of LNG carriers that have arrived for unloading. Additionally, delays in carrier arrival resulted in frequent plant shutdowns due to inventory shortfalls as a result of delayed replenishments. This eventually leads to reduced product send-outs, negatively impacting customer service quality. It is noted that a combination of the aforementioned domino effects amounted to a $350 million loss for the receiving terminal. We implemented feedback-control-based recovery actions (e.g., LNG emergency sourcing or spot procurement) that can minimize the losses due to the cascading effects of maritime disruption. Through this, we were able to achieve a recovery of around 70% due to the reduction in plant shutdown duration without the requirement of any policy intervention. The developed agent-based model of the LNG supply chain has been implemented in AnyLogic, a business simulation tool with a realistic case study involving a real-world operational facility of a specific company based in India. **We believe in its capability to serve as a robust decision support tool that helps evaluate and quantify the impacts of a variety of disruptions and also analyze the effectiveness of adopted countermeasures towards enhancing supply chain resilience.**

Keywords: LNG Supply Chain, maritime disruptions, domino effects, countermeasures, enhancing resilience, decision support tool

References

1. CNN, J.K. and B.M., n.d. See California's shipping gridlock in motion [WWW Document]. CNN. URL https://www.cnn.com/2021/10/23/business/20211023-california-ports-animation/index.html.
2. GIIGNL 2022 Annual Report -, URL https://giignl.org/document/giignl-2022-annual-report/.
3. Saadi, D., Anupam, A., 2021. Oman "almost back to normal" as Sohar port reopens after cyclone [WWW Document]. URL https://www.spglobal.com/commodityinsights/en/market-insights/latest-news/oil/100421-oman-almost-back-to-normal-as-sohar-port-reopens-after-cyclone.
4. The three lessons we learnt from the Suez Canal nightmare [WWW Document], 2021. URL https://supplychaindigital.com/logistics/three-lessons-we-learnt-suez-canal-nightmare.

5. United Nations Conference on Trade and Development, 2021. COVID-19 and Maritime Transport: Impact and Responses: Transport and Trade Facilitation Series, No. 15, Transport and Trade Facilitation Series. United Nations. https://doi.org/10.18356/9789216040123
6. US EIA: Maritime Chokepoints Are Critical To Global Energy Security [WWW Document], n.d. URL https://www.marineinsight.com/shipping-news/us-eia-maritime-chokepoints-critical-global-energy-security/.

428e Midstream Infrastructure Cybersecurity, Physical Security and Safety

Helen Lou, Tianxing Cai

The pipeline infrastructure, storage facilities and terminals are critical components of the energy infrastructure. The industry is increasingly relying on integrating information and communication technologies into operational technology for data analytics, automation, and remote operation. The advancements, although significantly improve operational efficiency and profitability, are prone to cyber-attacks.

It is clear both physical security, safety and cybersecurity are needed to ensure the resilience of the energy infrastructure. The relationships between safety, physical security, and cybersecurity are closely intertwined. Cyber-attacks can impact these facilities' safety and physical security, while safety and physical security measures can protect the facility against cyber threats. This presentation will discuss viable approaches to enhance physical security, safety and cybersecurity from different aspects. These efforts will assist the industry to identify, protect against, and respond to cyber threats and remediate undesired consequences of these attacks in a timely matter.

428f *Chemical Sector Risk Management Agency Resources for Enhanced Supply Chain Resilience and Security*

Ashley Pennington, Cheryl Louck, Janine Mason, Jacob Mehl

The US Department of Homeland Security's Cybersecurity and Infrastructure Security Agency (CISA) functions as the Sector Risk Management Agency (SRMA) for the chemical sector. The Chemical sector is one of the 16 critical infrastructure sectors(1) whose assets, systems, and networks, whether physical or virtual, are considered so vital to the United States that their incapacitation or destruction would have a debilitating effect on security, national economic security, national public health or safety, or any combination thereof. In order to have a resilient and sustainable supply chain, it is imperative for the critical infrastructure sectors to function at their full capability. Across CISA, teams work to support functions of the Chemical SRMA to ensure safe, secure, resilient, and sustainable functioning of the chemical supply chain. The impacts of the chemical supply chain span far beyond the chemical sector and influences the supply chain or function of all critical infrastructure sectors. The team will present an overview of the Chemical SRMA and its role in capacity building, training, publications, supply chain, voluntary programs, and no-cost resources available to the chemical sector and within the four industry segments (basic chemicals, specialty chemicals, agricultural chemicals, and consumer products). The Chemical SRMA focuses on coordination and collaboration across the sector and with cross-sector critical infrastructure partners, both at the headquarters level and at the individual site level through the CISA regional offices. Among the resources related to supply chain discussed will be an overview of the 2022 Effective All-Hazards Chemical Sector Response Playbook(2) and webinar series which provides a Standard Operating Procedure (SOP) to assist the Chemical Sector in preparing for, responding to, and recovering from an all-hazards emergency; as well as information pertinent to the supply chain of water treatment chemicals. The team will also discuss how to contact your regional office for interaction and resources, an overview of the Homeland Security Information Network – Critical Infrastructure (HSIN-CI) Chemical Portal information, attendance at classified and unclassified briefings, and how to receive pertinent Chemical SRMA updates will be covered.

- *(1) Critical Infrastructure Sectors | CISA ; (2) Chemical Sector Playbook | CISA*

565b Low Desolvation Energy Tailored Electrolyte for Enhanced Subzero Temperature Performance of Lithium-Ion Batteries

Ethan Adams, Jason Packard, Vilas Pol

The exploration of new territories in space and the Arctic has highlighted a growing demand for energy storage solutions that can withstand extremely low temperatures (LTs). The current lithium-ion batteries are not suitable for such conditions without significant heating or thermal protection, as the electrolyte can freeze and the desolvation energy barrier for the Li-ion is high. To address this issue, a new type of fluorinated electrolyte was designed on the basis of improving pre-existing fluorinated electrolytes to allow for both a lower freezing point (FP) and lower desolvation energy compared to the standard commercial electrolyte (COM). This work's electrolyte is comprised of a single lithium salt, lithium bis(fluorosulfonyl)amide (Li**F**SI), dissolved in a four-solvent mixture of a highly polar fluoroethylene carbonate (**F**EC), intermediately polar methyl (2,2,2-trifluoroethyl) carbonate (**F**EMC), weakly polar nonaflurobutyl methyl ether (**N**ONA) (FP= –135°C), and an additional solvent whose dielectric constant is slightly weaker than FEMC. This specific solvent mixture was chosen to allow for a greater incorporation of NONA, which serves as a nonpolar diluent to depress the freezing point and promote the incorporation of the salt anions into the solvation structure, by adding a fourth intermediately polar solvent like diethylene carbonate (**D**EC), ethyl methyl carbonate (**E**MC), or dibutyl carbonate (D**B**C). These three solvents were chosen due to their suitable freezing point and dielectric constants. The optimized F-FDFN electrolyte showed a superior ultralow temperature performance of 70% of its room temperature (RT) capacity at –50°C with a low desolvation energy of 60.1 kJ mol^{-1} compared to COM, which retained only 25% of its RT capacity at –25°C. Excellent LT performance for F-FDFN electrolyte in a Li||$LiFePO_4$ cell is attributed to the tailored solvation shell incorporating a higher fraction of lithium salt anions compared to other prepared electrolytes, which had more unassociated anions.

565c Solid State Electrolytes with Low-Temperature Plasticizer Optimization

Nestor Levin, Annie Sun, Ethan Adams, Soohwan Kim, Vilas Pol

Solid State Batteries (SSBs) are the next generation of commercial energy storage devices, as needs for battery innovation grow, owing to their safety, high energy density, wider potential stability, and high mechanical strength. However, SSBs have had historically poor performance at subzero temperatures, due to poor interfacial contact, charge transfer kinetics, and freezing of electrolyte. A solid-state electrolyte with good performance at low temperatures is of great interest to space and defense applications. PVDF-based polymer electrolytes have been recently explored as polymer matrixes for electrolytes due to their excellent stability, yet have been less explored compared to PEO[1,2,3]. Furthermore, it has been shown that a garnet-type ceramic, LLZTO, optimized at 5% wt. in a composite polymer electrolyte (CSPE), improves the ionic conductivity and thermal stability[4]. Here, we present a systematic mechanistic elucidation of a wide-range Quasi Solid-State Electrolyte (QSSE) optimized for low-temperature operation (-25 C to 25 C) with a binary solvent system via modification of a polar aprotic low-melting plasticizer as a wetting agent. Additionally, we provide insight into the transport phenomenon regulating the low temperature performance at varying concentrations of LiTFSI salt in the wetting agent and composite polymer film.

We show that a 2M LiTFSI/THF wetting agent introduced into the DMF/PVDF/LiTFSI/LLZTO composite electrolyte displays 146 mAh/g capacity in a Li|CSPE|LFP battery with 99% CE at room temperature. Low temperature studies show 95 mAh/g capacity at -20 C with 99.9% CE, and 85 mAh/g at -25 C with 96% CE. The ionic conductivity is reported at 1.2×10^{-3} s/cm at 25 C without ambient heating. We also report a transference number of 0.42. An in situ-safety study also demonstrated improved thermal safety with lower exothermic heat flux and more stable thermal degradation relative to traditional commercial liquid electrolytes. Additionally, we show the ionic conductivity profiles and performances at varying temperatures from 25 C to -30 C, providing activation energies and Arrhenius behavior. Finally, density functional theory (DFT) calculations and molecular dynamics (MD) simulations provide potential insight into stable configurations and mechanisms governing the complex system.

This work precedes further exploration into a low temperature QSSE, with exploration into alternative methods to polymer wetting such as in-situ polymerization and solution casting with higher boiling organic co-solvents, such as CPME. Ionic liquids are also a potential direction for nonvolatile solvents towards low temperature function. Further DFT and MD work and other transport characterization in these directions can magnify the research community's understanding of low temperature challenges for solid-state electrolytes.

1. Fang, L. *et al.* Quasi-Solid-State Polymer Electrolyte Based on Highly Concentrated LiTFSI Complexing DMF for Ambient-Temperature Rechargeable Lithium Batteries. *Eng. Chem. Res.* 61, 7971–7981 (2022).

2. Zhou, Z., Zou, R., Liu, Z. & Zhang, P. Deciphering the role of tetrahydrofuran residue in the poly(ethylene oxide)/LiTFSI hybrid used for secondary battery electrolyte. *Giant* 6, 100056 (2021).

3. X. Zhang, J. Han, X. Niu, C. Xin, C. Xue, S. Wang, Y. Shen, L. Zhang, L. Li, C.-W. Nan, *Batteries & Supercaps,* 2020, *3*, 876.

4. Zhang, S. *et al.* Room-temperature, high-voltage solid-state lithium battery with composite solid

polymer electrolyte with in-situ thermal safety study. *Chemical Engineering Journal* 400, 125996 (2020).

565e Can Fusion Help Decarbonize the Power Sector?

Amanda Farnsworth, Emre GenÃ§er

The power sector has been identified as the key to decarbonizing the global energy economy. As we work to electrify the other energy sectors (transportation, residential, commercial, and industrial), we must decarbonize our generator fleet. The majority of research focuses on conventional renewable resources, namely wind and solar. While these resources are economically competitive with fossil fueled generators based on levelized cost of electricity, their unpredictable intermittency poses problems at high penetration levels. Highly renewable systems often require expensive energy storage supplements.

Fusion is a prospective energy technology which is safe, low-carbon and firm, but also expensive. Fusion reactor create heat which can be used in conventional power conversion cycles to produce electricity. Fusion reactors produce very little nuclear waste, and are natural self-quenching, so there is no danger of a run-away reaction. This study explores the value of this technology or one with similar characteristics. Because this technology is not yet validated, there are very few studies which quantify its potential impact on the power sector. One such manuscript identifies capital cost ceilings that fusion must meet to be competitive with variable renewable energy (VRE) resources [1]. Another paper shows that fusion's potential is highest in regions where conventional low-carbon options are limited, such as Japan, Korea, or Turkey [2]. Another paper demonstrations that fusion can contribute to the decarbonization of the global energy economy, but that its capital cost and commercialization date will impact these results [3].

This analysis adds considerable value to the current collection of knowledge in three main ways. First, this study dives into fusion's impact both on the system's unit commitment, and its investment decisions. It is shown that certain technologies pair more complimentarily with fusion than others. Additionally, generator fleet buildout impacts land requirements, which is a more significant topic in regions of high population density. Secondly, we evaluate fusion integration in nine regions within the US: Atlantic, California, Central, North Central, Northeast, Northwest, Southeast, Southwest, and Texas. A profile is created for each region, so that national conclusions can be extrapolated to international locations. This analysis is meant to illuminate understanding on what regional characteristics encourage or discourage the adoption of this type of technology. Lastly, we complete an extensive capital cost sensitivity analysis. This is important to understand how sensitive investment strategies are to fusion's cost.

Preliminary results have been produced for the first two sections of this study. The model used to complete this analysis, Ideal Grid, is a greenfield linear capacity expansion model with an hourly timestep. Ten technologies are currently represented: wind, solar, run-of-river hydro, conventional hydro, fusion, natural gas combustion turbine, natural gas combined-cycle, natural gas combined-cycle with 90% carbon capture, pumped hydro storage, and lithium-ion batteries. It should be noted that fission is not included in this model due to current legislature and public opinion which are both preventing buildout. Both techno-economic analysis and life-cycle assessment methods are used to track costs and emissions from all stages of the life-cycle. For example, this means that solar panels are not considered carbon neutral. An extensive number of optimizations are conducted in order to fully capture the impact of fusion.

First, a case study is conducted in the Atlantic region. An imposed carbon ceiling is lowered from 50 gCO_2/kWh to 15 gCO_2/kWh. It is shown that fusion is not economically competitive until emissions intensity of 30 gCO_2/kWh or lower, but that it is heavily favored as regulations are further tightened. At the lowest emissions intensity, fusion satisfies 74% of electricity demand, while contributing 47% of emissions. Also, it can be seen that before fusion is introduced, the system installs slightly more wind resources than solar, but after fusion is introduced, more solar is installed than wind. Lastly, it should be noted that cost increases as fusion penetration increases because fusion is expensive, and system curtailment decreases because fusion operates at a higher capacity factor, so less generator capacity is required.

Fusion installation is tested in the aforementioned nine regions for comparison. It is shown that fusion is favored most heavily in the Californian and Southeast regions, and that the Northwest does not install any fusion resources. Further analysis is needed to understand the affinity of fusion in certain regions in contrast to others.

Preliminary results show that fusion has the potential to impact the power sector in a major way. Further analysis is being conducted to understand price sensitivity, regional characteristics that attract fusion, and fusion's impact on land requirements.

References

1. Bustreo, C., Giuliani, U., Maggio, D. & Zollino, G. How fusion power can contribute to a fully decarbonized European power mix after 2050. *Fusion Eng. Des.* **146**, 2189–2193 (2019).
2. Gi, K., Sano, F., Akimoto, K., Hiwatari, R. & Tobita, K. Potential contribution of fusion power generation to low-carbon development under the Paris Agreement and associated uncertainties. *Energy Strateg. Rev.* **27**, (2020).
3. Turnbull, D., Glaser, A. & Goldston, R. J. Investigating the value of fusion energy using the Global Change Assessment Model. *Energy Econ.* **51**, 346–353 (2015).

565f Electricity Generation from Hydraulic Fracturing Wastewater Using a Reverse Electrodialysis Stack

Arash Emdadi, Jamie Hestekin, Lauren F. Greenlee, Bruce E. Logan

Hydraulic fracturing (HF) wastewater is produced during the fracking process for oil and gas production. Due to the high salinity of produced wastewater, it is an environmental challenge with potentially severe impacts on soil and water resources. In this study, electricity generation from HF wastewater was examined by using a reverse electrodialysis (RED) stack. A 10-cell RED stack was used that contained pairs of cation exchange membranes (CEMs) and anion exchange membranes (AEMs). Different synthetic salt solutions ($NaCl+CaCl_2$, NaCl, $CaCl_2$, Na_2SO_4, $NaCl+Na_2SO_4$) and a real HF wastewater were employed to assess the effect of different divalent and monovalent ions on power production. Due to the temperature gradient between produced HF wastewater and feed solution, the effect of the temperature gradient between concentrated and diluted streams was also assessed. Membrane characterizations, including permselectivity and electrical resistance, were analyzed in the presence of the synthetic salt solutions and the HF wastewater over a temperature range of 25- 60 °C. Based on the results, the presence of calcium and sulfate ions drops the permselectivity of CEMs and AEMs by 50% and 35%, respectively. Results of electrochemical impedance spectroscopy (EIS) revealed a sharp reduction in the electrical residence of both AEMs and CEMs with increased temperature. Based on polarization data a maximum gross power density of 2.6 $W.m^{-2}$ was achieved when HF wastewater was used in a RED stack. Applying a temperature gradient further increased the power density to 3.7 $W.m^{-2}$ which is among the highest power densities reported for a RED stack with a waste stream.

565g Theoretical Analysis of Salt Selection for a Reverse Electrodialysis System

Arash Emdadi, Jamie Hestekin, Lauren F. Greenlee, Bruce E. Logan

Reverse electrodialysis (RED) is a membrane technology used for electricity generation when two solutions with different salinity mix. The difference in the Gibbs free energy of concentrated and diluted solutions is the driving force for ion transport between channels and electricity generation. The type and nature of the components dissolved in the streams play an important role in electricity generation by RED. Parameters, including the valence of ions, the charge density of ions, and the activity coefficient of ions, affect the diffusion process between concentrated and diluted streams. A model was developed for screening salt solutions, including 174 monovalent and multivalent ions. In the first step, the theoretical open circuit voltage (OCV_{th}) of 174 salt solutions, including mono and multivalent ions (A_xB_y; $x \in \{1,2,3,4\}$; $y \in \{1,2,3,4\}$) was calculated and compared with the reference salt (NaCl). The Bromley model was used for activity coefficients and OCV_{th} calculation of the RED cell. The effect of ionic properties, including charge density, ionic radius, and hydration energy, on the OCV_{th} was analyzed. In the second step, a model for hazard assessment was developed to select the salt solutions with minimum environmental, health, and physical hazards. Between the investigated salts, potassium formate ($KHCO_2$), sodium bromide (NaBr), potassium acetate ($KC_2H_3O_2$), and sodium acetate ($NaC_2H_3O_2$) were found to be promising salts for RED applications due to high theoretical OCV, low hazard potential, and low cost compared to NaCl.

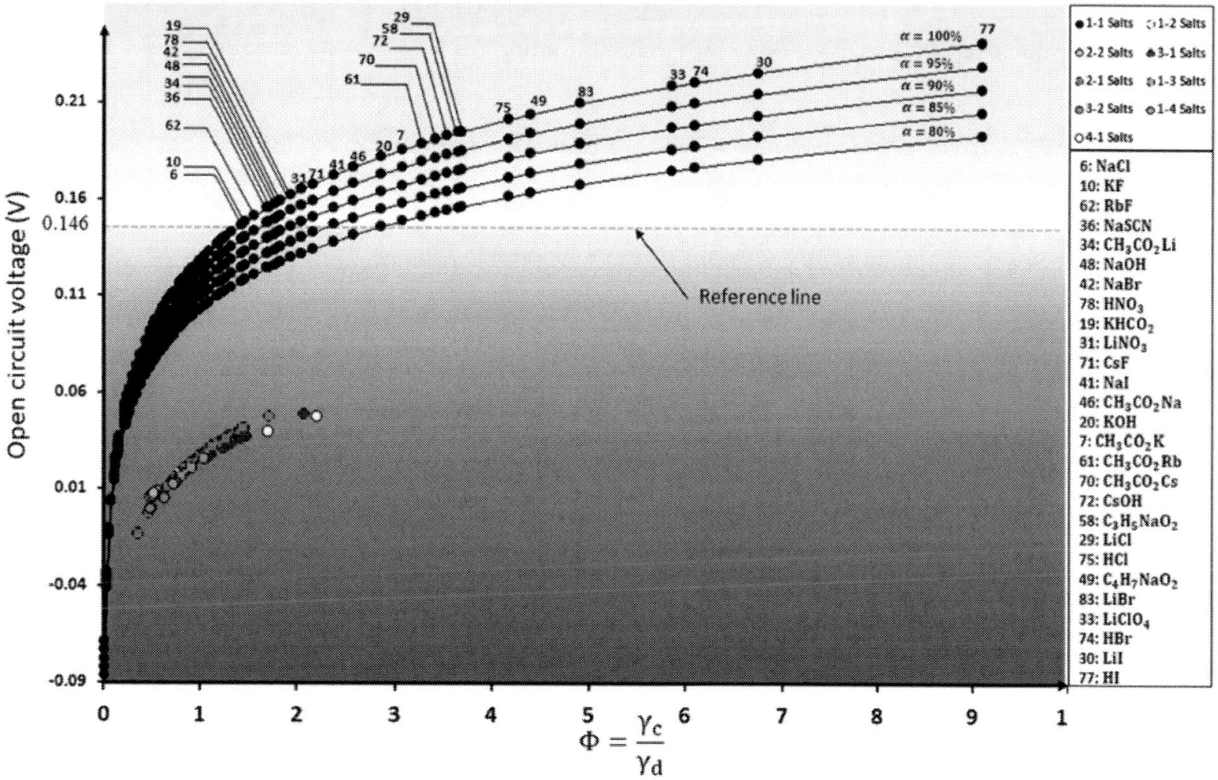

444b Rare-Earth Elements Recovery from Electronic Waste: Techno-Economic and Life Cycle Analysis

Esther Sanchez Moran, Mark Mba Wright

Rare earth elements possess unique magnetic, luminescent, and catalytic properties, making them a key role in our modern life. They are an essential component of many sectors of today's economy, including high-technological devices, industrial applications, defense, and a wide variety of clean energy technologies.

Electronic waste constitutes a rich source from which rare earth elements can be recovered. Recovering rare earth elements from electronic waste can lower the dependence on mining, stabilize rare earth elements market prices, and reduce the environmental impact of landfilling. Thus, researchers have been focused on developing technologies to recover rare earth elements from electronic waste.

This article presents a technical and environmental assessment of the steps involved in recovering didymium oxide from hard drive shreds through acid-free dissolution recycling. Although this recycling technology already minimizes the environmental impact of the recovery process, the goal of this study is to further improve the economic feasibility while making this process even more environmentally friendly than before. A facility with an annual processing capacity of 342.42 tons of hard drive shreds was modeled for this case study, leading to 2.53 tons of didymium oxide recovered yearly. The results of the techno-economic assessment indicated that the minimum selling price for didymium oxide was 130 \$/kg. Furthermore, as the 2022 average market price of didymium oxide were higher than the obtained minimum selling price, this means that this recovery process is economically profitable.

We thank DOE STTR for funding this project.

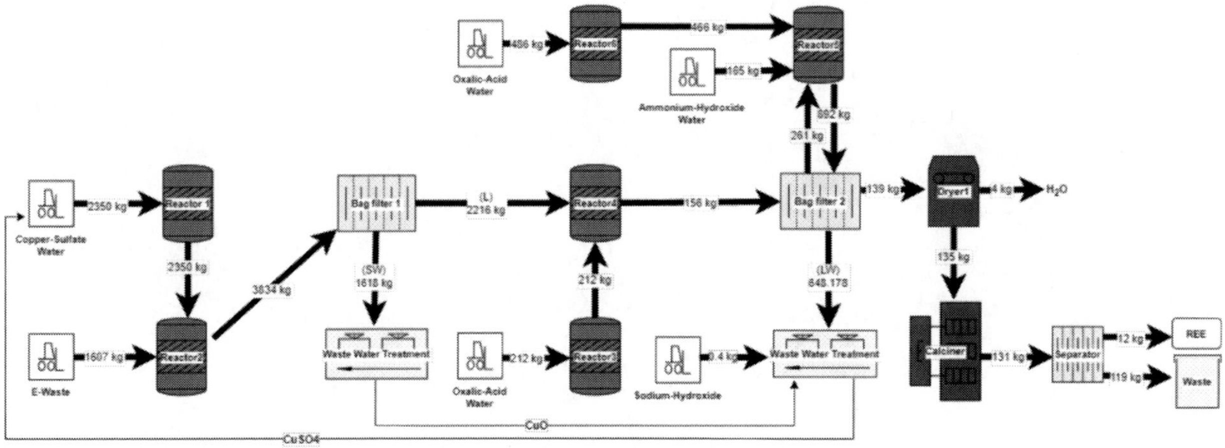

444d Development of a Predictive Model for in-Situ Resource Utilization of Space Exploration Waste

Ray Pitts, Joel Olson, Anne Meier

As we prepare for exploration-class human space missions, readily available, in-situ resources must be identified and prioritized. Through this lens, resource recovery from astronaut waste items becomes increasingly advantageous. With an approximate waste production rate of 1.8 kg per crewmember per day, considerable resources can be extracted from these waste materials as mission durations increase and sustainable on-planet surface architectures are developed. To estimate waste item production and resultant resource availability, an updated waste model for exploration was developed. This model incorporates up-to-date trash production estimates from the International Space Station, along with expected changes for deep-space exploration missions. To quantify the available resources outlined in the updated waste model, an elemental analysis of each major waste category was performed via CHONS elemental analysis and quantitation. Moisture content for current NASA waste models, along with user-inputted custom waste models, were also analyzed. Furthermore, NASA waste management technologies for exploration were integrated into the model to predict performance differences as the amount and composition of input waste feedstock varies. These technologies include Trash-to-Gas (thermal degradation of astronaut waste items into a syngas), Trash Compaction and Processing System (compaction of astronaut trash and recovery of water), and In-Space Manufacturing (3D printing of processed and recovered waste materials). Key performance parameters for each technology were assessed as a function of various space mission parameters – mission duration/location, crew size, trash feedstock, and trash production rate. An overview of the key findings from this study, including resource estimates from waste items for near-term human missions, is reported.

444e Optimization of Advanced Oxidation Processes for Removal of 1,4-Dioxane from Water Influents

Serenity Broussard, William E. Holmes, Rafael Hernandez, Daniel Gang, Mark E. Zappi

1,4-Dioxane is a synthetic cyclic ether that has been identified as a potentially harmful organic compound. As water quality testing procedures continue to advance, a growing number of valuable water resources around the world are being found to be contaminated with 1,4-dioxane. Historically, 1,4-dioxane was produced as a solvent stabilizer for the chemical degreaser 1,1,1-trichloroethane (1,1,1-TCA). In the early 2000s, 1,1,1-TCA was deemed an ozone depleting substance and its use was limited to essential applications. With the decline in 1,1,1-TCA production came the decline in 1,4-dioxane production. Unfortunately, 1,4-dioxane is formed as a by-product in several manufacturing processes used in the production of consumer goods. The polyester synthesis process and the ethoxylation process, which is utilized to manufacture surfactants found in detergents, shampoos, and cosmetics, are the two major processes that result in the formation of 1,4-dioxane as a by-product. 1,4-dioxane is considered a persistent organic pollutant as it does not readily degrade in the environment and instead accumulates. The USEPA has classified 1,4-dioxane as a Class 2B probable human carcinogen, but federal regulations for acceptable concentration levels in drinking water are not currently in place.

There is a critical need for effective treatment options to address water contamination caused by 1,4-dioxane. Due to its high solubility and affinity for water, adsorption processes have had challenges with treating 1,4-dioxane. Biotreatment methods are evolving, but more research is needed to yield a viable process. Surprisingly, little research has been applied toward the use of advanced oxidation processes for treating water contaminated with 1,4-dioxane. With 1,4-dioxane becoming an emerging pollutant of concern, more information is needed to characterize treatment process effectiveness and determine complimentary design parameters before these treatment systems can be properly applied in the field.

Advanced oxidation processes (AOPs) are a set of treatment processes that utilize hydroxyl radicals to degrade organic pollutants in water matrices. Various process configurations have been found to produce hydroxyl radicals. These configurations can be broken down into two main AOP groups: lighted (mainly UV photolytic-based) or dark (hydrogen peroxide and/or ozone based without photolysis). AOPs have a strong track record for treating a wide variety of organic pollutants in water matrices.

This study evaluated the comparative performance of various candidate AOPs (both lighted and dark) for the effective treatment of 1,4-dioxane within water matrices. Bench-scale AOP reactors were used to test the relative performance of the candidate AOPs for degrading 1,4-dioxane. Total organic carbon is used to evaluate the extent of mineralization along with the analytical tracking of key degradation products. Preliminary results indicated that 1,4-dioxane was easily degraded via oxidation. Efforts were then directed toward reducing oxidation system complexity and reagent cost without sacrificing treatment performance. Optimal dosing amounts and feed strategies have been optimized. Degradation kinetics coupled with completeness of oxidation reactions were also evaluated. Both lighted and dark systems were found effective with each type offering different benefits.

444f Respirator Cartridge Performance Testing for Chemicals of Potential Concern

Satish Nune, Michael Minette, Angela Melville, Stephen, D Davidson, Thomas Brouns, Michael Zabel, Eugene Morrey, Kevin D. Miller, Kenneth J Way

To accurately assess the performance of respiratory protection cartridges under varying environmental conditions, Pacific Northwest National Laboratory (PNNL) and Washington River Protection Solutions (WRPS) have jointly developed a laboratory test system to test respirator cartridges commonly used at Hanford tank farms in Richland, Washington, USA. The goal of this program is to understand respirator effectiveness across a wide range of vapors/gases and concentrations. The Hanford Tank Farms has over 170 nuclear waste tanks that have varying vapor mixtures. For workers operating around those tanks, cartridge respirators are used to reduce potential health risks associated with chemicals of potential concern (COPC) in complex vapor mixtures. Laboratory testing will provide industrial hygiene professionals an understanding of the performance of respirator cartridges with these varying mixtures of COPC vapors under a range of environmental conditions. Respirator cartridge performance was tested under humid conditions at room temperature (gas flow rate of 102.5 L/min +/- 5%, a temperature of 20°C +/- 2°C and a relative humidity of 70% +/- 5%), as humidity can have a major influence on breakthrough times. We will present our recent results on the experiments where the performance of the MSA OptiFilter TL multipurpose respirator cartridge was tested against a gas mixture with varying chemical vapor concentrations of Ammonia, N-Nitrosodimethylamine (NDMA), N-Nitrosodiethylamine (NDEA), N-nitrosomorpholine, N-Nitrosopiperdine and formaldehyde. The breakthrough times can be used for establishing cartridge change out schedules for improving worker safety and operating conditions.

286a Designing Chemical Value Chains for Net-Zero Emissions and Nature Positive Decisions

Amrita Sen, Ying Xue, Bhavik Bakshi

Adherence to the Paris Agreement necessitates the curbing of global average temperature rise to 1.5 degrees above pre-industrial levels. While direct contribution of the chemicals and materials industry (CMI) to global greenhouse gas emissions is relatively small, the growing importance of chemical energy carriers requires all chemical value chains to transition to a state of net-zero or net-negative emissions.[1] Biomass is an essential driver towards carbon-neutrality, especially in "hard to decarbonize" sectors such as the CMI which cannot be decoupled completely from carbon mobilization.[2,3] The availability of biomass and its fair allocation among polluting operations is crucial and still remains unaddressed in current relevant literature. Moreover, the impact of chemicals manufacturing on the natural environment is not limited to GHG emissions. Water use, critical mineral use and impact on bio-diversity are some considerations which are often overlooked in the scope of net-zero emission value chain design.[4] These can be quantified by the absolute environmental sustainability (AES) metrics which calculate the transgression of environmental impact of a product or process on nature's carrying capacity. Most existing literature in the area focusses on the net-zero objective and the economics associated. The few works that seek to quantify AES do not discuss the methodological gap in accounting for ecosystem services in LCA or incorporate spatially explicit datasets.[5,6] To bridge this gap, we design chemical and plastics value chains to study the implications of net-zero value chains on AES.

We postulate that while carbon sequestration by biomass is a driver of both carbon-negativity and AES, design for one objective might not necessarily satisfy the other. The biomass accounting for life-cycle and AES assessment methods operate on different principles. LCA treats biomass derived products as either inherently carbon neutral or expands the system boundary to incorporate carbon sequestration by biomass in the scope of emission calculations. For example, the inclusion of greater amounts of biomass derived feedstocks into chemical value chains may lower the overall greenhouse gas emissions of the value chain, when accounted for by the later principle. LCA thus implicitly assumes that the demand of biomass by the industry may be equitably fulfilled. However, AES assessment may allocate a different amount of biomass for sequestration of these emissions, depending upon the choice of sharing principle. Therefore, while a process network formulated to minimize emissions might seek to incorporate greatest permissible amount of biomass usage to leverage sequestration, a network formulated to maximize AES will instead lower the demand of ecosystem services in accordance with their available supply. Thus, the objectives of net-zero design and design for absolute environmental sustainability are not perfectly aligned.

To study the interplay of these objectives and determine technological solutions robust to the environmental objective, we formulate the steady state design of sustainable chemical value chains as a multi-objective optimization problem. The net emissions of the chemical process network are minimized using LCA practices for biomass accounting. We also seek to minimize the transgression level utilizing the best available regional information on chemical value chains, for the AES objective. We use our model of the global CMI developed earlier, as the baseline technological matrix.[7] Additionally, emerging technologies operating at all stages of the life cycle are incorporated. These include input side solutions such as renewable power, green hydrogen, biomass, captured carbon-dioxide; process upgradations such as process electrification, transformative technologies; output side solutions such as carbon capture, incineration, pyrolysis; and recycle options ranging from primary to tertiary recycling. Regional information on chemical production, usage, sourcing of inputs, and background emissions is used to estimate emissions and sequestration. The multiscale techno-ecological synergy (TES) based AES assessment method is used to quantify the AES metric. Integrated with biophysical models, this method brings in high geographical resolutions in assessing nature's carrying capacity. Private and public ownerships are considered while downscaling ecological thresholds at large scale which avoids subjectiveness and encourages stakeholders taking actions towards net-zero objectives[8].

The results of the optimization problem yield multiple Pareto optimal value chain solutions. We study the trade-offs between the most net-negative and most sustainable solutions. The net-negative solutions are disproportionately dependent on carbon sequestration to offset their emissions. The non biogenic emissions from this network reflect the same, with some fossil-based options being selected to offer greater sequestration credits. The most sustainable solutions on the other hand minimize the non-biogenic emissions from the network, relying on circular solutions like plastics recycling and carbon capture and utilization for carbon circularity. This helps us determine the technological solutions most robust to the subjective choice of environmental metrics, like carbon capture and utilization and plastics recycling. Our results caution us against dependence on one class of solutions in the quest to a net-zero world. While we recognize the importance of bio-based solutions to offset the hardest-to-abate emissions and their criticality towards net-negative operations, the abuse of such solutions raises issues of damage to bio-diversity and food-fuel conflict. This work provides a stepping stone towards consideration of multiple environmental stakeholders in planning the net-zero transition.

References

1. Wevers, J. B., Shen, L., & Van der Spek, M. (2020). What does it take to go net-zero-CO2? A life cycle assessment on long-term storage of intermittent renewables with chemical energy carriers. Frontiers in Energy Research, 8, 104.

2. Haus, M. O., Winter, B., Fleitmann, L., Palkovits, R., & Bardow, A. (2022). Making more from bio-based platforms: life cycle assessment and techno-economic analysis of N-vinyl-2-pyrrolidone from succinic acid. Green Chemistry, 24(17), 6671-6684.

3. Gabrielli, P., Gazzani, M., & Mazzotti, M. (2020). The role of carbon capture and utilization, carbon capture and storage, and biomass to enable a net-zero-CO2 emissions chemical industry. Industrial & Engineering Chemistry Research, 59(15), 7033-7045.

4. Persson, L., Carney Almroth, B. M., Collins, C. D., Cornell, S., de Wit, C. A., Diamond, M. L., ... & Hauschild, M. Z. (2022). Outside the safe operating space of the planetary boundary for novel entities. Environmental science & technology, 56(3), 1510-1521.

5. Bachmann, M., Zibunas, C., Hartmann, J., Tulus, V., Suh, S., Guillén-Gosálbez, G., & Bardow, A. (2023). Towards circular plastics within planetary boundaries. Nature Sustainability, 1-12.

6. Meng, F., Wagner, A., Kremer, A. B., Kanazawa, D., Leung, J. J., Goult, P., ... & Cullen, J. M. (2023). Planet-compatible pathways for transitioning the chemical industry. Proceedings of the National Academy of Sciences, 120(8), e2218294120.

7. Sen, A., Stephanopoulos, G., & Bakshi, B. R. (2022). Mapping Anthropogenic Carbon Mobilization Through Chemical Process and Manufacturing Industries. In Computer Aided Chemical Engineering (Vol. 49, pp. 553-558). Elsevier.

8. Xue, Ying, and Bhavik R. Bakshi. 'Metrics for a nature-positive world: A multiscale approach for absolute environmental sustainability assessment.' *Science of The Total Environment* 846 (2022): 157373.

286b Material Flow Assessment and Circularization of Waste in the Acetaminophen Production Network through Integration of a New Industrial System

Abhimanyu Raj Shekhar, Shweta Singh

The increasing demand for sustainable processes in the pharmaceutical industry has necessitated for an urgent need for the circularization of waste streams in the supply chain.[1] In this study, we present a comprehensive regional material flow assessment (MFA) of the acetaminophen production network, comprising of 6 industrial process systems classified into 4 NAICS sectors, using physical input-output tables (PIOTs) generated by the PIOT-Hub.[2] Our research aims to identify opportunities for circularizing waste flows through the integration of a new industrial system in the production network, and demonstrates the value of incorporating circular economy principles into the design and operation of chemical production systems in the pharma-industry.

The MFA will be conducted by generating PIOTs using the PIOT-Hub, a computational tool that employs bottom-up approach of leveraging mechanistic models of the industrial processes in various stages of the acetaminophen production network as the principal data input to generate material flow map in the form of PIOTs. This approach enables us to quantify waste streams, identify critical points of material losses, and map potential areas for process optimization. By utilizing state-of-the-art MFA techniques and the inherent capabilities of PIOT-Hub, we can evaluate different scenarios for waste reduction and circularization, leading to the selection of a promising new industrial system.[3,4]

This novel approach of conducting MFA, integrated with the proposed structure of the acetaminophen production network, is designed to repurpose waste streams as valuable inputs for other industries. By harnessing the principles of industrial symbiosis, our proposed system can transform waste products from the acetaminophen manufacturing network into high-value materials via inclusion of a new waste-sequestering industrial process.[5] Furthermore, this integrated approach not only reduces the environmental impact of acetaminophen production, but also generates new revenue streams and provides opportunistic pathways towards interdependent collaboration among different industrial sectors.

The implementation of our proposed industrial system demonstrates the potential for significant improvements in resource efficiency and waste reduction within the acetaminophen production network. The key findings of this research study emphasize on the importance of a comprehensive, mechanistic-driven MFA for identifying and addressing material flow inefficiencies in chemical processes categorized in the field of pharmaceuticals. By highlighting the benefits of circular economy principles and industrial symbiosis, this study can serve as a blueprint for the development of sustainable, resource-efficient processes in the pharmaceutical industry.

References:

1. Alshemari, A., Breen, L., Quinn, G. & Sivarajah, U. Can We Create a Circular Pharmaceutical Supply Chain (CPSC) to Reduce Medicines Waste? *Pharmacy* **8**, 221 (2020).
2. Vunnava, V. S. G., Shin, J., Zhao, L. & Singh, S. PIOT-Hub - A collaborative cloud tool for generation of physical input–output tables using mechanistic engineering models. *J. Ind. Ecol.* **26**, 107–120 (2022).
3. Krausmann, F., Schandl, H., Eisenmenger, N., Giljum, S. & Jackson, T. Material Flow

Accounting: Measuring Global Material Use for Sustainable Development. *Annu. Rev. Environ. Resour.* **42**, 647–675 (2017).

4. Fischer-Kowalski, M. *et al.* Methodology and Indicators of Economy-wide Material Flow Accounting: State of the Art and Reliability Across Sources. *J. Ind. Ecol.* **15**, 855–876 (2011).
5. Chertow, M. R. INDUSTRIAL SYMBIOSIS: Literature and Taxonomy. *Annu. Rev. Energy Environ.* **25**, 313–337 (2000).

286c Developing a Nitrogen Circular Economy By Recovering and Utilizing N_2o As a Feedstock for Phenol Production

Chayse Lavallais, Jennifer Dunn

Excess reactive nitrogen in the environment has led to a disruption of the global nitrogen cycle. A large fraction of that nitrogen ends up in wastewater treatment plants, where it is released back into the atmosphere as diatomic nitrogen or nitrous oxide (N_2O), a potent greenhouse gas. However, with an increasing emphasis on developing a nitrogen circular economy, various nitrogen recovery technologies are being investigated. One of these technologies is called the coupled aerobic-anoxic nitrous decomposition operation (CANDO). CANDO can convert ammonia in wastewater into N_2O, which is then co-combusted with methane for additional energy generation. However, N_2O is also a powerful selective oxidant that can be used for chemical production. Recovering N_2O from wastewater to produce industrially relevant chemicals could advance the nitrogen circular economy, but its environmental impacts have not been explored.

In this study, we investigate recovering N_2O from wastewater and utilizing it for phenol production, as traditional phenol production from the cumene process is energy intensive. By developing process models for both the N_2O recovery and the phenol production utilizing N_2O, a life cycle assessment was completed to compare the cumulative energy demand and the global warming potential (GWP) to the cumene process and other nitrogen recovery technologies. Results show that phenol production utilizing N_2O has lower energy demand and GWP than the cumene process. The results can encourage further exploration of using N_2O as an oxidant for the production of other industrial chemicals.

286d Integrated Anaerobic Membrane Bioreactor Platform for Resource Recovery from Swine Wastewater- Evaluation of Life Cycle Economic and Environmental Impacts

Prathap Parameswaran, Madison Kratzer, Vikas Khanna

In moving towards the goal of a circular economy, traditional waste streams are being considered for their potential for resource recovery. Recovering products such as energy, clean water, and nutrients from wastewater sources contributes to minimizing the extraction of virgin resources from the environment and its associated environmental impacts. In particular, phosphorus and biologically active nitrogen are nutrients of key importance because they are obtained using non-renewable resources with high life cycle environmental impacts. Concentrated animal feeding operations (CAFOs) are sources of high strength waste streams from large numbers of closely reared animals that contain significant amounts of nitrogen, phosphorus, and organics. We considered wastewater from hog CAFOs due to the high strength and relatively consistent wastewater matrix. A combination of a conventional activated sludge (CAS) and anaerobic digestion (AD) platform is typically used to treat these waste streams. This treatment platform requires high energy input for aeration of the activated sludge, and does not include recovery of resources including volatile fatty acids (VFAs), nitrogen, and phosphorus. We present an optimization based model of a novel treatment train configuration which uses an anaerobic membrane bioreactor (AnMBR). The AnMBR system couples anaerobic digestion with membrane separation which allows for long solids retention times and high carbon conversion within the system. Existing work in this domain has focused on evaluation of individual unit processes. A holistic systems approach to evaluating the environmental and economic performance of the treatment platform is needed to guide research and development and provide quantitative comparisons of the system to traditional treatment methods.

We present an optimization-based decision-making framework for the evaluation of economic and environmental impacts of the treatment of swine wastewater in an integrated resource recovery system. A physics-based model of unit operations including an An-MBR, VFA recovery, phosphorus recovery through coagulation and flocculation, nitrogen recovery through adsorption, and a final wetland polishing step was created. Parameters in individual unit operation pieces were tuned using available data from laboratory experiments and a pilot scale system. The model allows for the evaluation of a wide problem space rather than ad-hoc combinations for treatment trains. These include different configurations of parameters in unit operations, and different choices for unit operations to achieve separation and resource recovery. Optimization for the mixed integer nonlinear programing (MINLP) problem is conducted in Pyomo using the IPOPT algorithm. Optimization was also performed using multi-objective optimization to find a range of pareto-optimal solutions balancing environmental impacts with economic potential of the system. Environmental impacts were quantified using process-based life cycle assessment (LCA). Preliminary results show that there are solutions in which the AnMBR platform is comparable in economic cost to CAS and AD treatment while also having lower environmental impacts. This is achieved by offsetting high capital costs with the value of recovered nutrient and high value-added VFA products. Detailed sensitivity analysis was performed around key variables including energy for membrane cleaning, membrane replacement frequency, ambient temperature, removal of carbonate alkalinity from the system prior to phosphorus treatment, and the composition of the swine wastewater to quantify their impact on model results. The implications of the findings and challenges for developing circular economy solutions for wastewater streams will be discussed.

286e Towards a Circular Economy for PET Bottles in the US - a System-Dynamics Approach

Tapajyoti Ghosh, Alberta Carpenter

The United States generates the most plastic waste of any country. Along with that GHG emissions from the global plastic economy are expected to increase to 15% of the global carbon budget by 2050. It is imperative that plastic recycling is made a reality to reduce both plastic pollution in the environment and GHG emissions. A portfolio of end-of-life strategies must be implemented to minimize environmental impacts and retain valuable plastic material, but it is challenging to compare options that generate products with different utility and lifetime. Plastic use reduction, reuse and recycling are thus increasingly important, but making informed policy and research decisions within this space can be challenging given the diverse range of available solutions. The novel analysis framework, Plastic Parallel Pathways Platform (4P) has been equipped with consequential life cycle assessment, techno-economic analysis, and a plastic circularity indicator to estimate the greenhouse gas (GHG) emissions, circularity, and cost of polyethylene terephthalate (PET) down-cycling to lower-quality resin, closed-loop recycling to food-grade PET bottles, up-cycling to fiber-reinforced plastic (FRP), and conversion to non-plastic products (electricity, oil) on a United States economy-wide basis. Integrating system dynamics into this robust plastics model that already incorporates techno-economics, circularity, and environmental impacts will enable identification of key bottlenecks between manufacturers, waste sorters, and reclaimers that currently prevent rapid decarbonization of the plastics economy. System dynamics (SD) explore the evolution of activities and technologies based on changed macro parameters such as plastic demand and supply, market shifts, and cross-sectoral interactions. This project particularly aims to explore the interplay between, waste collection, plastic waste sorting, recycling, and manufacturing, as well as the effect of plastic bale quality and plastic reuse initiatives on the surrounding process stages. This functionality will facilitate combinatory analysis in which a portfolio of end-of-life pathways are assessed simultaneously, with the exact makeup of that portfolio affected by parameters such as technology scales, resource constraints, and waste mitigation efforts. Integrating SD with the 4P framework enables analyzing the effect of increased revenue and reinvestment into improving process efficiencies, sorting and collection quantities. Through that, market effects of increased recycled resin availability can be studied for the plastics systems model for the US. The results will help identify technical or economic bottlenecks that currently limit efforts to decarbonize the U.S. plastics economy.

286f Supply Chain Design and Optimization for Plastics at the End-of-Life: City-Based Study

Baibhaw Kumar, Natalia Cano-Londoño, Gerardo Ruiz-Mercado, Heriberto Cabezas, Csaba Deak

The world is transitioning to more efficient and sustainable waste management practices for promoting a circular economy. The management of Municipal solid waste (MSW) worldwide, particularly in developing nations, is challenging because of its ubiquitous pathways toward its end of life (EOL) as shown by life cycle assessments. The year 2019 witnessed almost 370 million tons of plastic produced globally, including 58 million tons produced in Europe. A closed-loop supply chain for plastics is formed when a basic forward logistics is connected with a reverse supply chain to reclaim that product from the start point. Reverse supply chain networks have been particularly used in plastic waste management, contributing to circular economy perspectives. Mathematical models aid in understanding the networks and have been used extensively by researchers to optimize supply chains. One such platform or approach is using the P-graph. The P-graph framework's fundamental algorithmic technique relies on structural configurations built on process graphs to design and optimize process networks graphically. Proper structuring is one facet of network designs frequently given insufficient consideration. As shown in the figure, the supply chain(s) for plastics at the end-of-life in Hungary is complex. This study aims to identify the gaps in the plastic supply chain framework, focusing on the city of Miskolc in "Borsod" County in Hungary. The objective is to create supply chain structures that make the reuse of plastic waste as profitable and environmentally sound. We will show results for several alternate supply chain structures that give decision makers in private firms and public utilities as much latitude as possible in making decisions. Philosophically, this approach is the core of the creative innovation that is necessary to manage an otherwise difficult challenge.

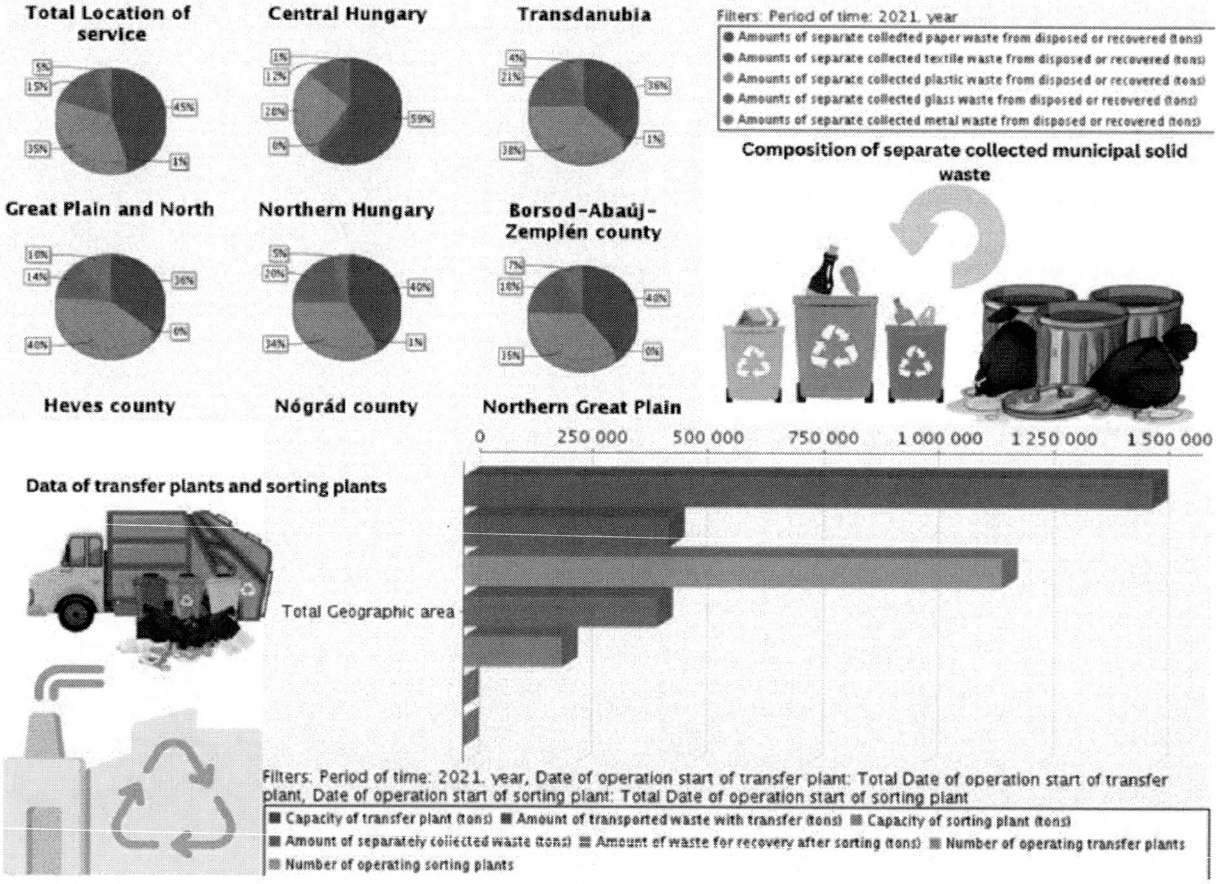

286g Computational Solvent-Based Extraction of Additives for Polymer Recycling and Upcycling

Pooja Bhalode, Sean Najmi, Pavel Kots, Montgomery Baker-Fales, Dionisios Vlachos

The ubiquitous use of polymers in varied industries has also brought to light the need to identify efficient and effective ways to recycle and upcycle polymers.[1] This is especially concerning for commodity polymers leading to an irreversible global catastrophe if not recycled appropriately.[1] A key challenge in the recycling and upcycling of end-of-use commodity polymers is separating the additives from the polymer. Such additives, if not isolated, can significantly challenge the catalytic decomposition of polymers and impact their circularity.[2,3] To address this challenge, it is imperative to introduce innovative strategies to isolate the additives, followed by polymer upcycling.

This work presents a computational framework focusing on high-throughput screening of solvents for investigating selective solvent-based extraction of additives from polymers aimed at reducing experimental efforts and enabling efficient polymer recycling and upcycling. The computational solvent selection framework is based on Hansen solubility parameter (HSP) method and a quantum chemistry-based equilibrium thermodynamics method - COnductor like Screening MOdel for Realistic Solvents (COSMO-RS).[4,5] HSP has been widely used to identify selective solvents based on the degree of dispersion, polarity, and H-bonding and is strategically designed to identify solvents for separating additives from the polymer. Following this, COSMO-RS is used to calculate thermodynamic properties and optimize process conditions. A process flowsheet is constructed to quantify the choice of solvents and overall procedure from a techno-economic (TEA) and life cycle analysis (LCA) perspective. Optimal solvents identified from this process are validated experimentally and the additives are then isolated prior to the polymer being used as a feedstock for valuable products.

References

1. Coates, G. W. & Getzler, Y. D. Y. L. *Nat Rev Mater* **5**, 501–516 (2020).
2. Ügdüler, S., Geem, K. M. V., Roosen, M., *et al. Waste Manage* **104**, 148–182 (2020).
3. Hinton, Z. R. *et al. Green Chem* (2022).
4. Hansen, C. M. CRC Press, Boca Raton, 2007.
5. Klamt, A. *Wiley Interdiscip Rev Comput Mol Sci* **1**, 699–709 (2011).

484a Relative Permeability of $scCO_2$ and Brine in Core from Potential CO_2 Reservoirs

Dustin Crandall, Thomas Paronish, Magdalena Gill, Johnathan Moore, Scott Workman

To predict the storage potential of CO_2 in geologic storage locations, we need to understand how CO_2 and brine flow in these reservoirs. This is difficult to do because CO_2 and brine flow differently under the high pressure and temperature conditions that exist in reservoirs. In addition, it is difficult to obtain accurate measurements of CO_2 and brine flow in cores that are representative of the formations being targeted for CO_2 injection.

The National Energy Technology Laboratory (NETL) is conducting research on the behavior of CO_2 and brine flow in reservoirs. NETL is using unsteady state relative permeability measurements to study CO_2 and brine flow, in addition to coupling this with non-destructive core analyses. The raw and processed data from these analyses are made publicly available on NETL's Energy Data eXchange (www.edx.netl.doe.gov) for researchers and developers to use in planning field scale operations.

NETL's research has shown that micro-scale fabric has an influence on the flow of CO_2 into brine saturated formations. These features are below the resolution of most well scale measurements, and below the resolution that would be expected in reservoir modeling efforts. Accounting for small scale features in the methods of upscaling flow parameters is an ongoing area of research.

The research being conducted by NETL is important for understanding the behavior of CO_2 and brine flow in reservoirs. This research will help us to predict the storage potential of CO_2 in reservoirs and to develop effective methods for CO_2 injection.

484c CCS-Analytics: Application of Artificial Intelligence for Carbon Dioxide Storage Systems

Shahab Mohaghegh, Vyacheslav Romanov, Guoxiang Liu, Derek M. Vikara

A well-known fact is that science and engineering are significantly improved through the correct implication of Artificial Intelligence. Science and engineering Application of Artificial Intelligence (SEA-AI) has the capability to reduce, curb, and even prevent Climate Change. There are no doubts that Artificial Intelligence can significantly help providing Climate Change solutions. Carbon Capture and Storage (CCS) can significantly contribute to the minimization of Climate Change, then CCS-Analytics (Application of Artificial Intelligence for Carbon Dioxide Storage Systems) that incorporates SEA-AI, is a serious technology for reducing and even preventing Climate Change.

The title of this article represents the fact that science and engineering application of Artificial Intelligence is being used to assist Carbon Capture and Storage. There are significant differences in using Artificial Intelligence to solve science and engineering problems versus general (non-engineering) problems. Since Carbon Dioxide Storage Systems is an engineering related problem then the correct version of the Artificial Intelligence must be used to enhance the solution. This article will provide explanations about how science and engineering application of Artificial Intelligence must be used in CCS-Analytics (Application of Artificial Intelligence for Carbon Dioxide Storage Systems).

Carbon storage and sequestration in the geological formations that are thousands of feet under the surface can keep the Carbon Dioxide in such reservoirs (porous media) for hundreds of millions of years. The research of Carbon Storage (Sequestration) will significantly contribute to the enhancement of this process and makes it as realistic and applicable as fossil energy production in the past several decades. It must be noted that Carbon Storage (Sequestration) requires domain expertise in reservoir engineering as CCS-Analytics requires application of Artificial Intelligence in reservoir engineering. Carbon Storage (Sequestration) can be done in several types of geological formations that are tens of thousands of feet below surface.

Two of the most important geological formations that are used for Carbon Storage are saline aquifers and oil and gas reservoirs that have already been emptied through the production of the hydrocarbons. Saline Aquifers are “Green Fields”, and Petroleum Reservoirs are “Mature (Brown) Fields”. The main difference between these two types of formations is the availability of the field measurements that can represent their geological characteristics. In many cases Geo-Analytics (AI-based Geological Modeling) can provide the required enhancement for CCS-Analytics. Some details of the Geo-Analytics will also be mentioned in this article.

CCS-Analytics (Application of Artificial Intelligence for Carbon Dioxide Storage Systems) can be applied in two different types of applications. The first application is developing “Smart Proxy Model” of the Numerical Reservoir Simulations that is used to model the Carbon Dioxide Storage Systems to generate pressure and saturation distribution of the Carbon Dioxide in space and time throughout the reservoir. Smart Proxy Model that is the Artificial Intelligence application in Proxy Modeling is far more different and orders of magnitude enhanced, exact, and better than the traditional proxy models (Reduce Order Models – ROM, and Response Surface Models - RSM). These traditional proxy models of the numerical simulation (ROM and RSM) have been used in the engineering related numerical simulations in the past several decades. In this article case study of Smart Proxy Modeling for the Carbon Dioxide Storage Systems will be presented.

The second CCS-Analytics (Application of Artificial Intelligence for Carbon Dioxide Storage Systems) is a purely AI-based (Top-Down) Reservoir Simulation and Modeling that only uses actual field measurements to model the pressure and saturation distribution of the Carbon Dioxide in space and time throughout the reservoir. However, this second application of CCS will be provided and presented as soon as such research projects are completed.

484d Enabling Gigaton-Scale CO_2 Mineralization and Utilization in Reactive Reservoirs

Ruoshi Cao, Katherine Muller, Quin Miller, Allie Nagurney, Emily Nienhuis, Eusebius Kutsienyo, Pejman Rasouli, Herbert T. Schaef

Basalt and ultramafic reservoirs provide the safest geologic carbon storage (GCS) solution through carbon mineralization. Commercial-scale deployment requires coupling the lab-based experiments with reactive transport modeling to derisk the process and fulfill permitting requirements. This work focuses on research and development needed to accelerate adoption of commercial-scale GCS in our mafic and ultramafic active drill sites, including the Columbia River Basalts. We report on storage resource analyses of carbon storage potential in basalt and peridotites, including the potential for critical mineral recovery. We also have continued to develop reactive transport reservoir simulators to evaluate well configurations, pressures, temperatures, flow rates, and predicted reservoir development, plume extent and migration, and fractional storage in supercritical, aqueous and mineral phases. The simulations are conducted to consider changes in reservoir porosity and permeability associated with geochemical effects; and to quantify changes in effective storage capacity, and storage/injection rate at the site. The results are guiding the downselection of priority areas and options for acquisition of new datasets to resolve key uncertainties. These collective projects are laying the groundwork to characterize and permit a multiple injection well, and to help develop the industrial regulatory and community partnerships needed to successfully demonstrate and upscale carbon storage and utilization technology in reactive mafic-ultramafic reservoirs across the globe.

484e Geologic Characterization and Fluid-Flow Modeling for Commercial-Scale CO_2 Injection and Storage in the Mt. Simon Sandstone Storage Complex

Oladipupo Babarinde, Fang Yang

Several large-scale carbon dioxide (CO_2) storage projects are currently underway in the Illinois Basin, with the potential to contribute to the commercial deployment of carbon capture utilization and storage (CCUS) across the US. To develop these projects, extensive efforts are required to characterize and model the geology and fluid flow of the subsurface geologic formations where CO_2 will be stored.

This study focuses on the geologic characterization and fluid-flow modeling conducted for the Carbon Storage Assurance Facility Enterprise (CarbonSAFE) Illinois storage corridor project, Phase 3. A stratigraphic test well was drilled and logged in eastern McLean County, Illinois, to evaluate the injectivity and storage resource of the Mt. Simon storage complex, which comprises the Mt. Simon Sandstone (stratigraphically subdivided into upper, middle, and lower), Eau Claire shale, and Argenta formation.

Using geophysical logs and formation cores from the test well, along with geophysical logs from two nearby wells, three geocellular models of the storage complex were developed with permeability estimations from three different methods (the difference in average permeability of Mt. Simon among the three models was within 20%). The "most-likely" model was chosen and calibrated to the local structural and reservoir properties of the formations. The depositional environment of the Mt. Simon Sandstone varies across its subdivisions vertically, from deltaic environment in upper Mt. Simon, fluvial dominated environment in middle Mt. Simon, to a fluvial-aeolian environment in lower Mt. Simon. As such, the properties in the model were heterogeneous both laterally and vertically. The calibrated geocellular model covers an area of 20 miles (32 km) by 20 miles (32 km) with an average thickness of 2977 ft (907 m). The storage unit, the Mt. Simon sandstone, is 1980-ft (604-m) thick with average porosity and permeability of 11% and 50 mD. The Arkose zone, the most porous and permeable 200-ft (61-m) thick interval at the bottom of the Mt. Simon Sandstone, has average porosity and permeability of 16% and 308 mD.

Fluid-flow modeling was conducted using the four geocellular models described above. A sensitivity study was completed to understand how various factors (including well inclination, the distance between wells, perforation interval, permeability, and porosity) affect maximum CO_2 injection rate, CO_2 plume size, and pressure front, thus determining an injection strategy that could meet the injection goal.

Sensitivity on well inclination showed that CO_2 could be injected at a maximum CO_2 injection rate of 1.2 Mt/yr to 2.5 Mt/yr via a vertical well by perforating the entire Arkose zone. A slant well with various vertical angles (45–78°, measured from vertical) going through the Arkose zone allowed an increase of 4–34% in the maximum CO_2 injection rate as the perforated length increased due to the slant well's angle. A horizontal well with a perforated length of 1000 ft (304.8 m) to 5000 ft (1524 m) allowed an increase of 22–200% in the maximum injection rate. When CO_2 was injected via a vertical well at a rate of 1 Mt/yr, the CO_2 plume radius was around 1.4–1.5 mi (2.3–2.4 km) at 12-year injection and 1.6–1.8 mi (2.6–2.9 km) at 20-year injection. The CO_2 plume radius increased by ~4% at 50-year post injection.

When multiple vertical wells were considered at the well site, the well locations were determined considering pressure interference among wells, distance to future monitoring wells, and land availability. Sensitivity on the distance between wells showed that shorter distances between the injectors yielded smaller CO_2 plume size but had little effect on the pressure front. In this study, the optimal distance between wells was 0.5–1 mi (0.8–1.6 km) to minimize the plume size while maintaining an injection rate of 1.5 Mt/yr.

Sensitivity on the perforation interval of a vertical well showed that perforating the Arkose zone yielded a smaller pressure front than injecting into the less porous and permeable 500-ft (152-m) thick interval in the upper part of the formation. Sensitivity on porosity and permeability showed that a 20% increase in porosity decreased the CO_2 plume radius by 9% at 20-year injection and 7% at 50-year post injection. A 20% change in permeability had little effect on the CO_2 plume size.

The findings from this study serve as a reference for early-stage commercial CO_2 storage site development. Future work will involve updating the models and incorporating additional geologic data to further refine the site evaluation process.

484f Benefits of Injecting Carbon Dioxide into Coal Seams

John Pope

This paper will share advances in sequestering carbon dioxide into underground coal seams. To date, approximately 375,000 metric tons of CO_2 have been injected into coal seams across the world. Over the past 30 years, the underlying science of CO_2 sequestration in coal has been investigated extensively, including several studies funded by NETL, and the technology has been performed repeatedly and successfully by many groups. CGC has advanced the technique to address issues around coal swelling and other formation damage that can sometimes accompany the process.

By using downhole Raman spectroscopy in CCS wells located in the US and Australia, the authors have been able to observe CO_2 gas and solution gas flowing through coal seams in a sustainable manner with no apparent damage to that coal. By mimicking the conditions under which those events were observed, the authors have been able to repeatedly and reproducibly sequester CO_2 into coal.

Those sequestration experiments have included:

- Sequestering CO_2 into one cubic centimeter chunks of coal charged with methane and water in high pressure laboratory cells while monitoring solution gas and head space gas using Raman spectroscopy;
- Sequestering CO_2 into a one meter long, 10 centimeter diameter tube filled with coal and water; and
- Sequestering CO_2 into a 30 meter thick coal located in a 160-acre field in the Powder River Basin of Wyoming.

It is well-known that coal preferentially absorbs CO_2 over methane and nitrogen. This characteristic of coal can be used to filter CO_2 from mixed gas streams including nitrogen. As a result, gas streams containing relatively low concentrations of CO_2 can be injected directly into coal seams and only the CO_2 is retained.

Technoeconomic analysis of the approach indicates that as a result, a CCS project using a coal sink vs. a traditional sink can reduce its capture and cleanup costs by 50% to 80%.

The paper will describe experimental design and results as well as the implications for global CCS.

484h Imaging the Dissolution of Chalk during HCl Injection

Peter Winkel Rasmussen, Anders Nymark Christensen, Nicolas Bovet, Benaiah Anabaraonye

Geological carbon storage is a promising climate change mitigation technology based on the injection of CO_2 into geological formations such as depleted oil and gas reservoirs, which prevents it from attributing to global warming as it is locked out of the atmosphere [1, 2]. In Denmark, depleted oil and gas fields are composed of chalk, which presents a challenge due to the chemical reactions initiated by CO_2-acidified brines and reservoir rocks [3]. The chemical dissolution of chalk could lead to the weakening of the reservoir, subsidence, and subsequently a failure to store the injected CO_2 [4]. Since the infrastructure for injecting and monitoring is already in place, a CS program in chalk could be economically viable [3]. It is, therefore, essential to study the dissolution of the chalk to determine its viability of it as a storage medium.

In this work, we present a systematic study of the dissolution of Danish North Sea chalk in the presence of acidic solutions. North Sea chalk is characterised by small pore spaces, low permeabilities, and high porosities. Experimental studies were performed in a novel triaxial flow cell that allows for the simulation of reservoir temperatures and pressure and the application of geomechanical tests; the central part of the cell is made of aluminium to maintain X-ray transparency. We used X-ray computed tomography (CT) to monitor the dissolution process across the chalk sample during core flooding. An example of a CT of a chalk sample can be seen in Fig. 1, along with a photo of the flow cell. The chemical compositions of the effluents were analysed using ion chromatography (IC) techniques, and the continuous measurements of differential pressures across the core provided information on the changes in permeability.

We present two case studies of chalk dissolution at two HCl concentrations. In the first study, HCl (7 x 10-4 mol/kg) was injected while in the second study, we injected a higher concentration of HCl (5 x 10-3 mol/kg). The concentration of HCl in the first experiment was chosen to simulate the pH of CO_2-acidified brine at reservoir conditions. In this experiment, we saw no discernible change in the permeability or porosity of the core after injecting over a short period indicating that chalk might be a suitable storage medium.

In the second experiment, with the more concentrated acid, we imaged dissolution of chalk as it occurred. We clearly resolved the dissolution pattern (see **Fig. 2**) and show the development of wormholes in the core. This figure only shows a tenth of the CT scans we performed during the experiment, which means we can track the development of the channels with a high temporal resolution. The development of tunnels seen in **Fig. 2** results in a 61% increase in permeability, despite the porosity only increasing by 0.3%. IC measurements shown in **Fig. 3** track the changes in of calcium, sodium, chloride and sulphate in the effluent during the saturation of the core sample using de-ionized water and the subsequent injection of HCl.

The combination of X-ray CT with effluent analysis and permeability and porosity measurements allows us to study chalk dissolution in detail. Furthermore, the triaxial capabilities of our flow cell makes it possible to quantify the weakening of chalk after dissolution. This is valuable data that feed into the risk analysis of CS and can be used as input for modelling and reservoir simulations.

Fig. 1: Left: An example of a CT scan of chalk sample.
Right: The triaxial flow cell inside a CT scanner.

(a) Dissolution patterns in chalk 65 PVs has been injected. **(b)** Dissolution patterns in chalk 100 PVs has been injected. **(c)** Dissolution patterns in chalk 140 PVs has been injected.

Fig. 2: Difference images between the scan of the saturated chalk sample and scans taking during injection of HCl.

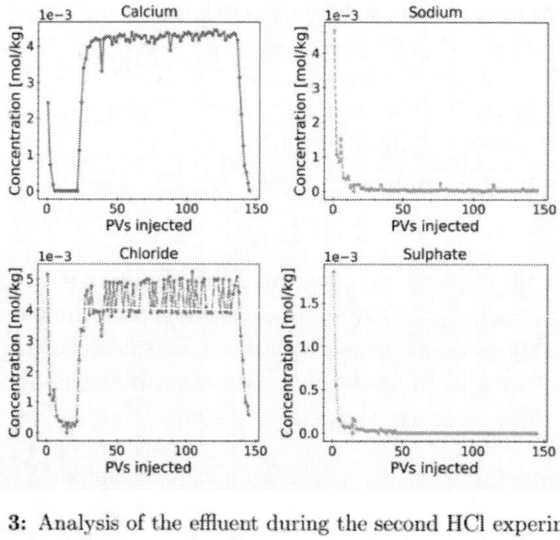

Fig. 3: Analysis of the effluent during the second HCl experiment.

438a Roadmap Towards Zero-Emission Shipping: Global Long-Term Planning of Green Corridors

Hanchu Wang, Prodromos Daoutidis, Qi Zhang

The International Maritime Organization is acting to regulate greenhouse gas emissions from marine shipping. The goal is to reduce emissions by at least 50% compared to 2008 by 2050 and continue to eliminate emissions from the shipping sector in alignment with the target set by the Paris Agreement [1]. Decarbonizing maritime transportation requires substantial investments in new technologies and infrastructure for producing, distributing, and utilizing alternative marine fuels. To accelerate the maritime energy transition away from fossil-based fuels, the concept of so-called green shipping corridors has been proposed [2]. In this context, a green corridor refers to a shipping route that is supported by low- and zero-emission maritime transportation solutions.

In a previous study, we conducted a global analysis of green shipping corridors by designing a network of alternative fuel production sites, transportation links, and bunkering ports that could meet a significant portion of the global marine fuel demand [3]. However, the study only considers green ammonia as an alternative fuel and is limited to shipping routes for bulk carriers and containerships. Moreover, it is a single-period analysis, which does not provide conclusive insights in terms of how such a network of green corridors should be established over time.

In this work, we propose an extended multiperiod supply chain planning model that allows us to optimize the development of a green corridor network over a given time horizon. We consider a larger set of vessel types, a finer distribution of shipping demands, and multiple alternative marine fuel options, including e-methanol and green ammonia. The results from our analysis show the optimal locations of alternative fuel production sites and bunkering ports as well as the interplay between different technologies over time, hence providing a feasible road map for the development of green shipping corridors in the future.

References:

[1] DNV (2022). Maritime Forecast to 2050.

[2] The Getting to Zero Coalition (2021). The Next Wave: Green Corridors.

[3] Wang, H., Daoutidis, P., & Zhang, Q. (2023). Ammonia-based green corridors for sustainable maritime transportation. Digital Chemical Engineering, 6, 100082.

438b The Role for Synthetic and Biofuel Pathways in an Integrated, Net-Zero Energy System

Jun Wen Law, Dharik Mallapragada

Integrated energy systems modeling is a key tool for identifying cost-effective, resilient, and equitable decarbonization pathways that account for various spatial, temporal, and technological interactions. While electrification of final energy is a key pillar for economy-wide decarbonization, the practical challenges associated with the electrification of certain end-uses as well as intermittency of variable renewable energy (VRE) electricity supply creates opportunities for a range of other vectors and technologies to achieve a net-zero emissions goal. These include the utilization of alternative energy carriers such as hydrogen (H2), biofuels and other synthetic fuels, as well as negative emission technologies (NETs) such as direct air capture (DAC) and bioenergy with carbon capture and sequestration (BECCS). The goal of this study is to quantify the drivers impacting the role for synthetic fuel and biofuel processes as part of the least-cost solution of a net-zero integrated energy system, while considering their interaction with supply chains for electricity, H2, and CO2, to satisfy specified final energy demands across end-uses in transportation, buildings, and industry.

Our approach relies on formulating and solving a capacity expansion model (CEM) that accounts for spatial and temporal variations in resources, supply and demand balances for each energy vector, as well as investment and operational cost characterization. The open-source modeling framework, *EnX,* builds on a previously developed open-source CEM for electricity and H2 infrastructure planning [1], and takes the perspective of a central planner to minimize operating and capital costs associated with the supply chain of each vector (electricity, CO2, H2, biomass, liquid fuels and methane). Uniquely, the EnX modeling framework allows for evaluating a range of technological options for meeting demand for liquid and gaseous fuels under net-zero emissions constraints while considering their interaction effects with the supply chain of other vectors. For liquid fuels and methane production, the model considers both synthetic fuel production from captured CO2 as well as biofuel production, along with the option of continuing to use fossil fuels in conjunction with NETs. For the biofuel and synthetic fuel processes, the model accounts for various liquid products (jet, gasoline, diesel) and their disposition along with the potential for sequestering or venting any unconverted CO2 in the process. In addition, the model emissions accounting boundary also incorporates energy needs and can incorporate non-CO2 greenhouse gases associated with production of primary energy sources like biomass and fossil fuels.

We apply the EnX model to explore net-zero energy system outcomes for the continental U.S. using a 9-zone spatial representation and considering 11 weeks of system operations at an hourly resolution to approximate temporal variability in energy supply and demand. We use the model to shed light on the value of alternative biofuel and other synthetic fuel processes under a net-zero emissions constraint, exogeneous final energy demands for H2, electricity, liquid and methane and varying assumptions about availability of biomass resources and CO2 sequestration sites. When CO2 sequestration and biomass resources are highly constrained, we find that synthetic fuels processes with high carbon conversion efficiency are favored for liquid fuel production, where the H2 is produced primarily from electrolysis and CO2 sourced from DAC. On the other hand, when both CO2 sequestration and biomass resources are less constrained, there is greater reliance on NETs, primarily via BECCs, to offset emissions from using fossil fuels to meet liquid demand.

[1] See DOLPHYN model: https://github.com/macroenergy/DOLPHYN

438c Multi-Period Joint Capacity Expansion Planning of Renewable Electricity and Hydrogen Production: An Application to South Korea

Yechan Choi, Jay Hyung Lee, Seongmin Heo

The increasing concerns over greenhouse gas emissions and the consequential climate change have driven the search for sustainable alternatives to conventional fossil fuels. One such alternative is green hydrogen, generated from the electrolysis of water using renewable electricity and free from greenhouse gas emissions during combustion. In order to meet the increasing demand for hydrogen and achieve cost-effective hydrogen production, the capacity expansion of both renewable energy systems and hydrogen supply chain must be jointly planned while considering the dynamics between the two systems. However, despite the interdependence between these two systems, research on their joint planning is rather scarce [1, 2].

One study proposed a linear continuous truck routing model for hydrogen transportation within a joint planning framework [3], where a robust optimization was employed to address uncertainties. Another study investigated various technologies associated with the generation and storage of both electricity and hydrogen [4]. Specifically, hydrogen production through electrolysis was compared against that from steam methane reforming with or without carbon capture and storage. However, these studies did not account for the long-term dynamics of the hybrid system or the lifetimes of different facilities, which have significant implications for long-term planning. On the other hand, there exist several works which addressed long-term, multi-period planning for the hydrogen and renewable power production by considering facilities with varying lifetime [5, 6]. However, in these studies, the focus was on the planning of either hydrogen production or renewable power generation, while the other was assumed to have a fixed infrastructure, i.e., joint planning was not explicitly considered.

To this end, in this talk, a multi-period joint capacity expansion planning framework is proposed for renewable electricity and hydrogen production to meet the long-term future demands. The proposed framework determines the optimal size and location of production, storage, transmission/transportation, and operation of the hybrid supply chain network by solving a cost minimization mathematical programming problem. This problem is a multi-time scale decision making problem, where detailed operational decision constraints at an hourly level are combined with investment planning decisions over a few decades. The effectiveness and advantages of the proposed framework will be illustrated by a comprehensive case study for the hydrogen economy of South Korea, taking into account a range of policy and technology scenarios.

[1] Riera, J. A., Lima, R. M., & Knio, O. M. (2023). A review of hydrogen production and supply chain modeling and optimization. *International Journal of Hydrogen Energy.*

[2] Farrokhifar, M., Nie, Y., & Pozo, D. (2020). Energy systems planning: A survey on models for integrated power and natural gas networks coordination. *Applied Energy*, 262, 114567.

[3] Wang, S., & Bo, R. (2021). Joint planning of electricity transmission and hydrogen transportation networks. *IEEE Transactions on Industry Applications*, 58(2), 2887-2897.

[4] Bødal, E. F., Mallapragada, D., Botterud, A., & Korpås, M. (2020). Decarbonization synergies from joint planning of electricity and hydrogen production: a Texas case study. *international journal of hydrogen energy*, 45(58), 32899-32915.

[5] Li, C., Conejo, A. J., Liu, P., Omell, B. P., Siirola, J. D., & Grossmann, I. E. (2022). Mixed-integer linear programming models and algorithms for generation and transmission expansion planning of power systems. *European Journal of Operational Research*, 297(3), 1071-1082.

[6] Yoon, H. J., Seo, S. K., & Lee, C. J. (2022). Multi-period optimization of hydrogen supply chain utilizing natural gas pipelines and byproduct hydrogen. *Renewable and Sustainable Energy Reviews*, 157, 112083.

438d Spatial Trade-Offs in Designing a Sustainable Cellulosic Biofuel Supply Chain: Insights from High Resolution Modelling

Eric O'Neill, Caleb Geissler, Christos Maravelias

Cellulosic biofuel produced from dedicated energy biomass requires crops to be established on marginal lands, new facilities to be constructed, and large supply systems to be designed to connect the upstream and downstream portions of the process. The cellulosic biofuel supply chain (SC) is highly interconnected and the design and operational decisions, all the way from field to product, need to be considered simultaneously and at a high spatial resolution to account for the trade-offs among different components of the system.

In this work, we use a high resolution mixed-integer linear programming model to analyze the spatial trends that arise when designing a system to produce liquid cellulosic biofuel in the US Midwest. Because greenhouse gas (GHG) mitigation is one of the primary goals of producing liquid biofuel, we use the epsilon constraint method to find pareto-optimal solutions for a range of biofuel demands. Upstream, model decision variables determine biomass establishment, fertilization, and harvesting at the field level which influence the yield and soil carbon sequestration. Downstream, we consider 3 primary conversion pathways at the biorefinery including fermentation to ethanol, pyrolysis, and gasification and explore the option to install carbon capture and storage (CCS) technologies to capture and sequester CO_2 from a variety of process streams with a range of CO_2 concentrations.

The results of our analysis show that landscape design, choosing where to establish biomass crops and how to manage them, has a large impact on the GHG balance of the system especially at restrictive epsilon constraints. We show that the technology portfolio at the biorefineries is dependent on the location of each facility, the spatial trade-offs between the local price and GHG impact of electricity, the bioenergy landscape design, and the cost to capture, transport, and sequester CO_2. Finally, we offer insights into the SC network design and operational strategies to meet a range of biofuel demands under different epsilon constraints.

438e Industry Perspective on Green Ammonia: Current Status and Challenges in Synthesis, Transportation and Utilization

Shantanu Patwardhan, Kiran Chaudhari

Commercial scale ammonia is synthesized by utilizing the Haber-Bosch process, where hydrogen and nitrogen are exothermically reacted over a catalyst at high temperature and pressure. Nitrogen is sourced directly from air and purified via an Air Separation Unit (ASU). The source of hydrogen, which is more complex and requires energy intensive processes, determines the sustainability of ammonia production. For example, hydrogen produced from steam reforming of light hydrocarbons or gasification of coal to produce synthesis gas typically has an intensive carbon footprint i.e., they release large quantity of carbon dioxide to atmosphere and therefore, the generated hydrogen is brown (coal gasification) or grey (steam methane reforming (SMR) or similar). In contrast, green hydrogen can be produced by using renewable electricity in an electrolyzer to split water into hydrogen and oxygen, where the net carbon dioxide emission is close to zero. One more example of green hydrogen is when it is produced from synthesis gas generated by biomass gasification. Although, this type of process releases carbon dioxide to air, it is that carbon dioxide which is already absorbed into the biomass from air via photosynthesis, creating a carbon dioxide cycle and therefore generating close to net zero carbon dioxide. The corresponding ammonia synthesis using green hydrogen is termed as green ammonia. It should be noted though, that the color-coding of the synthesized ammonia is simply the result of source of hydrogen generation processes that have varying carbon footprints.

The presentation will focus on the electrolyzer-based green ammonia synthesis process (renewable electricity generation used to produce hydrogen in electrolyzers which, along with nitrogen from ASU, feeding to ammonia synthesis), ammonia storage, transportation and its utilization. The discussion will be around the challenges associated with the green ammonia life cycle.

The cyclic nature of the availability of the renewable electricity and difficulties in synchronizing the ammonia synthesis with it, is one example of challenges during synthesis. Water, being the main feedstock for the process, may compete with limited water resources that are typically used for human consumption. The industry normally looks at green ammonia as the carrier of the renewable energy, where it is expected to use ammonia as is, to produce energy or convert it to hydrogen by cracking process and then use the hydrogen to produce energy. The technologies to convert ammonia to energy are still not fully commercialized. The green hydrogen process is capital intensive – especially the electrolyzers. Added to that are the challenges associated with electrolyzer procurement, which makes the green ammonia synthesis process economically unattractive. The need of co-locating renewable electricity source, water availability and easy connectivity to transportation hub may pose hurdles for the location choice of the green ammonia synthesis facility.

All the above issues with green ammonia synthesis will be elaborated and discussed. It is also intended to propose probable solutions or recommend alternatives.

438f Comprehensive Analysis of Various Types of Hydrogen Refueling Stations : Economics, Environmental, and Safety

Ga-Young Lee, Joo-Sung Lee, Chul-Jin Lee

Efforts to reduce carbon emissions are being made in all sectors for global carbon neutrality. In the transportation sector, which accounts for 15% of greenhouse gas emissions worldwide, a shift from traditional fossil fuels to electric or hydrogen-based fuels is being pursued to overcome direct carbon emissions. Among these, Fuel Cell Electric Vehicles (FCEVs) using hydrogen as a fuel are considered advantageous for their short refueling time, high fuel efficiency, and long driving range.

As of the end of 2021, there were 685 hydrogen refueling stations worldwide, but to meet the increasing demand for FCEVs, about seven times more hydrogen refueling stations are needed. Therefore, it is crucial to determine the type of hydrogen refueling station (HRS), as large-scale new infrastructure needs to be built. HRSs can be classified into two types: on-site, which produce hydrogen directly on site (e.g., water electrolysis, steam methane reforming, and ammonia cracking), and off-site, which store hydrogen transported in a compressed or liquefied form from other regions.

This study comprehensively analyzed the performance in terms of the economic, environmental, and safety aspects for five different kinds of HRSs with high commercialization potential: gaseous, liquid, steam methane reforming, ammonia cracking, and electrolysis. In addition, we analyzed their performances by reflecting on the scenario of commercialization of renewable energy in 2050 to identify the potential of each refueling station.

As a result, in 2020, we found that gaseous HRS (GHRS) was the most economical and environmentally friendly hydrogen refueling station due to its simple process configuration and low cost and global warming potential of materials. However, in terms of safety, electrolysis had the lowest risk because the water is used as the raw material. In 2050, assuming a decrease in investment costs due to the introduction of renewable energy and the commercialization of the electrolysis system, GHRS still has good economic and environmental performance while the electrolysis HRS has a comparable carbon intensity to GHRS.

438g Economic and Environmental Potential of Green Hydrogen Carriers (GHCs) Produced Viareduction of Amine-Captured CO_2

Changsoo Kim, Younggeun Lee, Hyunjoo Lee, Ung Lee, Kyeongsu Kim

Hydrogen is deemed as a crucial component in the transition to a carbon-free energy system, and researchers are actively working to realize the hydrogen economy. While hydrogen derived from renewable energy sources is a promising means of providing clean energy to households and industries, its practical usage is currently hindered by difficulties in transportation and storage. Due to the extreme operating conditions required for liquefying hydrogen, various hydrogen carriers are being considered, which can be transported and stored at mild operating conditions and provide hydrogen at the site of usage. Among various candidates, green hydrogen carriers obtained via carbon dioxide utilization have been proposed as an economically and environmentally feasible option.

Herein, the potential of using methanol and formic acid as green hydrogen carriers are evaluated regarding various production and dehydrogenation pathways, within a hydrogen distribution system including the recycle of carbon dioxide. Recent progress in carbon dioxide utilization processes, especially conversion of carbon dioxide captured in amine solutions, have demonstrated promising results for methanol and formic acid production. This study analyzes seven scenarios that consider carbon dioxide utilization-based thermocatalytic and electrochemical methanol and formic acid production, as well as different dehydrogenation pathways, and compares them to the scenario of delivering liquefied hydrogen. The scenarios are thoroughly analyzed via techno-economic analysis and life cycle assessment methods. The results of the study indicate that methanol-based options are economically viable, reducing the cost up to 43% compared to liquefied hydrogen delivery. As for formic acid, only the electrochemical production method is profitable, retaining 10% less cost compared to liquefied hydrogen delivery. In terms of environmental impact, all of the scenarios show higher global warming impact values than liquefied hydrogen distribution. However, results show that in an optimistic case where wind electricity is widely used, electrochemical formic acid production is competitive with liquefied hydrogen distribution, retaining 51% less global warming impact values. This is because high conversion can be achieved at mild operating conditions for the production and dehydrogenation reactions of formic acid, reducing the input of utilities other than electricity. This study suggests that while methanol can be a short-term solution for hydrogen distribution, electrochemical formic acid production may be a viable long-term option.

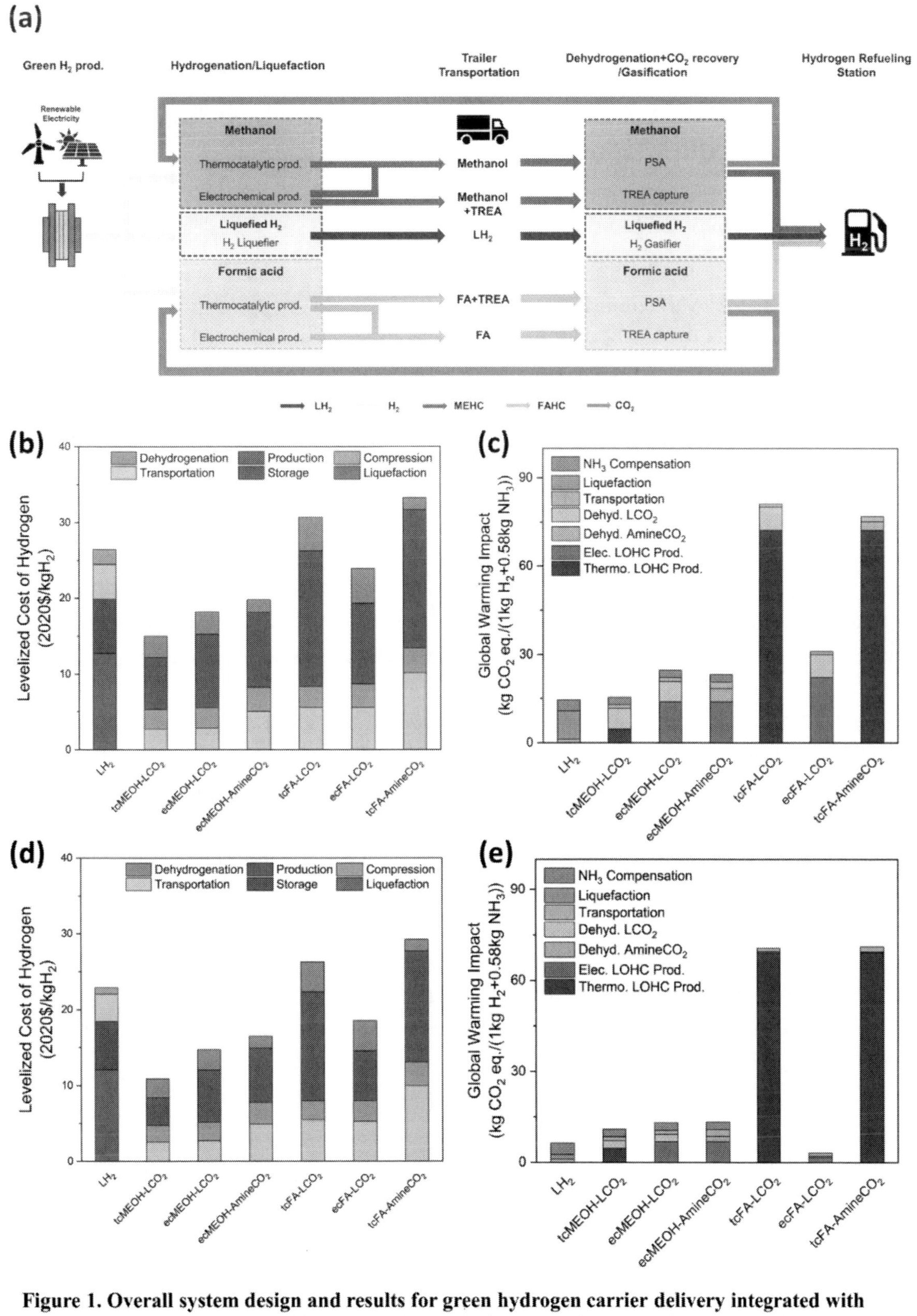

Figure 1. Overall system design and results for green hydrogen carrier delivery integrated with carbon dioxide utilization. (a) The proposed green hydrogen carrier system integrated with carbon dioxide utilization, the techno-economic analysis results for the (b) base case and (d) optimistic case, and the life cycle assessment results for the (c) base case and the (e) optimistic case. Results show that in the current situation methanol production is the most economical, while with future cost reductions considering green hydrogen and renewable electricity production, electrochemical formic acid production provides an economically and environmentally feasible option.

533a *In-Situ* CO_2 Capture-Conversion in Methane Dry Reforming for Direct Synthesis of Hydrogen over Bifunctional Structured $Ce_{1-x}Co_x$nio/Ca perovskites-Type Oxides Monoliths

Khaled Baamran, Shane Lawson, Ali Rownaghi, Fateme Rezaei

The need for CO_2 utilization is rising on account of its high abatement cost and low market value. Bifunctional materials (BFMs) comprised of adsorbents and catalysts have shown promise as a pathway to capture CO_2 at high temperature and subsequently utilize the contaminant as a feedstock. [1, 2] However, previous efforts in the development of BFMs have been limited, with efforts put forth towards structuring such materials into facile contactors being narrower still. In this study, we therefore expanded the use of 3D printing technology to previously unexplored BFMs – comprised of a CaO adsorbent and trimetallic Ni/Co/Ce perovskite catalysts – for the application of combined CO_2 capture and methane reforming. Three honeycomb monoliths comprised of 50% adsorbent and 50% catalyst with varied ceria/cobalt ratio were 3D printed to assess the role of cobalt on catalytic properties and overall performance. The samples were vigorously characterized using X-ray diffraction (XRD), energy dispersive spectroscopy (EDS), N_2 Physisorption, X-ray photoelectron spectroscopy (XPS), H_2-TPR, *In-Situ* CO_2 adsorption/desorption XRD, and NH_3-TPD. From these, it was determined that the ceria/cobalt ratio did not influence crystallinity, textural properties, CO_2 adsorption/desorption reversibility, or metallic dispersion, but did cause reductions in acidity, CO_2/CaO bond strength and reducibility at elevated cobalt loading. Assessing the samples via combined CO_2 capture and methane reforming from 500-700 °C revealed that such differences in physiochemical properties lowers H_2 and CO yield at higher cobalt loading, leading to the best performance in the $Ce_{0.75}Co_{0.25}$NiO/Ca sample that achieved 52% CO_2 conversion, 94% CH_4 conversion, 61% H_2 yield, and 2.30 H_2/CO ratio at 700 °C. The stability of this sample was assessed across five adsorption (T = 600 °C)/reaction (T = 700 °C) cycles, showing only marginal losses in H_2/CO yield. Thusly, these findings successfully expand the use of 3D printing to unexplored perovskite BFMs and demonstrate an important proof-of-concept for their use in combined CO_2 capture and utilization for methane reforming processes.

533b Tandem CO2 Electrolysis System for Chemical Production

Feng Jiao

As our society faces the pressing challenges of climate change and global warming, driven in part by increasing atmospheric CO2 levels, reducing CO2 emissions has become a critical mission in the pursuit of a sustainable future. Traditional chemical industry processes often rely on fossil fuels, which inevitably emit substantial quantities of CO2. In response, electrochemical processes have garnered interest for their potential to be more environmentally friendly and exhibit a smaller carbon footprint when powered by renewable energy sources.

Our research group is currently dedicated to the development of CO2 electrolysis devices that convert CO2 into value-added chemicals and fuels through innovative electrocatalyst design and reactor engineering. In this presentation, we will showcase our recent work on a two-step tandem CO2 electrolysis system. We have reported an internally coupled purification strategy that significantly enhances acetate concentration and purity in CO electrolysis. This approach employs an alkaline-stable anion exchange membrane with high ethanol permeability and a selective ethanol partial oxidation anode to regulate the CO reduction product stream.

We successfully demonstrated a stable 120-hour continuous operation of the CO electrolyzer at a current density of 200 mA cm-2 and a full-cell potential of less than 2.3 V, consistently producing a 1.9 M acetate product stream with a purity of 97.7%. This performance is among the best reported in the literature [1]. The ability to convert CO2 into acetate has opened up the possibility of developing an electrochemical-biological hybrid approach to produce food from CO2, offering much higher efficiency than natural photosynthetic pathways [2].

References:

1. Overa, B. Crandall, B. Shrimant, D. Tian, B. H. Ko, H. Shin, C. Bae and F. Jiao*. Enhancing acetate selectivity by coupling anodic oxidation in carbon monoxide electroreduction. Nature Catalysis5, 738-745 (2022). 10.1038/s41929-022-00828-w
2. C. Hann, S. Overa, M. Harland-Dunaway, A. F. Narvaez, D. N. Le, M. L. Orozco-Cardenas, F. Jiao* and R. E. Jinkerson*. A hybrid inorganic-biological artificial photosynthesis system for energy-efficient food production. Nature Food3, 461 (2022). 10.1038/s43016-022-00530-x

533c Production of Carbonates from Industrial Waste Via Carbon Mineralization Towards a Circular Economy

Ning Zhang, Danny Huang, Aaron Moment

Carbon mineralization is a process to fix CO_2 into solid carbonates for permanent storage. This process has been proposed as an efficient approach to achieving the goal of net-zero greenhouse gas emissions. From a thermodynamic perspective, mineralization occurs spontaneously, however, the kinetics are very slow and can be accelerated through engineering. Industrial wastes (e.g. demolition & construction materials, coal fly ash, steel slags, etc.), which contain a high concentration of alkaline earth metals (i.e. calcium) can be used as sources to produce calcium carbonates instead of landfilling them.

Herein, we present studies focusing on the utilization of alkaline industrial wastes to produce different polymorphs of calcium carbonates for different industrial applications. We propose a multiple-stage process to treat the feedstock to obtain high purity carbonates. First wastes are dissolved in an acidic solution to extract out Ca, and followed by an impurities removal (Fe, Al and partial Si) via pH swing process, and finally a crystallization reaction is occurred by bubbling CO_2. We also investigated the effects of various parameters on different metals extraction, such as pH values, temperatures, acid dosing speed, slurry densities, etc. We are optimizing the process by applying life cycle assessment and techno economic assessment to evaluate it from environmental and economic perspectives. Salt electrolysis is used to produce acid and base to offset chemical costs and the recycled salty solution will be the parent solution for this process. Finally, we scaled up this process to a 30 L reaction with mild and simple reaction conditions. These produced carbonates and solid residues can be used as supplementary cementitious materials or fillers to achieve a circular economy.

86f Techno-Economic and Life Cycle Assessment of Reactive CO2 Capture to Renewable Natural Gas

Alvina Aui, Hannah M. Goldstein, Nathan C. Ellebracht, Wenqin Li, Simon H. Pang

The global atmospheric concentration of carbon dioxide (CO2) has increased to 400 ppm over the last few years, mainly due to the generation of energy from fossil sources. The alarming increase in CO2 levels has led to a global need to simultaneously reduce CO2 concentration and deploy renewable energy sources. A strategy to achieve this is through a novel process called Reactive CO2 Capture (RCC). RCC refers to a process that integrates the capture and conversion of CO2 into a value-added product via fewer unit operations than separate capture and conversion. In our study, RCC utilizes and integrates two existing and developed processes – direct air capture and the Sabatier reaction. Through RCC, CO2 can be captured and directly converted into methane using renewable hydrogen. Methane, also known as natural gas, is a large source of energy carrier; used in hydrogen, electricity and heat production, and even as a transportation fuel. The natural gas produced via RCC is also considered a renewable source when the hydrogen used in the Sabatier reaction is renewable, and thus, this new form of renewable energy can be deployed and used in existing infrastructure.

To systematically compare this novel technology, a techno-economic analysis (TEA) and life cycle assessment (LCA) was performed. The TEA evaluates the potential energetic and economic savings of this integrated process, while the LCA analyzes the total carbon intensity of the production of renewable natural gas (RNG) via RCC. RCC is also being compared to a separated process and other conventional and existing technologies in producing the same product. Observed in the preliminary TEA and LCA is a 10-15% reduction in the minimum selling price of RNG and the carbon intensity of RNG. This is mainly due to increased energy efficiency, process intensification and heat integration, reduced capital expenditure, and the ability to channel CO2 from the air as feedstock into a product resulting in potentially net-negative emissions. Finally, a sensitivity analysis of an array of experimental parameters and its effect on the TEA and LCA is also being studied to better inform and understand the opportunities and limitations of the technology.

533f Feasibility Study of Alkaline Metal Oxide Modified Cu-ZnO-Al_2O_3 for Reactive Capture of CO_2 to Methanol

Chae Jeong-Potter, Martha A. Arellano-Treviño, Wilson McNeary IV, Anh To, Daniel Ruddy

Rising global temperatures associated with increased anthropogenic emissions of CO_2 to the atmosphere have put CO_2 capture, utilization, and storage (CCUS) technologies at the forefront of all the scenarios for climate change mitigation. Reactive carbon capture (RCC) technologies, in which capture and conversion of CO_2 occur in a single reactor, are particularly energetically and economically attractive by avoiding the need to purify, compress, and transport the captured CO_2. To this end, the dual function materials (DFM) have been developed to enable CO_2 capture and subsequent conversion to methane. Dual function materials (DFMs) have recently been developed as a promising technology for RCC.

DFMs are comprised of basic CO_2 adsorption sites co-located with catalytic hydrogenation sites on the same high surface area carrier. They have demonstrated stable performance over simulated realistic direct air capture (DAC) and post-combustion capture conditions with subsequent production of methane upon exposure with renewably sourced H_2. While renewable CH_4 would be an excellent transition fuel, it is economically noncompetitive compared to fossil methane and minimally adaptative as a platform molecule for upgrading to higher C-containing compounds. This requires design and investigation of DFMs that enable CO_2 capture and conversion to more valuable and more useful C1 products like methanol. Herein, we propose a DFM consisting of an alkali-modified Cu-ZnO-Al_2O_3 (CZA), the state-of-the-art commercial methanol synthesis catalyst, for CO_2 capture from a CO_2-rich waste gas stream and in-situ conversion to methanol (Figure 1). We explore the hydrogenation activities of alkali modified CZA over a variety of temperatures (150-300°C) and pressures (0.6-4 MPa) at stoichiometric 3:1 H_2:CO_2 ratio. The DFMs are characterized via CO_2 TPD, CO_2 chemisorption, and DRIFTS. Proof of concept of cyclic CO_2 capture and hydrogenation to methanol at the selected temperature and pressure conditions is also presented.

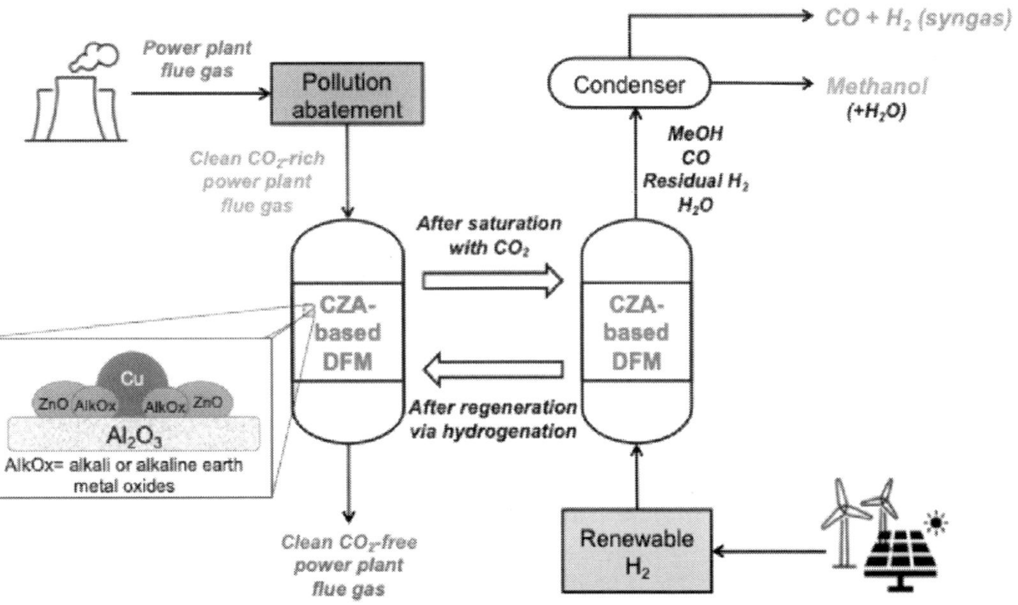

Figure 1. Simplified representation of process flow diagram for reactive CO_2 capture (RCC) to methanol using renewable H_2 and dual function materials (DFMs). The DFMs are comprised of an alkali modified commercial $Cu\text{-}ZnO\text{-}Al_2O_3$ (CZA) catalyst.

533g Accelerated CO2 Mineralization of Acid Mine Drainage Waste

Hamid Radfarnia, Kourosh Zanganeh, Laleh Dashtbankenari, Sanaz Mosadeghsedghi, Konstantin Volchek

Mining sector is one of the largest industrial CO_2 emitters in Canada. To support Canada's commitment for the transition to net zero emissions by 2050, several regulations and approaches are recently taken into effect to this sector, trying to reduce CO_2 emissions and increase energy efficiency in mining operations. One approach to achieve this goal is the implementation of negative emissions technologies that can permanently store CO_2. One such carbon capture and sequestration strategy is mineral carbonation which involves reacting the CO_2 with an alkaline source to produce solid carbonate minerals. Application of the alkaline wastes have become increasingly more attractive than natural minerals, owing to their higher reactivity with the CO_2.Canada's mining industry produces tonnes of mining wastes such as mine tailing and acid mine drainage (AMD) yearly, which are a potential secure source for CO_2 mineralization deployment. These wastes are rich in magnesium and calcium compounds, which are the promising candidates for sinking CO_2 through carbonation processes to produce solid stable carbonate products. On the other hand, the later source, AMD, is primarily one of the most hazards to ground and underground water sources on the mining sites and is characterized by low pH and high concentrations of heavy metals and sulfates. These contaminants can be easily leached out, having devastating effects on surrounding environment. On mining sites, AMD remediation is commonly achieved by the addition of lime that neutralizes the solution, consequently, makes the treated ADM solution rich in calcium species.

The current ex-situ solution-based CO_2 mineralization processes suffer from either low reaction rates or a need for expensive thermal pre-treatment that negatively affect the economy and environmental burdens of the system. Therefore, acceleration of involved reactions is a promising approach that could effectively enhance such CO_2 carbonation technology competitiveness. In this respect, the main goal of this work is to present a new mineralization process assisted by ultrasonic treatment that accelerates CO_2 reaction with active species of Canadian AMD solution, that is, magnesium and calcium compounds. Ultrasonic processing is a well established technology that can improve chemical reaction kinetics that are inherently slow towards gaining a higher product yield. Successful application of ultrasonic processing for producing precipitated calcium carbonate (PCC) and acceleration of calcium dissolution from the industrial slags has been addressed in the literature [M. Altiner et al, Ultrasonics Sonochemistry 72 (2021) 105421; A. Said et al. / Chemical Engineering and Processing 89 (2015) 1–8].

It is speculated that AMD-based CO_2 mineralization reaction integrated with ultrasonic processing is featured with a better CO2 gas dissolution and kinetics of carbonate precipitation which is slow in its nature. To this end, we developed a CO_2 mineralization test set-up, enabled us to apply different operation conditions for CO_2 carbonation study. Industrial AMD solution was subjected to CO_2 carbonation trials. The impact of solution pH, temperature, CO_2 partial pressure and the dosage of the alkaline agent were investigated to find the optimum conditions of operation for achieving a higher carbonation yield. The precipitate samples were characterized, and CO_2 carbonation efficiencies with and without ultrasonic processing were determined.

264a Dynamic Model of a Solar Fuel System for the Thermochemical Production of Syngas from H_2o and CO_2

Remo Schäppi, Aldo Steinfeld

We report on a dynamic model of a solar fuel system for the solar thermochemical production of syngas – a specific mixture of CO and H_2 – from CO_2 and H_2O via a ceria-based redox cycle using concentrated solar energy.

The system is driven by concentrated solar process heat to effect the simultaneous co-splitting of CO_2 and H_2O via a ceria-based thermochemical redox cycle. The solar reactor for effecting this redox cycle consists of a cavity-receiver containing a reticulated porous ceramic (RPC) foam structure made of pure CeO_2. The characteristic redox cycle is operated under a temperature/pressure-swing mode, consisting of three phases: 1) The reduction phase, during which the solar reactor is heated with concentrated sunlight up to the desired reduction end temperature of up to 1500°C to release O_2 from CeO_2, assisted through lowered total pressure by a vacuum pump and an Ar sweep gas flow. 2) A cool-down phase under atmospheric pressure during which the solar reactor, re-pressurized by injecting CO_2, cools down to the oxidation start temperature. 3) The oxidation phase, during which CO_2 and H_2O are co-injected into the reactor's cavity, react with the reduced ceria to form syngas. An in-depth experimental parametric study of the system has previously been carried out and the presented dynamic model allows further parameter analysis and offers additional insight to the system performance.

The dynamic reactor model is a grey box model based on energy and mass conservation equations within predefined model domains and considers mass and energy transfer between the domains and over the system boundaries. Model domains contain solid, gaseous and porous volumes with the redox reaction taking place in the porous volumes representing the RPC structure inside the cavity. The model follows a lumped parameter approach to describe the mass and energy reservoirs and is implemented using MATLAB. Unknown system parameters are determined empirically through comparing numerically simulated results to experimental data collected over multiple characteristic cycles with varying process parameters. Model validation is done by comparing simulated results to a different data set of reactor cycles.

Apart from the model description, we present the parameter identification and validation as well as first model applications. We look at system behaviour when linking two or more reactors together in continuous operation and study the reactor behaviour under intermittent solar conditions.

264b Continuum Model for Transport Phenomena and Thermochemical Reactions in Oxidation Reactors for High-Temperature (>1000 °C) Heat Extraction

David Korba, Michael Hayes, Philipp Schimmels, Kelvin Randhir, James F. Klausner, Nick AuYeung, Like Li

A transient three-dimensional heat and mass transfer continuum model was developed to simulate the transport phenomena coupled with exothermic oxidation of magnesium-manganese-oxide particles in a high-temperature moving-bed reactor for thermochemical energy storage. The reactor included direct high-temperature heat extraction at temperatures greater than 1000 °C. The model was applied to simulate a corresponding experimental setup, the results of which are used for detailed model validation. The predicted temperature and oxygen concentration profiles were in good agreement with experimental data. A follow-up power budget and efficiency comparison between the model and experiments demonstrated excellent agreement, with a predicted steady-state efficiency of 29.4% for a baseline set of operating parameters. The performance of the reactor was further investigated through two parametric studies (varied particle flow rate and gas extraction ratios), which demonstrated a strong dependence of reactor efficiency (varied between 15% – 40%) on both operating parameters. The present heat and mass transfer model will be extremely useful in further reactor design and scale-up studies, as well as for selection of operating conditions to maximize efficiency.

264c Thermal Energy Transport Using the Sorption-Assisted Boudouard Reaction

Alexandre Yokochi, Riley Choquette, Benjamin Phillips, Annette von Jouanne

High-quality process heat from clean sources is necessary for decarbonizing the industrial sector. Thermal energy may be transported long distances with high exergetic efficiencies using the set of reversible chemical reactions described below

$2CO + MO \leftrightarrow MCO_3 + C$ ΔH = -350 kJ ~ -440 kJ

where M represents a Group 2A element. The reactions combine the well-known Boudouard equilibrium with the decomposition of metal carbonate groups [1-5], which combined we have denominated the sorption-assisted Boudouard reaction (SABR). Because CO is kinetically stable at ambient temperatures, thermal energy can be stored indefinitely, transported, and released with modeled roundtrip exergetic efficiencies as high as 80% or at temperatures over 1000°C. A plot of roundtrip heat transport exergy efficiency and optimal heat release temperature as a function of process pressure is presented in the attached figure.

In this presentation, the round trip exergy efficiency, reaction kinetics as determined from DSC/TGA data, and an initial technoeconomic model estimating the cost of integrating the technology into existing industrial applications will be shared. In addition, the exergetic efficiency of steampipe heat transport and the SABR process will also be presented.

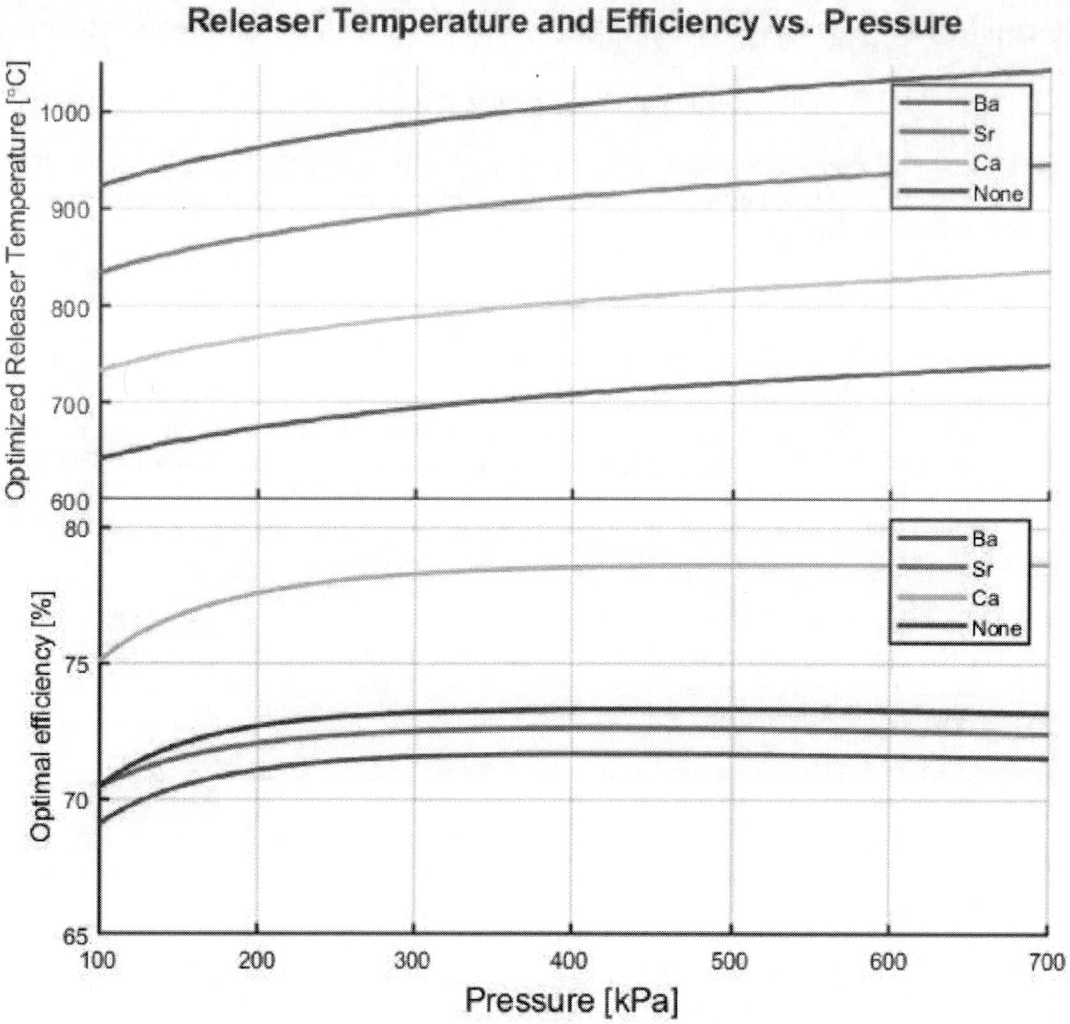
Releaser Temperature and Efficiency vs. Pressure
Optimized Releaser Temperature [°C]
1000
900
800
700
600
Ba
Sr
Ca
None
Optimal efficiency [%]
80
75
70
65
100
200
300
400
500
600
700
Pressure [kPa]

264d Heat Transfer in Directly-Irradiated High-Temperature Particle–Gas Flows for Solar Particle Receiver Applications

Jingjing Chen, Apurv Kumar, Joe Coventry, Wojciech Lipiński

Solar particle receivers (SPRs) are pursued as a promising technology for conversion of concentrated solar radiation for power generation and chemical processing. Particle receivers use absorbing particles as heat transfer and thermal energy storage media. Falling particles form a particle curtain inside a receiver to absorb solar irradiation, and can attain temperatures above 1000°C. A thorough understanding of heat transfer addressing radiative heat transfer in polydisperse particles is crucial to advance the solar particle receiver technology. A complete heat transfer model involving hydrodynamics of particle–gas multi-phase flows, radiation in the particle phase, interfacial convection, and conduction in the gas phase is required to understand the physical problems encountered in SPRs.

We present simulations of particle–gas two-phase flows using the multi-phase particle-in-cell method. The method allows for capturing high-fidelity solid–gas flow characteristics with increased computational efficiency by combing the advantages of both Eulerian and Lagrangian methods. Radiative transfer in polydisperse particles is investigated using a novel radiative transfer model. The radiative transfer model considers polydisperse particles as a multi-component system and is implemented using the Monte Carlo ray-tracing method. A three-dimensional computational fluid dynamics model is developed for directly-irradiated high-temperature particle–gas flows laden with polydisperse particles by coupling particle–gas hydrodynamics of particle–gas flows, radiative heat transfer in non-grey absorbing, emitting and anisotropically-scattering multi-component media, conduction in the gas phase, and interfacial convective heat transfer. The computational model is employed to study thermal performance of a free-falling particle receiver.

264e Challenges in Using Particulate Materials As Heat Transfer Fluid or Reactant in Solar-Chemical Processes

Martin Roeb

Concentrating solar technology can replace fossil fuels not only in electricity generation, but also via generation of fuels and commodities in the industrial and transportation sectors.

To advance such solar applications, new technologies are needed: new receivers in which absorbs solar radiation, heat transfer media, high-temperature heat exchangers and storage. Particles made of ceramics or metals have proven in recent years to be an interesting alternative to the previous liquid heat transfer media. They can be used both in receivers, where they directly absorb solar radiation, and as storage materials. In some cases even the particles themselves represent the material to be treated and converted to the desired product like in the case of solar lime or cement production (Moumin et al 2019, Meier et al 2006).

A major advantage is that they can be used for temperatures over 1000 °C. At the same time, they are often cheaper than liquid heat transfer media and non-toxic. Various solar particle receivers have already been successfully tested. However, working with particles at high temperatures and under high solar concentrations comes along with a number of challenges like particles degradation and attrition, erosion of particles ducts, heat transfer and temperature homogeneity in the particle bed, radiation absorptivity of the particles, flowability of particles.

The present contribution will deal with two challenges specifically. One is related to the transport of hot particles, the second one deals with directly heated reactors which require a quartz window which if often prone to failure because of deposition of material on its surface.

Handling the particles is still a major challenge, because the hot particles have to be transported further from the solar receiver to the next process components, such as the storage tank or the heat exchanger.

A project, carried out at DLR, deals with the transport of particles for solar thermal and solar thermochemical applications. Within the project a high temperature transport and lock system for particles at temperatures up to 1000 °C was develop and qualified. For this purpose, a protype setup was developed to demonstrate the controlled handling and low-heat-loss transport of particles. The particles are heated in a closed rotary receiver by concentrated radiation from a solar simulator and then automatically conveyed through the system. The system transports the particles to a storage tank from where the particles are returned to the receiver, thus closing the particle cycle.

For some of the key components the use of innovative fibre-enforced materials was investigated, which can withstand the high temperatures. They turned out chemically inert and suitable for frequent transfer of particles.

A quartz window is often the bottleneck of directly irradiated receiver-reactors due to the deposition of material on its surface, which causes overheating and failure of the window. Several systems analysed in the past to overcome the window problem have been developed based on aerodynamic protection methods such as air curtains or vortex flows (e.g. E. Koepf 2015, A. Chinnici 2016). However, the high amount of gas required by these systems considerably reduces the efficiency of the reactor, and in several cases the window could not be completely protected from particle deposition.

We intend to present an unexplored method to avoid the window problem in a solar reactor. The novel system developed uses electrostatic precipitation of particles (ESP) to prevent the migration of solid particles to the window. The application of ESP in a solar reactor is analysed in depth. An experimental setup was developed to demonstrate and proof the performance of the system at representative conditions. The precipitator showed its best performance using positive corona discharge. Compared to comparison with negative discharge this mode is characterised by lower current densities, enabling a wider operation range with more stable corona discharge, and lower energy consumption. The window protection system turned out to be able to keep the window free from particle contamination when using particles of limestone and cement raw meal.

References:

1. Moumin, S. Tescari, P. Sundarraj, L. de Oliveira, M. Roeb, C. Sattler. (2019) Solar treatment of cohesive particles in a directly irradiated rotary kiln, Solar Energy 182, 480-490
2. Meier, E. Bonaldi, G.M. Cella, W. Lipinski, D. Wuillemin. (2006) Solar chemical reactor technology for industrial production of lime. Solar Energy 80, 1355–1362.
3. Koepf, W. Villasmil, and A. Meier. (2015) High Temperature Flow Visualization and Aerodynamic Window Protection of a 100-kWth Solar Thermochemical Receiver-reactor for ZnO Dissociation, Energy Procedia, vol. 69, pp. 1780–1789, doi: 10.1016/j.egypro.2015.03.148.
4. Chinnici, M. Arjomandi, Z. F. Tian, and G. J. Nathan. (2016) A Novel Solar Expanding-Vortex Particle Reactor: Experimental and Numerical Investigation of the Iso-thermal Flow Field and Particle Deposition, Solar Energy, vol. 133, pp. 451–464, doi: 10.1016/j.solener.2016.04.006
5. P. Rincon Duarte, D. Kriechbaumer, B. Lachmann, S. Tescari, T. Fend, M. Roeb, C. Sattler. (2022) Solar calcium looping cycle for CO2 capturing in a cement plant. Definition of process parameters and reactors selection," Solar Energy, vol. 238, pp. 189–202, doi: 10.1016/j.solener.2022.04.031

264f Development and Testing of a Novel Combined Moving-Fluidized Bed Oxidation Reactor for Efficient High-Temperature Discharge of Thermochemical Energy Storage Particles

Juvenal Ortiz-Ulloa, Lucas Freiberg, Fuqiong Lei, Owen Ramsey, David Korba, Michael Hayes, Philipp Schimmels, Kelvin Randhir, Like Li, Joerg Petrasch, James F. Klausner, Nick AuYeung

Thermochemical energy storage (TCES) using metal oxide particles has emerged as a promising technology to enable renewable energy dispatchability. The basic principle of TCES is to store energy as chemical bonds in metal oxide particles for short or long-term periods, which can then be released as heat when needed. In addition to providing energy storage, TCES offers several other advantages, including high energy density, decoupled charging and discharging steps for long-term storage, and delivery of high temperature heat (~1000 °C). However, there are inherent mass and heat transfer limitations associated with the oxidation reaction that takes place during the discharge process due to the interaction of solids and gases. While the design space for oxidation reactors offers numerous possibilities, including solid-gas contacting patterns like moving beds, fixed beds, and fluidized beds, there is currently no effective decoupled discharge reactor design that can accommodate a heat transfer fluid different than the process gas or fluidizing gas.

To enable effective extraction of high-temperature heat from TCES, a reactor concept has been proposed in which a counter-flow fluidized bed with heat exchanger is located between two moving beds. The operation of the reactor involves introducing reduced magnesium manganese oxide particles into the reactor, which undergo an exothermic oxidation reaction under air. The moving beds enable sensible heat exchange between gas and solids to completely preheat and cool down gas and particles, whereas the fluidized bed facilitates the oxidation reaction and heat extraction. The heat transfer fluid (HTF) destined for the high-temperature application passes through the heat exchanger, removing the reactive heat from the fluidized bed. The proposed design offers several advantages, including feeding and removal of particles at ambient temperature, enabling the integration of different HTFs such as air or sCO2, preventing the carriage of fines downstream with the HTF, and enhancing heat transfer between reactive particles and HTF.

Low order mathematical models were developed to aid in the design and to estimate the effect of operating parameters such as solids and gas flowrates, particle size, final extent of oxidation, and HTF type on the performance of the moving beds, fluidized bed, and heat exchanger plates. The design calculations indicate that a heat output between 0.8W to 1.2W from the particles undergoing oxidation can be produced with a particle flowrate between 1 to 2 g/s, achieving conversions between 70% to 90% using air as process gas. Furthermore, due to the counterflow nature of the operation, the top and bottom moving beds have the potential to completely preheat from 25 °C to 1000 °C the particles and the incoming process gas respectively in a short distance. A finned, narrow channel heat exchanger is employed to accommodate different heat transfer fluids and to maximize heat transfer from the particle bed to the HTF to achieve high temperatures at the outlet.

In addition, a 1kW reactor has been constructed using high-temperature resistant materials such as superalloy Hastelloy X for the heat exchangers and fluidized bed, and non-porous alumina for the top and bottom moving beds. The ceramic-metal joints are achieved by using a combination of ceramic based insulating putty and a high-temperature high-thermal expansion adhesive. The reactor is modular, enabling easy access for maintenance and potential scalability to test larger reactive beds. The system includes an 'L-valve' pneumatic conveyance mechanism for particle flowrate control.

Initial testing using N_2 as a heat transfer fluid indicates that HTF outlet temperatures between 800 °C to 1000 °C can be obtained in the heat exchangers from ~1kW of heat produced on the particle side. The reactor concept has the potential to be used in several applications such as concentrated solar power plants storing chemical energy during the day, and another potential application is in industrial processes, where excess heat from a process can be stored and reused at a later time, leading to energy savings and reduced carbon emissions.

The proposed reactor design has several advantages over existing TCES technologies. One major advantage is the decoupled charging and discharging steps for long-term storage. The proposed reactor design separates these two steps, allowing for greater flexibility and efficiency in energy storage and retrieval. The use of a counter-flow fluidized bed with a heat exchanger also allows for the integration of different heat transfer fluids such as air or supercritical CO_2. This makes the system adaptable to different types of applications and allows for the optimization of heat transfer and energy efficiency. Additionally, the use of a narrow-channel heat exchanger with fins maximizes heat transfer between the particle bed and the HTF, leading to higher temperatures at the outlet.

In conclusion, the presented reactor design for TCES using metal oxide particles represents a potential advancement in the field of energy storage. The implementation of the system in various thermal applications is enhanced by the decoupled charging and discharging steps, integration of different HTFs, and the ability to feed and remove solids and gas at ambient temperature. With further development and refinement, this reactor concept has the potential to advance the way we store and utilize energy from renewable sources.

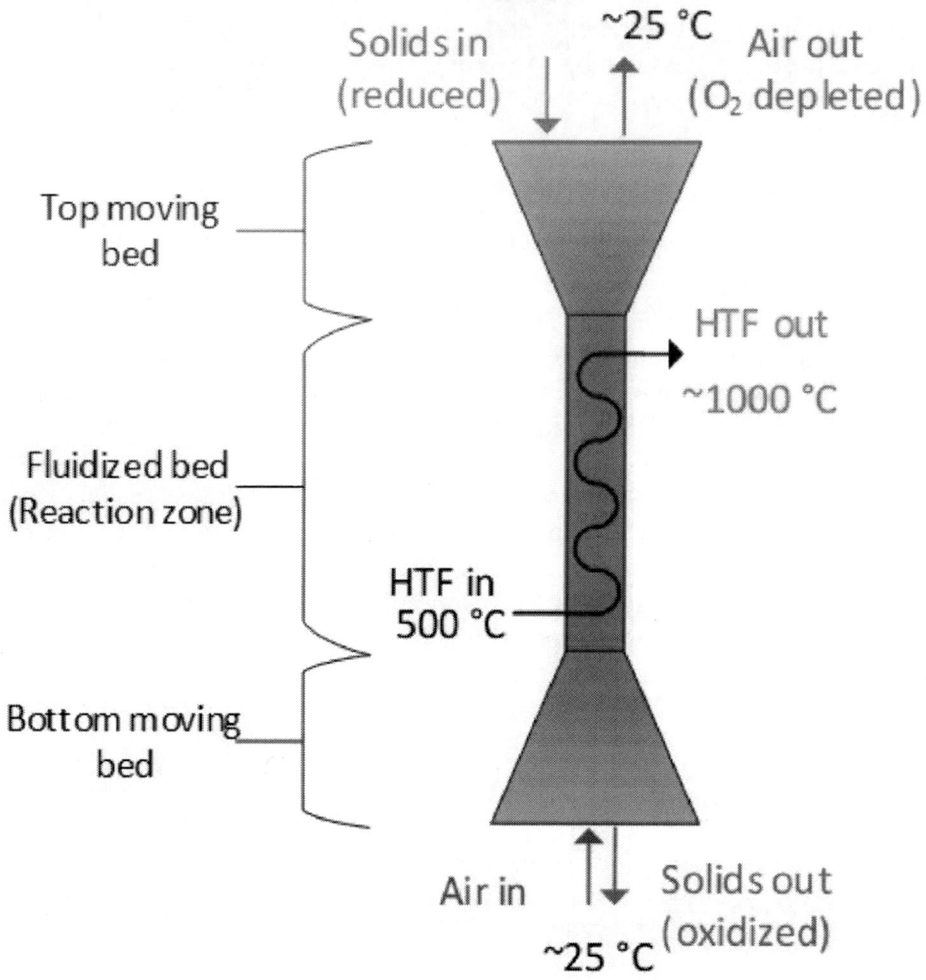
Solids in
(reduced)
~25 °C
Air out
(O_2 depleted)
Top moving
bed
HTF out
~1000 °C
Fluidized bed
(Reaction zone)
HTF in
500 °C
Bottom moving
bed
Air in
Solids out
(oxidized)
~25 °C

325a Experimental Demonstration of Heat Recovery in a Solar Redox Reactor with an Integrated Thermal Energy Storage

Alon Lidor, Sven Hellmüller, Philipp Haueter, Aldo Steinfeld

We consider the production of solar fuels via a thermochemical redox cycle for splitting H_2O and CO_2, driven by concentrated sunlight. The product is a specific mixture of H_2 and CO – syngas – which can be further processed downstream by the established Fischer-Tropsch synthesis to liquid hydrocarbons such as kerosene (jet fuel). The entire solar fuel process chain has been successfully demonstrated with a solar reactor concept based on a cavity-receiver containing reticulated porous ceramic (RPC) ceria structures. The key performance metric of the solar reactor is its solar-to-fuel energy efficiency, which is strongly dependent on the ability to recover heat during the temperature-swing redox cycle.

In previous work, we reported on an experimental investigation of a novel heat recovery method based on coupling the solar reactor with two thermocline energy storage (TES) units made of a packed bed of alumina spheres and using an inert heat transfer fluid (HTF). With this arrangement, the heat rejected from the solar reactor during the switching from the reduction to the oxidation step can be recovered, stored in the TES units, and recuperated back to the solar reactor for the switching from the oxidation to the reduction step. Experimental campaigns with two versions of the setup were able to demonstrate efficient heat extraction from the reactor following reduction, with HTF temperature over 1300°C and up to 70% of the rejected heat transferred to the HTF. However, severe heat losses in the piping manifold between the reactor and TES units prevented storing heat at sufficiently high temperature for the recuperation stage.

In this work, a new system has been developed, combining the solar receiver-reactor and a single TES unit into an integrated assembly, thus removing the need for high-temperature piping and valves. In a preliminary experimental demonstration, using a solar receiver with an inert porous absorber material, successful heat recovery was demonstrated, for the first time, for this type of cavity receiver-reactors, to the best of the authors' knowledge. The new system and the experimental results will be presented.

325b Experimental Dataset and Numerical Model Validation of a Lab Scale Solar Volumetric Receiver for High Temperature Industrial Heating

Aysha Melhim, Fathya Salih, Insha Shaikh, Konstantinos Kakosimos

Concentrated solar technologies (CST) have the potential to revolutionize chemical industry due to their ability of delivering irradiance directly on the transformed chemicals and achieving very high temperatures. Unlike photovoltaics', where energy efficiency is hindered by the intermediate energy conversion steps, the CST converts the light into thermal energy, directly, achieving higher thermal efficiencies. At the same time, few of the main challenges are the intermittent character of solar energy and lack of a true volumetric absorption of the delivered power. Concerning the latter, multiple solar receiver designs were suggested to improve the volumetric effect, therefore minimizing the radiation losses and improving the thermal efficiency. Nevertheless, in most cases the criteria of achieving a volumetric behavior differ and available experimental datasets are limited to validate modeling works. Therefore, first objective of this work is the generation of an open experimental dataset for the validation of numerical models. Second objective is to demonstrate the use and application of the dataset for the validation of a 3D numerical model employing ray tracing for the incident light and coupled momentum and energy equations for the heat transfer modeling. Finally, a third objective is to build a 1D model heat transfer to capture the apparent phenomena in the receiver, and investigate the possible dimensionless correlations. In brief, the experiments were conducted using a 6 kWe powered xenon arc lamp with an ellipsoidal reflector, with irradiance fluxes ranging from 150 – 600 kW/m^2 Ambient air was flowed from front to back into the receiver with varying flowrates of 4 – 17 lpm. The tested receivers comprise ceramic components commercially available i.e., 20x20 mm^2 SiC particle filter, 20mm in ID Al_2O_3 tube, and 20mm in ID SiC tube. Transient temperature profiles were collected at three different axial positions at the center of the receiver and on the side walls, including the inlet and exit gas temperatures. Finally, the irradiance levels on the receiver were collected using a calibrated heat flux gauge. The data and uncertainly analysis will be made available on a public repository upon completion of the work.

325c Solar-Driven Biomass Chemical Looping Gasification for Production of Syngas and H_2: Thermodynamic Analysis and Proof-of-Concept Experiment

Dequan Xu, Bo Wang, Xian Li, Yanjun Dai, Chi-Hwa Wang

A novel conceptual solar-driven biomass chemical looping gasification (BCLG) system was theoretically and experimentally investigated. The air and fuel reactors in the state-of-the-art BCLG systems were replaced with a steam reactor and a directly irradiated solar fuel reactor in the proposed system, respectively. Metal oxides were circulated between the steam reactor and the solar fuel reactor as the oxygen carrier. The solar-driven BCLG exhibits two advantages over its conventional counterparts: (1) the syngas yield is raised as the high-temperature process heat is provided by high-flux solar radiation rather than biomass combustion, and (2) two streams of fuels, hydrogen and syngas, can be mixed to achieve a desirable carbon-to-hydrogen ratio for carbon-neutral liquid fuel synthesis via Fischer–Tropsch process.

A zero-dimensional thermodynamic model was built to determine the theoretical solar-to-fuel energy conversion efficiency. First law and second law analyses were conducted to examine the effect of various pivotal parameters, including the reaction temperatures in both reactors, the ratio of oxygen carrier to biomass, the solar concentration ratio, and the heat recovery effectiveness, on the overall performance of the proposed system. A proof-of-concept experiment was also conducted in a cavity-type packed-bed solar reactor under concentrated solar radiation emitted by Singapore's first 28 kW_e high flux solar simulator. The results indicated that the solar chemical looping gasification of biomass is a viable method for producing syngas and hydrogen in separate streams, which can be subsequently combined in a specified proportion for downstream chemical processing.

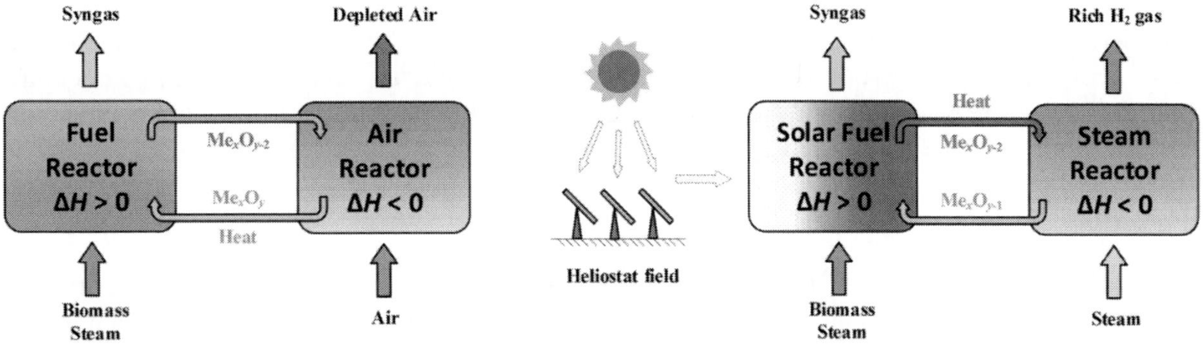

(a) Chemical looping gasification of biomass (b) Solar chemical looping gasification of biomass

325d Techno-Economic Assessment of Heat Integration from Concentrated Solar Thermal Energy into Solid Oxide Electrolysis for Green Hydrogen Production

Martin Janse Van Vuuren

Background

In the Paris Agreement of 2015, 196 countries agreed to take action to mitigate climate change and limit global warming to below 2°C. To achieve this goal, countries aim to decarbonize their energy sectors as much as possible. The use of green hydrogen may play a vital role in reducing the emissions of 'hard to abate' sectors like fertilizer production, ore processing and long-distance transport. It could also be used in long term energy storage to mitigate the variable nature of renewable energy sources.

Problem statement

Green hydrogen can be produced with water electrolysis that utilizes renewable electricity to split water into hydrogen and oxygen. The cost of green hydrogen electrolysis is currently still significantly higher than steam methane reformation which produces a lot of carbon dioxide. The highest contribution to the levelized cost of hydrogen is the cost of renewable electricity. The electricity demand for electrolysis can be reduced if heat energy is supplied by vaporizing water before it enters the electrolyzer. High-temperature solid oxide electrolysis cells (SOECs) that uses steam at 700-1000°C has the highest electrical efficiency compared to other low temperature water electrolyzer technologies. The high efficiency is a result of the faster reaction kinetics due to the higher temperature. This work investigates the technoeconomic impacts of integrating concentrated solar thermal (CST) energy into a hypothetical SOEC plant to determine if this integration could produce more cost-effective green hydrogen compared to using only electric heating.

Motivation

Technoeconomic analyses on the integration of SOEC plants with concentrated solar thermal energy have been previously documented in literature. However, limited attention has been given to the direct technoeconomic comparison between solar heat integration and electric heating. To the authors' best knowledge, there is no published literature on the direct comparison with using transient weather data to simulate annual production of hydrogen for both cases.

Methodology

This work modelled a 100 MW hypothetical SOEC plant with wind farm, photovoltaic plant, and a CST plant. Two systems are compared: one where heat is supplied via CST and the other where electric heating is used. Both systems source electricity from a photovoltaic and wind farm. A mathematical model of the SOEC has been created based on the work of previous published literature that uses area specific resistance to characterize the SOEC. A balance of plant model around the SOEC was built in Ebsilon Professional software that includes the compression and purification of the produced hydrogen to 30 bars. Both systems were modelled with one year of hourly variable weather data for a predefined location to determine annual hydrogen production.

Expected results

The primary question that will be answered in the presentation is whether CST makes sense in this system or not. The presentation will cover the economic comparison based on the levelized cost of produced hydrogen of the two systems. Additionally, the presentation explores the effects of changing operating conditions of the SOEC at both cell and balance of plant levels and their implications on heat integration and the levelized cost of hydrogen.

SELF-CONTROLLED ELECTRIC POWER SUPPLY SYSTEM FROM SOLAR CELL USING HYDROGEN STORAGE

Satoshi Wada, Katsuhiko Tsuno, Takeharu Murakami, Takayo Ogawa, Katsushi Fujii
RIKEN Center for Advanced Photonics, Wako, Saitama 351-0198, JAPAN

Introduction

The increase of global warming gases including carbon dioxide in the atmosphere due to the massive consumption of fossil resources and the production of foods associated with human activities causes global warming and sea acidification. Therefore, there is an urgent requirement to make such human activities in a state close to natural cycles as possible. Conversion of the energy sources to natural energy is one of them. Thus, energy production as electricity using solar cells and wind power generation is growing rapidly.

The electricity cannot be stored in its form, and nature originated electricity is not possible to generate the power with a user-on-demand balance because solar and wind energy fluctuate depending on nature. As a result, it is impossible to control the balance between the generation and demand like conventional power generation systems. Thus, a balancing system must be installed between the renewable power generation and the consumption. Currently, existing power generation like gas turbine generation and pumped storage power generation, which is usually greater than the amount of power generated based on renewables, are used to balance the two. When the amount of power generated from renewables is too large to achieve this balance, the renewable electricity is disconnected from the grid to maintain the balance of the grid. In other words, the establishment of an energy management system (EMS) to achieve this power balance, that is not based on conventional power generation methods, is an urgent requirement for a large amount of renewable energy installation.

Controlling Method of Electric Management System (EMS)

Many ideas for EMS control methods were reported [1-8]. Our proposed control method for EMS, which was modified from the previous methods, is introduced in this section. In this control method, high-speed power fluctuations are managed by the Main Voltage Controller (MVC), which is an energy storage device that can handle high-speed fluctuations, such as rechargeable batteries, while mass storage of energy is controlled by the Sub Voltage Controller (SVC), which is a storage method suitable for mass storage, such as hydrogen. Since commonly installed renewable energy generators like solar cells are DC output, we proposed a system with a DC-Bus line based on a specific voltage that controls everything with DC [9,10]. This control method is called "Modified DC Bus Signaling". It is characterized that it can be connected to the power grid and can also be used as an independent self-controlled system.

In the proposed method, the voltage of the DC-Bus line is actively controlled. A deviation from the reference voltage is an amount corresponding to the time integration of the power (current) including the direction of absorption or emission at the MVC. The SVC absorbs or emits power (current) following the amount of deviation from the reference voltage. In other words, when there is excess power in the DC-Bus line and power flows into the MVC, power absorption is performed to the SVC accordingly, and vice versa. The details of this control protocols are shown in Fig. 1. Theoretical analysis showed that this control method exhibits a first-order delay system and is stable in a steady state. Details of the control operation are found elsewhere [9, 10]. Another feature of this method is that the amount of digital signal traffic can be reduced because it is an analog-like control method in which the DC bus voltage is used as the signal.

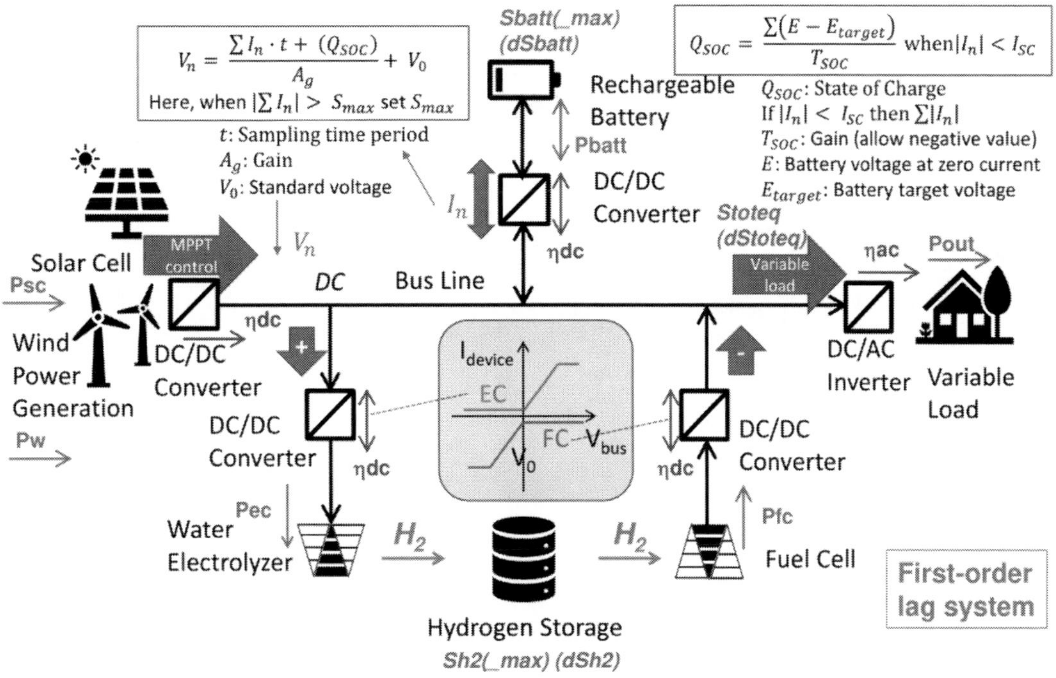

Figure 1. Schematic diagram of proposed Modified DC Bus Signaling System. The control details and abbreviations of power, storage, and efficiency are also shown.

Estimation of Basic Equipment Size for a Stand-Alone EMS

In regions prone to natural disasters, island areas, and low population density with dispersed residential areas, it is better to install a stand-alone and self-controlled system. This system installation is desired rather than installing a power grid because of the high cost of installing and maintaining a grid power system. The proposed controlling method can also be applied to such as an independent system.

On the other hand, it is not easy to estimate the scale of the system required. Many conditions affect the size of the system, including the method of power generation, the weather conditions of the region where the system is installed, the amount of electricity used and its time trend, the system connecting or not to the grid, and the ratio of the using electricity from the system and the grid. Here, we examine a method for estimating the size of the independent and self-controlled energy management system proposed here [11].

EMS Estimating Conditions

In the case of independent self-controlled energy system, energy fluctuations are automatically controlled by the system. Thus, the system size for long-term system operation can be designed by considering the state of the energy balance of a certain time period in accordance with the energy cycle as a unit. Here, this unit of energy balance can be considered as one day. Based on this approach, when the maximum power supply (P_{out_max}) and daily power supply (S_{toteq_n}: "_n" indicates daily changing

Unit period: 1-day per period: y-days energy storage per period
"d" at the front indicates the changing amount during one period

<<Power>>

Calculation INPUT:			
Average power output per unit period	(kWp)	Pout	
Maximum power output	(kW)	Pout_max	
Calculation OUTPUT:			
Rating power from solar cell	(kWp)	Psc	at 1 SUN
Rating power from wind power	(kWp)	Pw	at rating wind speed
Maximum power from/to MVC (battery)	(kW)	Pbatt	
Maximum power to SVC (EC)	(kW)	Pec	
Maximum power from SVC (FC)	(kW)	Pfc	

<<Storage>>

Calculation INPUT:		
Time for supply energy per Ti	(h)	Ts = y (days) × 24 (hrs)
Solar power converted to maximum solar cell output		
Time for solar irradiation	(h/day)	Ti
Calculation OUTPUT:		
Battery storage amount	(kWh)	Sbatt
Hydrogen storage amount	(kWh)	Sh2
Converted required energy	(kWh)	Stoteq

Constants

<<efficiency / cost>>	Conversion efficiency (max 1)			Estimated Cost as 2020		
Solar cell	energy	ηsc	0.20	Csc	($/kW)	1500 (incl. setting)
DC/DC converter	energy	ηdc	0.955	Cdc	($/kW)	1000
DC/AC inverter	energy	ηac	0.955	Cac	($/kW)	1000
Battery	input/output	ηbatt	0.90	Cbatt	($/kWh)	2000
Electrolyzer (EC)	energy	ηec	0.80	Cec	($/kW)	1000 (1/10 of market)
Fuel cell (FC)	energy	ηfc	0.50	Cfc	($/kW)	500 (1/10 of market)
Hydrogen storage				Ch2	($/kWh)	500

Figure 2. Parameters for calculations and constants with abbreviations and used values.

variables) are specified by the user at the supply destination, the system can be designed.

In designing the system, the following points should be further considered.

A. Assume solar and wind power as natural energy sources.
B. Natural energy generation should be consumed as soon as possible since storing the energy reduces energy efficiency due to the conversion.
C. Consider two cases: one is to store energy in rechargeable batteries, and the other is the proposed controlling way. That is, to use rechargeable batteries to absorb rapid fluctuations and to store a large amount of energy as hydrogen using water electrolyzers (EC) and fuel cells (FC).
D. The system is assumed to supply the maximum required amount of electricity from both rechargeable batteries and fuel cells.

A list of variables and constants used in these calculations is shown in Fig. 2.

System Size Estimation

Overall system balance: The overall system balance does not require detailed control because an independent and self-controlled energy system is used. Therefore, it can be considered the energy balance in one cycle (one day in this study). In regional optimization, energy storage for y days needs to be set in this single cycle.

A. Input (solar cells and wind power): The amount of electricity generated by wind power generation is proportional to the third power of the wind speed, and the rated output is obtained at the rated wind power. Therefore, wind power cannot be considered as a primary energy source in areas where wind speed is not constant. Solar cells are considered the main energy source, and wind power is used as a supplemental source in this study.
B. Output (rechargeable battery or FC): When the energy storage device is a rechargeable battery, the output from the rechargeable battery should be greater than the maximum system output, and when hydrogen is used, the outputs from the FC and rechargeable battery should be greater than the maximum system output. The output needs to take into account the efficiency of conversion devices during the electrical path.
C. Storage (rechargeable battery or/and EC): the input to the rechargeable battery when the energy storage is a rechargeable battery, or to the EC and rechargeable battery when the energy storage is hydrogen, must be greater than the maximum input. The input needs to take into account the efficiency of the conversion devices during the electrical path.

To summarize these all (b: batt, H: h2, EC: ec, FC: fc),

$$dS_{toteq} = (1/\eta_{ac})P_{out}(T_s - T_i), \quad (1)$$

$$dS_{b\_or\_H} = \left(1/\left(\eta_{dc}\eta_{b\_or\_FC}\eta_{ac}\right)\right)dS_{toteq}, \quad (2)$$

$$P_{b\_or\_EC} = dS_{b\_or\_EC}/\left(\eta_{b\_or\_EC}T_i\right), \quad (3)$$

$$P_{sc} = \left(1/\left(\eta_{dc}\eta_{dc}\eta_{b\_or\_EC}\right)\right)(dS_{batt}/T_i) + (1/(\eta_{dc}\eta_{ac})P_{out}), \quad (4)$$

$$dS_{H\_or\_b} = 0. \tag{5}$$

Real System Size

In this section, the calculation was performed with average conditions in Japan as an example. The unit cycle (T_s) is 1 (day) = 24 (hrs), as described above. For the input, the used solar radiation time (T_i) is 1000 (hrs/year) = 3 (hrs/day), which is measured as the time when solar cells can generate electricity at their rated capacity in Japan. Wind power is considered a supplemental power source, with $P_w = 5$ (kW) at a rated wind power of 12 m/s, but is not considered as an input to the basic system. For the output, $P_{out} = 1$ (kW) is assumed as the average output during the cycle (one day). This means that the power is 24 kWh per day. The maximum output is set to be considered to be $P_{max\_out} = 5$ (kW).

Assuming the same conditions of sunshine every day, the required converted output is $dS_{toteq} = 21$ (kWh). For rechargeable batteries, the sizes are $dS_{batt} = 28.85$ (kWh), $P_{batt} = 9.58$ (kW), $P_{sc} = 11.72$ (kW). For hydrogen, the sizes are $dS_{h2} = 46.54$ (kWh), $P_{ec} = 19.39$ (kW), $P_{sc} = 22.59$ (kW). In other words, to make a 1 kW constant power output energy system under standard conditions for rechargeable battery storage, 11.72 kW of solar cells, 9.58 kW of rechargeable batteries, and 28.85 kWh of stored energy for one day is required. For hydrogen storage, those are 22.59 kW of solar cells, 19.39 kW of water electrolysis cells, and 46.54 kWh of hydrogen storage energy for one day. When solar cells with a conversion efficiency of 20% are used in this system, the area of the solar cells is approximately 60 m^2 for the battery storage and 110 m^2 for the hydrogen storage.

The total device cost of the system as a function of energy storage ($S_{b\_or\_H}$) is shown in Fig. 3. Rechargeable battery storage is less expensive when the energy storage volume is small, but hydrogen storage is less expensive when the energy storage volume is large. This is because the energy storage volume greatly affects the cost, and the unit cost of energy storage is set to be lower when hydrogen storage is used.

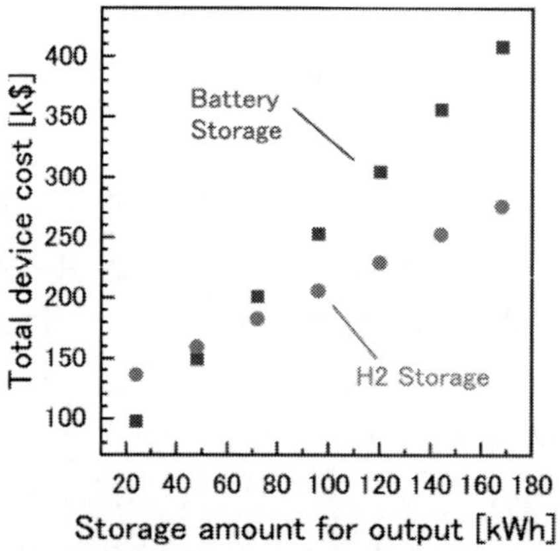

Figure 3. Total device cost of system compared between battery and hydrogen storage with the constants and parameters given in Fig. 2.

Regional Dependence Consideration

Standard Condition and Regional Dependence

When every day has the same conditions of sunshine, the input/output size was discussed in the previous section, however, the reality of the required energy storage is different since weather conditions are not the same every day. In other words, when considering regional dependence, the local weather conditions especially sunshine and wind power have a significant impact on the system size, even assuming the output is the same. The required energy storage depends on the number of consecutive days when sunshine is not expected since wind power is considered as a supplement. Since the number of these days varies greatly from year to year, we considered the standard energy storage capacity for four days. In other words, the amount of converted energy required for output (S_{toteq}) is 24 (kWh) × 4 (days) = 96 (kWh). Considering the efficiency of devices, we set the storage capacity of 100 kWh for rechargeable batteries and 200 kWh for hydrogen storage, which is equivalent to about 4 days of energy storage.

Regional dependence is evaluated by fixing the standard output conditions (maximum 5 kWp at 1 kW average output) and estimating input power and energy storage. The evaluated condition is this output is uninterrupted throughout the year without input from other sources. The actual calculations were performed for several selected Japanese regions. Since the annual variations are very large, the 30-year averages from 1980 - 2010 for sunshine and temperature were used. For wind speed, the actual values from 2019 were used due to the lack of average data [12].

Calculation Method for Regional Dependence

The regional dependences are considered as follows.

A. Wind power: The effect of the introduction of wind power was studied. The average daily output (P_{av}) with the introduction of wind power, under the rated wind speed $W_r = 12$ (m/s), the rated output $P_w = 5$ (kW), and the average daily wind speed (W_{av}),

$$P_{av} = (W_{av}/W_r)^3 P_w \qquad (W_{av} < W_r) \tag{6}$$

$$P_{av} = P_w \qquad (W_{av} \geq W_r).$$

B. Solar Cells: The following three points were considered for solar cells.
The relative time of solar power generation to the actual sunshine hours differs due to the effect of the altitude of the Sun. This difference was estimated from actual results to be 0.55 as a coefficient.
Assuming that solar power generation increases by 20% over standard conditions in the summer solstice approaching and decreases by 20% over standard conditions in the winter solstice approaching, the power generation differs throughout the year approximate expression set as a sine

curve.

The temperature dependence of the energy conversion efficiency of solar cells was assumed to be -0.45%/deg.

These three influences are collectively defined as *factor*.

C. Daily energy balance of the system: The daily energy balance is obtained by the following (6) and (7), which are variants of (1) and (2), using Psc and Pec obtained from the standard system scale estimation.

$$dS_{h2} = (\eta_{dc}\eta_{ec})(\eta_{dc}\eta_{sc}P_{sc} \times factor - (1/\eta_{ac})P_{out}) + 24P_{av} - (24 - T_i) \times \left(1/\eta_{ac}\eta_{dc}\eta_{fc}\right)P_{out} \tag{6}$$

$$dS_{toteq} = \eta_{ac}\eta_{dc}\eta_{fc}dS_{h2} \tag{7}$$

The difference between the energy obtained from natural energy and the energy used throughout the year (excess or deficiency of energy throughout the year) can be obtained by totaling these daily energy balances over the year.

Since an actual system cannot store energy infinitely, the standard conditions set as 100 kWh for rechargeable batteries and 200 kWh for hydrogen storage were used as a reference. Overloading of solar cells, which is less expensive, was evaluated as the highest priority, and when the overloading of solar cells was not enough, energy storage was increased to satisfy the demand to satisfy the energy output.

Calculation Results of Regional Dependence

The following four sites were evaluated.

Shinjo: An area where neither sunlight nor wind is expected in winter.

Aikawa (Sado Island): An area where winter sunlight is not expected, but wind is expected.

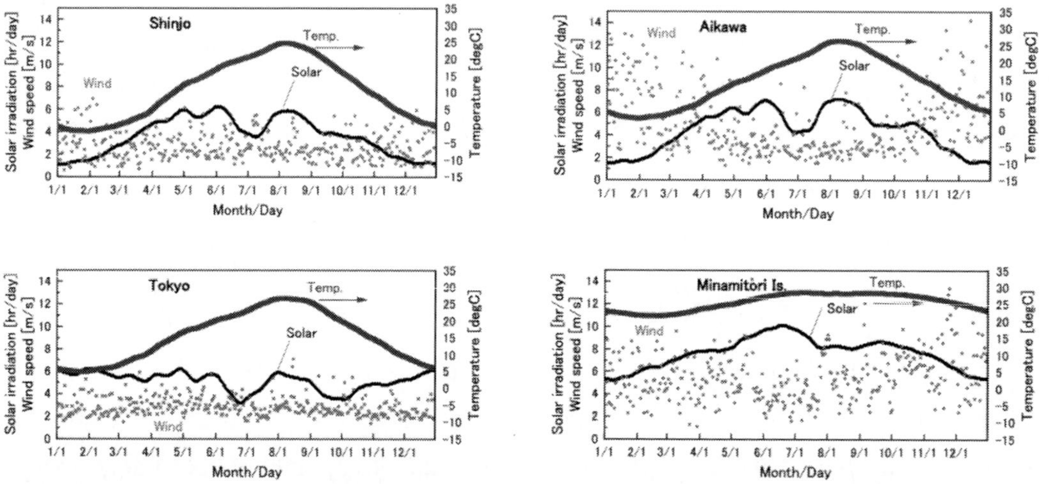

Figure 4. Solar irradiation, temperature (1980 – 2010 average) and wind speed (2019) observation data at Shinjo, Aikawa, Tokyo, and Minami-Torishima island.

Tokyo: No wind power is expected, but relatively stable sunlight is available.

Minami-torishima Island: Sunlight is available throughout the year, and little difference in the duration. The data on sunshine, temperature, and wind power are shown in Fig. 4.

An example of this energy balance and storage under standard conditions in Aikawa is shown in Fig. 5. Under standard conditions, periods of energy shortage exist, especially in winter when there is little sunlight, and this tendency is more pronounced when there is no assistance from wind power generation. Therefore, we attempted to eliminate this energy shortage by optimizing the solar cell size and energy storage. The results are summarized in Table 1.

From these results,

A. System size varies greatly depending on local weather conditions, and wind assistance is particularly effective in areas where fast wind speeds are expected,

B. The shortage of solar radiation, especially during the winter in the Northern Hemisphere, requires larger solar cell sizes than standard especially when no assistance exists by wind power generation,

C. The energy storage capacity needs to be considerably increased when a prolonged period of inadequate solar radiation exists, but the periods tend to be only for a limited period of time.

These results indicate that it is important to make the various devices into unit, and to facilitate their installation and removal for the system when making a system using renewable energy as its energy source. Since the "Modified DC Bus Signaling" system proposed here uses the voltage of the DC bus as a signal, the devices both MVCs and SVCs can be easily increased or decreased in number as a unit.

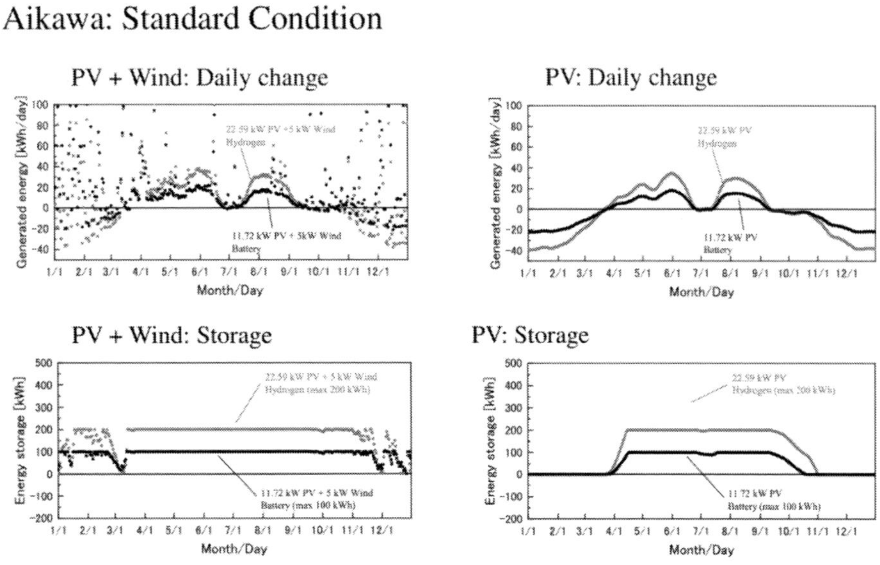

Figure 5. Daily energy increase/decrease of energy and the energy storage change with maximum limit at Aikawa under the standard conditions. The cases of solar cell + wind power generation input and solar cell input are shown. Red: hydrogen storage, black: battery storage.

Table 1. Optimization of solar cell and storage size to eliminate the energy shortage. The cases of solar cell + wind power generation input and solar cell input are shown.

Place	Characteristics	Solar +Wind Solar cell Battery / Hydrogen (kW)	Storage Battery / Hydrogen (kWh)	Solar Solar cell Battery / Hydtogen (kW)	Storage Batttery / Hydrogen (kWh)	Remarks
Standard		-	-	11.72	100	
		-	-	22.59	200	
Shinjo	Winter: A few sunshine and wind	30	500	35	500	Both solar cell and storage enlargement required
		60	1000	70	1000	
Aikawa (Sado)	Winter: A few sunshine but exist wind	15	100	30	400	Wind power is effective
		30	200	60	800	
Tokyo	Relatively mild condition	17	100	17	100	No wind power assist required
		30	200	30	200	
Minami-Torishima Island	A lot of sunshine	10	25	15	25	Smaller storage is enough
		20	50	30	50	

Summary

"Modified DC Bus Signaling" System that uses DC bus voltage as a signal, which can handle both fast fluctuations and large storage capacity, as an independent self-controlled energy system that can be used renewable energy as its energy source. Since this method is an analog-like control method of the DC bus voltage used as a signal, the amount of digital signal communication can be reduced. In addition, the optimization of system configuration for such independent energy systems based on natural energy sources is highly dependent on the region. Therefore, it is an advantage to the system design to make each device a unit and to increase or decrease its number. Since the "Modified DC Bus Signaling" system discussed is easy to increase or decrease the number of devices, regional optimization is also considered to be easy.

References

1. Thounthong P, Chunkag V, Sethakul P, Sikkabut S, Pierfederici S, Davat B. (2011), "Energy management of fuel cell/solar cell/supercapacitor hybrid power source", J. Power Sources 196, 313 – 324.

2. Gao D, Jin Z, Zhang J, Li J, Ouyang M. (2016), "Development and performance analysis of a hybrid fuel cell/battery bus with an axle integrated electric motor drive system", Int. J. Hydrogen Energy 41, 1161 – 1169.
3. Gavriluta C, Candela JI, Rocabert J, Luna A, Rodriguez P. (2015), "Adaptive droop for control of multiterminal DC bus integrating energy storage", IEEE Trans. Power Deliv. 30, 16 – 24.
4. Li Q, Wang T, Dai C, Chen W, Ma L. (2018), "Power management strategy based on adaptive droop control for a fuel cellbattery-supercapacitor hybrid tramway", IEEE Trans. Veh. Technol. 67, 5658 – 5670.
5. Yang H, Li Q, Zhao S, Chen W, Liu H. (2019), "A hierarchical selfregulation control for economic operation of AC/DC hybrid microgrid with hydrogen energy storage system", IEEE Access. 7, 89330 – 89341.
6. Maeda T, Ito H, Hasegawa Y, Zhou Z, Ishida M. (2012), "Study on control method of the stand-alone direct-coupling photovoltaic water electrolyzer", Int. J. Hydrogen Energy 37, 4819 – 4828.
7. Schonberger J, Duke R, Round SD. (2006), "DC-bus signaling: a distributed control strategy for a hybrid renewable nanogrid", IEEE Trans. Ind. Electron. 53, 1453 – 1460.
8. Qu D, Wang M, Sun Z, Chen G. (2015), "An Improved DC-bus signaling control method in a distributed nanogrid interfacing modular converters", In: Proceedings of the international conference on power electronics and drive systems, vol. 184.
9. Yamashita D, Tsuno K, Koike K, Fujii K, Wada S, Sugiyama M. (2019), "Distributed control of a user-on-demand renewable-energy power-source system using battery and hydrogen hybrid energy-storage devices", Int. J. Hydrogen Energy 44, 27542 – 27552.
10. Tsuno K, Koike K, Fujii K, Ogawa T, Wada S. (2023), "User-on-demand renewable energy supply system using modified DC-bus signaling", Preprints of the 22nd IFAC World Congress, 9800 – 9805.
11. Fujii K, Koike K, Tsuno K, Wada S. (2021), "Regional optimization guideline for small-sized energy system with hydrogen storage based on natural energy source (in Japanese)", J. Jpn. Institute Energy 100, 45 – 54.
12. Japan Meteorological Agency, Past data download page (in Japanese) , https://www.data.jma.go.jp/gmd/risk/obsdl/index.php (last access 2023.10.30)

325f Potential of Monetizing Stranded Natural Gas through Process Intensification and Green Energy

Abdullah Al-Aboosi, Fadhil Al-Aboosi, Mahmoud El-Halwagi, Wei Zhan

The crucial shift in global resource use and technological innovation are significant for achieving a smooth transition to a low-carbon economy and climate-resilient development. Despite the current abundance of natural gas in the worldwide market, there are considerable gas resources that remain unusable for economic and physical reasons. This unexploited gas resource is known as a stranded gas. Consequently, developing feasible alternatives for monetizing stranded natural gas are essential to attract investors' attention. The most appropriate alternatives should offer significant security and stability of investment along the life cycle of a project to overcome market fluctuations, technological advances, and political conflicts or trade bans. This study presents a new design framework for an integrated system. The proposed conversion system of natural gas to valuable products shows a significant capital cost reduction while preserving energy and improving sustainability through finding optimal natural gas technologies, integrated process flowsheets, and techno-economic feasibility.

AUTHOR INDEX

Aasberg-Petersen, Kim 150
Abdelsalam, Emad 185
Ackermann, Simon 77
Adams, Ethan 235, 236
Adams, Shelby 152
Adelan, Samantha 178
Ahmed, Mahdi 88
Ajayi, Oyelayo 102
Al Ketife, Ahmed M. D. 6
Al Lagtah, Nasir 4
Al Moinee, Abdullah 164
Al-Aboosi, Abdullah 306
Al-Aboosi, Fadhil 306
Alam, Md. Monjurul 130
Alareeqi, Seba 38
Allen, Zachary 152
Al-Mohannadi, Dhabia 87, 110
Almomani, Fares 6, 185
Alnawafah, Hamza 185
Alomía, Victoria 42
Amador, Carlos 35
Amar, Vinod 173
Amini, Shahriar 151
Amundsen, Ted 197
Anabaraonye, Benaiah 263
Anastas, Paul T. 3, 126
Arellano-Treviño, Martha A. 278
Ashna, Nizar 4
Askari, Zohreh 89
Aston, John E. 101
Atiyeh, Hasan 32
Aui, Alvina 277
Auyeung, Nick 282, 288
Avraamidou, Styliani 17
Baamran, Khaled 274
Babarinde, Oladipupo 260
Bademi, Apoorva 133
Badger, Nicholas 151
Bagdonas, Davin 98
Bahamon, Daniel 38
Baker-Fales, Montgomery 255
Bakshi, Bhavik 78, 188, 227, 246
Balch, Robert 93
Baldea, Michael 66
Baltrusaitis, Jonas 58
Baral, Nawa 30
Barbera, Elena 168
Bayzid, Kazi 130
Beltrán, Katty 42
Bennett, Gabrielle 152
Berezniovas, Edvinas 152
Berglund, Noah 99, 103
Bertucco, Alberto 168
Bhalode, Pooja 255
Bhattacharjee, Tiasha 99, 101, 103, 104
Bhattacharya, Moumita 149
Bhattacharyya, Debangsu 111
Bhosale, Rahul 152
Biegler, Larry 62
Bishop, Taylor 152
Biswas, Pratim 120
Bommarius, Andy 123
Bonnema, Michael 152
Bovet, Nicolas 263
Boylu, Rahim 151
Brennecke, Joan 66
Briones, Paola Alejandra Munoz 17
Brouns, Thomas 245
Broussard, Serenity 244
Burke, Emily 31
Burli, Pralhad 104, 107, 108
Burton, Gareth 110
Caballero, Daison Manuel Yancy 68
Cabezas, Heriberto 23, 254
Cai, Hao 29
Cai, Tianxing 233
Caisaluisa, Oliver 42
Cañada, Louise M. 66
Cano-Londoño, Natalia 254
Cao, Ruoshi 259
Carman, Carl 91
Carpenter, Alberta 12, 253
Cavalcante Jr., Celio 179
Chacón, Julio 42
Chakrabarti, Dhurjati Prasad 113
Chanona, Antonio Del Rio 84
Chaudhari, Kiran 270
Chaudhari, Utkarsh 83
Che, Fanglin 8
Chen, Jingjing 285
Chen, Qiang 122
Chen, Wu 22
Cheng, Ming-Hsun 176
Cheng, Yoke Wang 37
Chipkar, Sarvada 31
Choi, Yechan 267

Choquette, Riley 283
Chou, Joseph 147
Christensen, Anders Nymark 263
Clutter, Sequoia 152
Cordiner, Joan 88
Coventry, Joe 285
Cox, Philip 60
Crandall, Dustin 256
Cronk, Briana 31
Cui, Zhe 51
Dai, Yanjun 293
Dantas, Beatriz 195
Dantzler, Alex 51
Daoutidis, Prodromos 7, 265
Dashtbankenari, Laleh 280
Davidson, D 245
De Hert, Sergio Carrillo 52
De Luna, Francisco Murilo Tavares 179
Deak, Csaba 254
Debrauske, Derek J. 31
Desveaux, Jason 51
Dewey, Oliver 51
Diemler, Nathan 61
Diwekar, Urmila 23
Dosso, Cheick 62
Dunn, Jennifer 16, 251
Dupuis, Dan 100
Dutta, Abhijit 28
Ebrahimi, Parisa 139
Eisa, Mohamed 58
El-Halwagi, Mahmoud 306
El-Kaseeh, George 93
Ellebracht, Nathan C. 277
Emdadi, Arash 240, 241
Emerson, Rachel 104
Erguvan, Mustafa 151
Eriekat, Jihad 185
Errington, Ethan 127
Escobar, Lourdes Orejuela 42
Evans, Bergen 180
Evans, Caria 123
Fagan, Ryan 152
Fang, Jian 178
Fantke, Peter 14
Farnsworth, Amanda 238
Farzana, Sumaiya 98
Fath, Brian D. 227
Fenske, George 102
Ferguson, Alex 60
Ferguson, Steven 52
Fermeglia, Maurizio 168
Filho, Paulo Roberto C. F. Ribeiro 179
Finley, Tim 22
Frailey, Scott M. 91
France, Stefan 123
Freiberg, Lucas 288
Freitag, Joerg 223
Fu, Wenming 37
Fujii, Katsushi 296
Fujita, Masayuki Fujita Masayuki 221
Fushimi, Chihiro 223
Gager, Elizabeth 74
Gallardo, Yuliana Pullas 42
Gang, Daniel 244
Geissler, Caleb 269
Genã§er, Emre 81, 238
Ghosh, Tapajyoti 12, 253
Gibbons, William 72
Gilcher, Elise B. 3
Gilcher, Elise 126
Gill, Magdalena 256
Goldstein, Hannah M. 277
Greenlee, Lauren F. 240, 241
Guerrero, Víctor H. 42
Gunukula, Sampath 33, 109
Guo, Miao 127
Guzman, Luisdomingo 187
Gypakis, Antonis 169
Halabrin, Jordan 152
Hamed, Ahmed 101
Hamilton, Stephen 31
Handler, Robert 83
Hanumante, Neeraj 23
Haque, Md Emdadul 111
Hartley, Damon 83, 107, 108
Hartmann, Michael 119
Hashimoto, Jayni 73
Hastings, Elizabeth 178
Hatton, T Alan 71
Hatzell, Marta 123
Haueter, Philipp 291
Hawkins, Troy 132
Hayashi, Yusuke 226
Hayes, Michael 282, 288
Hellmüller, Sven 291
Heng, Jerry 127
Heo, Seongmin 267
Hernández, Borja 191
Hernandez, Rafael 244
Hernandez-Betancur, Jose 20
Herrera, Nelson 42
Hestekin, Jamie 240, 241
Hirano, Nanami 223
Hobbs, Mason 152

Holguín, Eduardo....42
Holmes, William E....244
Hopkinson, David....61, 62
Hori, Hiroshi....223
Hossain, Samira....129, 130
Hossain, Tasmin....83
Hu, Jianli....111
Huang, Danny....174, 276
Huang, Yinlun....19, 21
Huhnke, Raymond L....32
Hunt, Daniel....152
Huynh, Khang....186
Ichikawa, Takayuki....220
Ilsa, Kristiina....28
Ilojianya, Valentine....151
Ino, Isshin....125
Irvin, Glen....51
Ismail, Mohammad....130
Iv, Wilson McNeary....278
Ivarra, Miguel....152
Jadidi, Yasheen....135
Jeong-Potter, Chae....278
Jiang, Yuan....29
Jiao, Feng....275
Jiao, John....96, 98
Jin, Wencheng....101, 103
Johnson, Anne....83
Johnson, Matthew....96
Johnson, Sydney....81
Joo, Taigyu....59
Jordan, Mattigan....152
Joshi, Leela....31
Junaedi, Christian....137
Kakosimos, Konstantinos....292
Kansha, Yasuki....125, 193
Kar, Saurajyoti....132
Karakitsios, Spyros....169
Karakoltzidis, Achilleas....169
Karimi, Iftekar....183
Karunanithi, Pooja....152
Karunarathne, Sampath....109
Keene, Tony....52
Keiser, James....102
Keller, Austin N....66
Kenney, David....54
Khan, Irfan....22
Khanna, Vikas....252
Khatun, Mst. Lucky....130
Khosravi, Mansour....89
Khraisheh, Majeda....139
Kim, Changsoo....272
Kim, Kyeongsu....272
Kim, Soohwan....236
Kim, Wookyung....220
Kirtania, Kawnish....129, 130
Kishi, Naoya....222
Klausner, James F....282, 288
Klinger, Jordan....99, 101, 103
Kokosis, Antonios....55
Kono, Masafumi....223
Korba, David....282, 288
Kots, Pavel....255
Kratzer, Madison....252
Krynock, Michelle....147
Kumar, Anand....139
Kumar, Anikesh....183
Kumar, Apurv....285
Kumar, Baibhaw....254
Kumar, Kundan....176
Kuns, Miranda....102
Kusuma, Victor A....61, 62
Kutsienyo, Eusebius....259
Lacey, Jeffrey A....102
Lail, Marty....149
Lameh, Mohamed....87, 110
Lamers, Patrick....12
Landazuri, Andrea....42
Lane, Mary Kate....3, 126
Lau, Raymond....1
Lavallais, Chayse....251
Law, Jun Wen....266
Lawson, Shane....274
Lecaro, María Fernanda....42
Ledwith, Jessica....178
Lee, Chul-Jin....271
Lee, Ga-Young....271
Lee, Hyunjoo....272
Lee, Jay Hyung....267
Lee, Joo-Sung....271
Lee, Tae Hoon....59
Lee, Uisung....27
Lee, Ung....272
Lee, Younggeun....272
Lei, Fuqiong....288
León, Marco....42
Levin, Nestor....236
Li Sr., Wenzhen....172
Li, Like....282, 288
Li, Shuyun....27, 29
Li, Wenqi....106
Li, Wenqin....277
Li, Wenzhen....165
Li, Xian....293
Lidor, Alon....291

Liebold, Aaron 197
Lima, Fernando 195
Lin, Guiying 37
Lin, Yingqian 107, 108
Linke, Patrick 87, 110
Lipinski, Wojciech 285
Lipscomb, Glenn 61
Liu, Dupeng 30
Liu, Guoxiang 257
Loebick, Codruta 137
Loehde-Woolard, Hailey 180
Logan, Bruce E. 240, 241
Lopez, Aurora Del Carmen Munguia 17
Lopez, Gary 105
Lopez-Rodriguez, Rafael 52
Lou, Helen 178, 233
Louck, Cheryl 177, 234
Lu, Hong 63
Lu, Yongqi 63
Luciani, Christian 42
Luo, Guangsheng 122
Maddipudi, Bharathkiran 186
Maghfuri Sr., Abdullah 172
Maier, Jacob 123
Mallapragada, Dharik 266
Mana, Arun Sankar Eenhakkattu 229
Manousiouthakis, Ioannis 26
Manousiouthakis, Vasilios 26
Maravelias, Christos 117, 269
Martin, Mariano 9, 35, 191
Mason, Janine 177, 234
Matsumura, Yukihiko 220
Matuszewski, Michael S. 68
McCord, Dylan 74
McDaniel, Anthony H. 74
McDonald, Molly 117
McLaughlin, Jonathan 'Fred' 96, 98
McPherson, Brian 93
McReynolds, Julianna 152
Mehl, Jacob 177, 234
Meier, Anne 243
Melhim, Aysha 292
Melville, Angela 245
Mendez, Alonzo 73
Meramo, Samir 14
Meyer, Pimphan 27
Miller, Eric 72
Miller, Kevin D. 245
Miller, Quin 259
Minette, Michael 245
Mio, Andrea 168
Mirabá, Pierina 42
Modi, Sujit 120
Mogollon, Daniel 149
Mohaghegh, Shahab 257
Moment, Aaron 174, 276
Mondal, Mriganka 52
Moni, Sheikh 147
Moore, Johnathan 256
Moran, Esther Sanchez 242
Morrey, Eugene 245
Mortensen, Peter M. 150
Mosadeghsedghi, Sanaz 280
Mosalpuri Sr., Manish 172
Mosalpuri, Manish 165
Muhich, Christopher 73
Mukarakate, Calvin 28
Muller, Katherine 259
Murakami, Takeharu 296
Murillo, Herman 42
Nagurney, Allie 259
Najmi, Sean 255
Nakashimada, Yutaka 220
Nazemi, Farshid 227
Nemoto, Kosuke 226
Niebieskikwiat, Darío 42
Nielsen, Paul 63
Nienhuis, Emily 259
Nikiforou, Fotini 169
Nino, Juan C. 74
Ninomiya, Yoshihiko Ninomiya Yoshihiko 221
Nisal, Apoorva 23
Nishida, Keiya 220
Nune, Satish 245
Núñez, Sofía González 35
Nye, Charles 98
O'Connell, Margaret 16
Oestberg, Martin 150
Ogawa, Takayo 296
Ogunniyan, Opeyemi 111
Okumura, Yukihiko 222
Okwen, Roland 91
Olarte, Mariefel V. 29
Olson, Joel 243
Oluleke, Comfort 61
Oluleye, Gbemi 84
Omell, Benjamin P. 68
O'Neill, Eric 269
Ong, Rebecca 31
Ooi, Amanda Whai Shin 182
Ortiz-Ulloa, Juvenal 288
Ossai, Stephanie 33
Ou, Longwen 29
Packard, Jason 235

Panagakos, Grigorios 61, 62
Pang, Simon H. 277
Papathanasiou, Maria 113
Parameswaran, Prathap 252
Park, Ah-Hyung 174, 175, 182
Park, Jimin 123
Paronish, Thomas 256
Pasquali, Matteo 51
Passuto, Eleonora 14
Patwardhan, Shantanu 270
Paul, Brandon 68
Paul, Mahasweta 1
Pavan, Alessandro Massi 168
Pennington, Ashley 177, 234
Peralta, Alejandra 83
Petrasch, Joerg 288
Philippidis, George 2
Phillips, Benjamin 283
Phillips, Steven 27
Phillpot, Simon 74
Pirasaci, Tolga 2
Pitts, Ray 243
Poddar, Tuhin 132
Pol, Vilas 235, 236
Polychronopoulou, Kyriaki 38
Ponce, Sebastián 42
Pope, John 262
Posso, Juan Fernando 42
Prakash, Smrithi 229
Prócel, Luis M. 42
Qu, Jun 102
Quillinan, Scott 96, 98
Radfarnia, Hamid 280
Rahbar, Nima 54
Ramakrishnan, T. S. 219
Ramasamy, Karthikeyan 29
Ramsey, Owen 288
Randhir, Kelvin 282, 288
Rasmussen, Peter Winkel 263
Rasouli, Pejman 259
Rautenbach, Marené 150
Reck, Barbara 83
Restrepo-Florez, Juan Manuel 187
Reza, Toufiq 129, 130
Rezaei, Fateme 164, 274
Ring, Valerie 152
Ritchie, Stephen 151
Robertson, G. Philip 31
Roeb, Martin 286
Romanov, Vyacheslav 257
Rony, Asif Hasan 106
Roufogali, Nefeli 55
Rousseau, Joseph 110
Rownaghi, Ali 164, 274
Roy, Abhishek 51
Roychoudhury, Subir 137
Ruddy, Daniel 278
Ruiz-Mercado, Gerardo 20, 254
Sadeek, Sherard 113
Saha, Chayan Kumer 130
Saha, Nepu 101, 103
Saha, Swarna 129
Sakai, Yuka 125
Salih, Fathya 292
Salih, Hafiz 63
Sampat, Apoorva 22
Santamaría, Diego 191
Santosa, Daniel 29
Sarigiannis, Dimosthenis 169
Sarkar, Amrita 124
Sato, Trey 31
Sato, Yuki 125
Savage, Tom 84
Schaef, Herbert T. 259
Schäppi, Remo 281
Scheffe, Jonathan 74
Schimmels, Philipp 282, 288
Schmidt, Andrew J. 29
Scown, Corinne D. 30
Sedaghatnia, Kamand 83
Sen, Amrita 246
Shaikh, Insha 292
Shamsuzzaman, Farooq 183
Sharma, Richa 219
Shastri, Yogendra 23
Shekhar, Abhimanyu Raj 249
Shende, Rajesh 186
Shi, Jian 105
Shonnard, David 83
Siddiqui, Abdurrafay 19, 21
Sievers, Carsten 123
Singh, Gita 52
Singh, Shweta 133, 249
Sloan, Arthur 51
Small, Mark 101
Smith, Sam 152
Smith, William A. 106
Smith, Zachary P. 59
Snowden-Swan, Lesley J. 27, 29
Spanu, Leonardo 51
Sridhar, Preethi 183
Srinivasan, Rajagopalan 230
Stadtherr, Mark 66
Starr, John 167

Steinfeld, Aldo 77, 281, 291
Stephen 245
Storhaug, Vincent 60
Sugiyama, Hirokazu 226
Suiseeya, Kimberly Marion 16
Sukumara, Sumesh 14
Sultana, Al Ibtida 129
Sun, Annie 236
Sun, Lili 61
Sun, Ning 30
Sunol, Aydin 2, 41
Suzuki, Ryuta X. 223
Swaminathan, Venkataramanan 230
Sykes, David 197
Taft, Kendall 89, 92
Talan, Deniz 195
Talero, Gabriel 193
Tan, Eric 28
Tanner, Ralph S. 32
Tanoue, Ken-Ichiro 221
Teixeira, Andrew R 54
Tello, Rodrigo Buitrago 27
Thacker, Pranav J. 66
Thelen, Kurt 31
Thompson, David 101, 107, 108
Thompson, Vicki 83, 99, 101, 102
Thorson, Michael R. 29
Thunuguntla, Rahul 32
Timko, Michael T. 54
To, Anh 278
Trost, Jenna 16
Tsolas, Spyros 22
Tsuno, Katsuhiko 296
Ukawa-Sato, Ryu 223
Umehara, Nao 221
Van Vuuren, Martin Janse 294
Van Walsum, Gerard 33
Van Walsum, Peter 109
Vega, Lourdes 38
Vega-Montoto, Lorenzo 104
Vela, Nicolás 42
Venkatasubramanian, Venkat 229
Vibbert, Hunter B. 182
Vikara, Derek M. 257
Vlachos, Dionisios 255
Volchek, Konstantin 280
Von Jouanne, Annette 283
Wada, Satoshi 296
Wahlen, Bradley 99, 106
Walker, Devin M. 132
Wang, Bo 293
Wang, Chi-Hwa 37, 293
Wang, Hanchu 265
Wang, Haotian 71
Wang, Huamin 29
Wang, Shuai 54
Wang, Ximeng 74
Wang, Yujun 122
Wang, Yuxin 111
Ward, Keeran 113
Warren, Kent J. 180
Watkins, David W. 83
Way, Kenneth J. 245
Weimer, Alan 180
Wendt, Lynn 101
West, Jonathan 152
Wheeler, Clayton 33, 109
Wi, Sungsool 59
Williams, C. Luke 101, 109
Williams, Jonah 175
Workman, Scott 256
Wright, Mark Mba 165, 172, 242
Wu, Wan-Ni 59
Xia, Yidong 99, 101, 103
Xiao, Gang 75
Xu, Dequan 37, 293
Xue, Ying 78, 246
Yancey, Neal 99, 101
Yang, Fang 260
Yang, Kexuan 11
Yang, Xiaokun 40
Yao, Yuan 131
Ye, Qing 63
Yelvington, Paul 197
Yépez, Susan K. 42
Yokochi, Alexandre 283
Yu, Ying 96, 98
Zabel, Michael 245
Zaimes, George G. 132
Zakir, Hossain 222
Zamarripa, Miguel A. 68
Zanganeh, Kourosh 280
Zappi, Mark E. 244
Zavala, Victor 17
Zhan, Wei 306
Zhang, Mengli 220
Zhang, Ning 174, 276
Zhang, Qi 265
Zhao, Lin 22
Zhen, Xinyu 188
Zhu, Lingxiang 61, 62
Zhu, Peng 71
Zimmerman, Julie B. 3, 126
Zuber, Mario 77

AIChE
120 Wall Street, FL 23
New York, NY 10005-4020

ISBN 978-1-7138-9300-4

CURRAN ASSOCIATES INC.
proceedings.com